# Lecture Notes in Civil Engineering

Volume 477

**Lecture Notes in Civil Engineering** (LNCE) publishes the latest developments in Civil Engineering—quickly, informally and in top quality. Though original research reported in proceedings and post-proceedings represents the core of LNCE, edited volumes of exceptionally high quality and interest may also be considered for publication. Volumes published in LNCE embrace all aspects and subfields of, as well as new challenges in, Civil Engineering. Topics in the series include:

- Construction and Structural Mechanics
- Building Materials
- Concrete, Steel and Timber Structures
- Geotechnical Engineering
- Earthquake Engineering
- Coastal Engineering
- Ocean and Offshore Engineering; Ships and Floating Structures
- Hydraulics, Hydrology and Water Resources Engineering
- Environmental Engineering and Sustainability
- Structural Health and Monitoring
- Surveying and Geographical Information Systems
- Indoor Environments
- Transportation and Traffic
- Risk Analysis
- Safety and Security

To submit a proposal or request further information, please contact the appropriate Springer Editor:

- Pierpaolo Riva at pierpaolo.riva@springer.com (Europe and Americas);
- Swati Meherishi at swati.meherishi@springer.com (Asia—except China, Australia, and New Zealand);
- Wayne Hu at wayne.hu@springer.com (China).

**All books in the series now indexed by Scopus and EI Compendex database!**

Babu T. Jose · Dipak Kumar Sahoo ·
Eun Chul Shin · Deepankar Choudhury ·
Anil Joseph · Rahul R. Pai
Editors

# Proceedings of the Indian Geotechnical Conference 2022 Volume 2

## Geotechnics: Learning, Evaluation, Analysis and Practice (GEOLEAP)

Springer

*Editors*
Babu T. Jose
Cochin University of Science and Technology
Kochi, Kerala, India

Dipak Kumar Sahoo
Cochin University of Science and Technology
Kochi, Kerala, India

Eun Chul Shin
Department of Civil and Environmental Engineering
Incheon National University
Incheon, Korea (Republic of)

Deepankar Choudhury
Department of Civil Engineering
Indian Institute of Technology Bombay
Mumbai, Maharashtra, India

Anil Joseph
Geotechnical and Structural Consultant
Geostructurals Pvt. Ltd.
Ernakulam, Kerala, India

Rahul R. Pai
Department of Civil Engineering
SCMS School of Engineering and Technology
Ernakulam, Kerala, India

ISSN 2366-2557 ISSN 2366-2565 (electronic)
Lecture Notes in Civil Engineering
ISBN 978-981-97-1740-8 ISBN 978-981-97-1741-5 (eBook)
https://doi.org/10.1007/978-981-97-1741-5



This Springer imprint is published by the registered company Springer Nature Singapore Pte Ltd.
The registered company address is: 152 Beach Road, #21-01/04 Gateway East, Singapore 189721, Singapore

# Preface

The Indian Geotechnical Society, Kochi (IGS-Kochi) Chapter, in association with Cochin University of Science and Technology (CUSAT), India, organized the Indian Geotechnical Conference (IGC-2022) at Kochi during 15–17 December 2022. The main theme of the conference was "**GEOLEAP—GEO**TECHNICS: **L**EARNING, **E**VALUATION, **A**NALYSIS AND **P**RACTICE".

The sub-theme of the conference includes:

1. Behaviour of Soils and Soil—Structure Interaction
2. Soil Stabilization, Ground Improvement and Land Reclamation
3. Shallow and Deep Foundations
4. Geotechnical, Geological and Geophysical Investigation
5. Rock Engineering, Tunnelling and Underground Structures
6. Slope Stability, Landslides and Liquefaction
7. Earth Retaining Structures and Deep Excavations
8. Geosynthetics Engineering
9. Geo-Environmental Engineering, Sustainable Geotechnics and Landfill Design
10. Geohydrology, Dam and Embankment Engineering
11. Earthquake Geotechnical Engineering
12. Transportation Geotechnics
13. Forensic Geotechnical Engineering and Retrofitting of Geotechnical Structures
14. Offshore Geotechnics, Marine Geology and Subsea Site Investigation
15. Computational, Analytical and Numerical Modelling
16. Reliability in Geotechnical Engineering.

The proceedings of this conference consists of selected papers presented at the conference, organized into five volumes. The keynote and theme lectures presented during IGC 2022 are being published as a special issue of the *Indian Geotechnical Journal*.

We sincerely thank all the authors who have contributed their papers to the conference proceedings. We also thank all the theme editors and reviewers who have been instrumental in giving their valuable inputs for improving the quality of the final papers. Finally, we thank the Springer team for their support and full cooperation for publishing these five volumes of IGC-2022 proceedings.

Kochi, India

Dr. Babu T. Jose
Chairman IGC-2022

# Contents

# About the Editors

**Dr. Babu T. Jose** completed his B.Tech. from NIT Warangal in 1967 with the Nehru Memorial Gold Medal, M.Tech. from IIT Madras in 1972 and Ph.D. from Cochin University of Science and Technology in 1990. He was a lecturer in NIT, Calicut, for 13 years (1967–1980). Currently, he is an emeritus professor at Cochin University of Science and Technology since 2004. Dr. Jose has commendable experience in research. He has published more than 150 research papers, and on the consultancy front, he has successfully completed more than 500 projects. On the administrative side, he was a principal at the School of Engineering, Cochin University, during 1980–1995 and director till 2004. Dr. Babu T. Jose is the founder chairman of IGS, Kochi Chapter. He was the chairman of the organizing committee of IGC in 2011 and the chief editor of its proceedings.

**Dr. Dipak Kumar Sahoo** has been a full-time faculty member since 2001 and currently a professor since 2009, in the School of Engineering, Cochin University of Science and Technology, Kochi. Prior to academics, he was working in the Fertilizers and Chemicals Travancore Limited, a Central Public Sector Company, in India for more than ten years as a consulting engineer to many public and private sector companies for setting up their green-field petrochemical complexes in different parts of India. He received his B.Sc. Engineering (Civil) from CET Bhubaneswar in 1987, M.Tech. in Construction Engineering and Management from Cochin University of Science and Technology in 1998, Ph.D. in Structural Engineering from the Indian Institute of Technology Roorkee in 2009 and LL.B. from the Cochin University of Science and Technology in 2015.

**Prof. Eun Chul Shin** is currently a professor emeritus at the Incheon National University, South Korea. He graduated from Chungbuk National University in Korea. He obtained his M.S. degree from the University of Colorado at Boulder, USA, in 1987 and a Ph.D. degree from the Southern Illinois University at Carbondale, USA, in 1994 under the supervision of Prof. Braja M. Das. Since graduation, he has been teaching geotechnical engineering at the Incheon National University, Korea, where he obtained a full professorship in 2005 and served as a dean of the Urban

Science College until 2016. He is an immediate past vice president of International Society for Soil Mechanic and Geotechnical Engineering in Asia region (2017–2022). He has published five books on the geotechnical, geo-environmental and civil engineering-related subjects as well as more than 100 journal papers and 150 conference papers.

**Prof. Deepankar Choudhury** *FNAE, FNASc, FASc, F.ASCE, FIGS, FIE, Humboldt Fellow of Germany, JSPS Fellow of Japan, TWAS-VS Fellow of Italy,* is a Prof. T. Kant Chair professor and Head of Civil Engineering department of IIT Bombay, Mumbai, India. Always 1st Class 1st Gold Medalist, Professor Choudhury did his B.E. (Civil) from Jadavpur University, Kolkata; M.E. (Geotechnical Engineering) from IISc Bangalore and Ph.D. (Civil Geotechnical Earthquake Engineering) from IISc Bangalore. Professor Choudhury worked as a lecturer at IIT Kanpur during 2002–2003 before moving to IIT Bombay as an assistant professor in 2003 and elevated to full professor in 2012. He also mentored new IITs at Gandhinagar in 2009 and Dharwad in 2017 on deputation from IIT Bombay. He is the author of one major textbook, several edited books, more than 180 peer-reviewed reputed journal papers and more than 130 full peer-reviewed conference papers and supervised already 35 PhDs. He executed several national and international field projects of repute as an expert consultant to various organizations of India and abroad. Prof. Choudhury organized two major national examinations, IIT-JEE (Adv)-2015 and GATE-2021 as National Organizing Vice-Chairman and Organizing Chairperson respectively. He is Chairman of Disaster Resilience & Capacity Building of SDMA, Govt. of Maharashtra, member of High-Rise Committee (HRC) of BMC and member of BIS code committee CED-43. He is the Editor-in-Chief of Indian Geotechnical Journal, Springer and Associate Editor of ASCE-International Journal of Geomechanics.

**Dr. Anil Joseph** completed his graduation and post-graduation in Civil Engineering from N.I.T. Surat. He was awarded Ph.D. from NIT Calicut, for his work on "Precompression of Soft Marine Clays". He is the managing director of Geostructurals (P) Ltd., a leading foundation and structural consultancy firm and has provided consultancy services for more than 3000 high rise structures in India and abroad. He is also the managing director of CECONS (P) Ltd., a construction firm specialized in the execution of pile foundations, and a director of Engineers Diagnostic Centre (P) Ltd., a firm specialized in geotechnical investigation and retrofitting works. He is currently serving as the national president of Indian Geotechnical Society and chairman of Civil Division Board, Institution of Engineers India.

**Dr. Rahul R. Pai** is currently a faculty in the Civil Engineering Department of SCMS School of Engineering and Technology, Kerala, India. He has an experience of around 12 years in the field of academics as well as industry. He completed his B.Tech. in Civil Engineering from M. G. University, Kerala, India, and M.Tech. in Soil Mechanics and Foundation Engineering from S. V. National Institute of Technology, Surat, India. He also did his Ph.D. from S. V. National Institute of Technology,

Surat, in Transportation Geotechnics. The Ph.D. research was a funded project on waste management technology sponsored by the Department of Science and Technology, India. His key research areas include utilization of sustainable and innovative materials in highway infrastructure, structural evaluation of flexible pavement with falling weight deflectometer and quality control aspects of construction of pavements with industrial wastes.

# Shallow and Deep Foundations

# Bearing Pressure of Foundation on Strong Layer Overlying Weak Clay Accounting for Compressibility

**Joga Himavarsha, J. Y. V. Shiva Bhushan, and Madhira R. Madhav**

## 1 Introduction

Soft soils undergo excessive settlements due to compressibility causing engineering structures to collapse or undergo serious or detrimental damage. Ground profile consists usually of several strata, which contribute to the complex behavior when subjected to various loads. Layered soft clays are commonly encountered in Geotechnical Practice. The top layer often of even normally consolidated sedimented soil is subjected to atmospheric and climatic conditions forming a crust or a desiccated layer that is stiffer and stronger than the lower one.

Button [1] proposed the bearing capacity factors for strip footing resting on two-layer cohesive soil by limit equilibrium analysis, which depends on the relationship between strength ratio and ratio of thickness of upper cohesive layer to width of footing. The effect of non-homogeneity in shear strength on the ultimate bearing capacity of a strip footing on two-layered clay was given by Siva Reddy and Srinivasan [2]. Model experiments on circular and strip footings on layered clays were carried out by Brown and Meyerhof [3] and proposed modified bearing capacity factors. Failure of the foundation happens by punching through the top layer of the rigid layer overlying the soft layer. Merifield et al. [4, 5] investigated the undrained bearing capacity of a rigid footing on two-layer clay deposit. Papadopoulou and Gazetus [6] modified the coefficients for strength ratio and normalized thickness, for the estimation of undrained bearing capacity of footings on two-layered clays.

Sunil and Sheetal [7] studied on bearing capacity of surface square footing resting on strong over weak clay using PLAXIS 2D and stated that the bearing capacity factor increases with an increase in thickness of top layer. Benmebarek [8–10] studied the

J. Himavarsha · J. Y. V. Shiva Bhushan (✉) · M. R. Madhav
Department of Civil Engineering, VNR Vignana Jyothi Institute of Engineering and Technology, Hyderabad, India
e-mail: shivabhushan_jyv@vnrvjiet.in



B. T. Jose et al. (eds.), *Proceedings of the Indian Geotechnical Conference 2022 Volume 2*, Lecture Notes in Civil Engineering 477,
https://doi.org/10.1007/978-981-97-1741-5_1

effect of two-layered clays on bearing capacity of foundations using numerical FLAC software. The modified bearing capacity factor is shown to be dependent on relative thickness of top layer with respect to width of footing, the strength ratio of the soils, and the degree of non-homogeneity.

The load-carrying capacity of the layered strata is also influenced by the soil's stiffness properties. McMahon et al. [11] determined the load–settlement behavior of a linear-elastic perfectly plastic soil based on the ellipsoidal cavity expansion model. Shiva et al. [12] determined the effect of compressibility on undrained bearing pressure of homogeneous soft ground.

It has been observed that none of the above works considers the effect of compressibility of soils on undrained bearing pressure—settlement response of footings on two-layered soil system. The present study considers a strong and stiff layer overlying weak and soft clay accounting for compressibility of the two layers.

## 2 Problem Definition

A circular foundation of diameter, $D = 2$ m resting on two-layered clay, was considered (Fig. 1). Strong and stiff top clay layer of thickness, $H$, possessing deformation modulus, $E_1$, and undrained shear strength, $c_{u1}$, overlies a weaker and softer clay with an undrained shear strength, $c_{u2}$, of 10 kPa and deformation modulus, $E_2$, of 3 MPa. Undrained behavior of clays was simulated by considering Poisson's ratio of 0.495. Table 1 lists the values of soil parameters considered for the study.

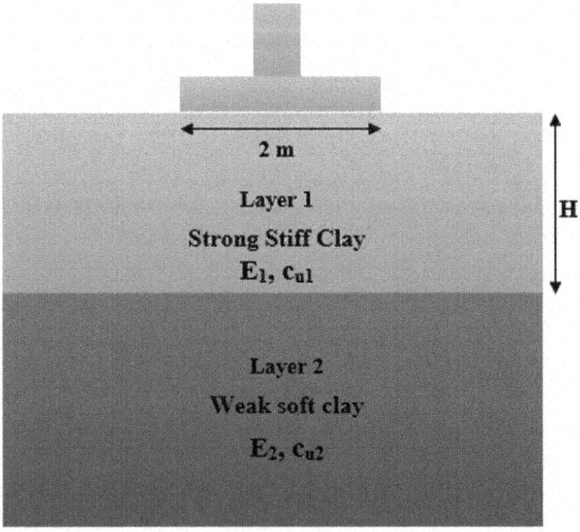


**Fig. 1** Sketch describing the problem definition

**Table 1** Properties of soil

| Soil property | | Values |
|---|---|---|
| Soil type | Stiff layer 1 | Weak layer 2 |
| Undrained strength, $c_u$ (kN/m$^2$) | 25 | 10 |
| Poisson's ratio, $\nu$ | 0.495 | 0.495 |
| Unit weight of soil, $\gamma$ (kN/m$^3$) | 18 | 15 |
| Deformation modulus, $E$ (MPa) | 6, 15, 30 | 3 |

## 3 Methodology

Numerical study was performed using finite element program PLAXIS 2D. Axisymmetric modeling was chosen for circular footing. The soil mesh was generated using 15-noded triangular elements. An undrained drainage type was chosen in Mohr–Coulomb soil model. The lateral boundaries in the radial direction were positioned at a distance of 6 times the diameter of the footing and the bottom one was fixed in both vertical and radial directions (Fig. 2).

The study is confined to a strength ratio, $c_{u1}/c_{u2}$, of 2.5 and analyzes the effect of compressibility and relative thickness of the top layer on bearing pressure–settlement response of two-layered clay. The study is limited to a vertical displacement of 20 cm, i.e., 10% of diameter of the footing.

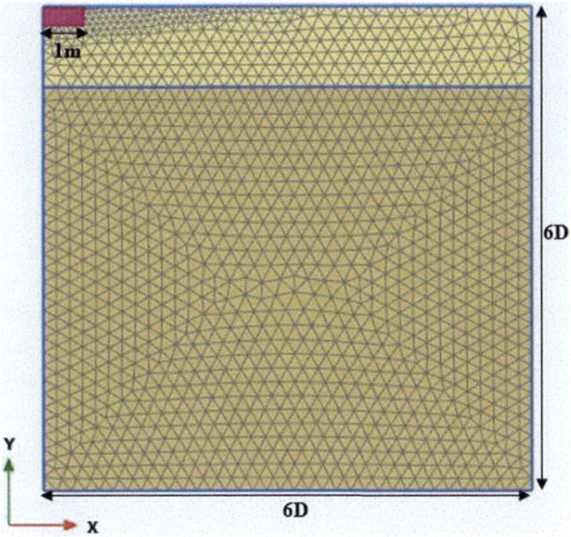


**Fig. 2** Meshing of finite element model with uniform displacement

## 4 Results

Undrained bearing pressure, $q$,—settlement, $s$, responses for the considered cases are obtained. Settlement is normalized by diameter of the footing and is defined as settlement ratio, SR = 100 $s/D$. The bearing pressure is normalized with undrained strength of the top layer and defined as bearing pressure factor, $N_{cf}$ ($= q/c_{u1}$). For studying the effect of compressibility on $N_{cf}$ of this two-layered soil, for considered cases, SR is used as reference. Ultimate bearing capacity factors given by Merifield and Nguyen [5], for different thicknesses of top layer, are compared with $N_{cf}$ values obtained at a settlement of 10% of footing diameter (Table 2). The predicted values agree closely with those of Merifield and Nguyen [5] with maximum deviation of about 1%.

Variation of bearing pressure factor, $N_{cf}$,—settlement ratio, SR curves for different values of $R_E$ are depicted in Fig. 3, for a strength ratio ($R_{cu} = c_{u1}/c_{u2}$) of 2.5. Figure 3a–d show the curves obtained for layered soil with thickness, $H$, of top strong and stiff clay of 0.5 m, 1.0 m, 2.0 m and 3.0 m corresponding to $H/D$ ratios of 0.25, 0.5, 1.0 and 1.5 respectively. Bearing pressure factor ($N_{cf}$) increases significantly with increase in settlement but marginally with stiffness ratio of $R_E$ for all the relative thickness ratios ($H/D$) implying increase in carrying capacity with increase in relative stiffness and relative thickness of the upper clay at all settlements.

For a thin top layer with $H = 0.25D$, $N_{\text{cf}}$ at SR = 1% increases by 3.9 and 6.7% for the deformation modulus of top layer increasing from 2.5 to 5 and 10 times the value of the bottom layer. For an intermediate relative thickness ratio of 1, Ncf value at SR = 1% increases by 24% and 34% for RE increasing by 4 folds. For a relatively thick top layer with $H$ = 1.5D, bearing pressure factors increase by 33% and 49% signifying that the effect is more on bearing pressure at larger thicknesses of the top layer. The effect of RE on bearing pressure factor is much less at larger settlement say at SR = 5%, with increments of 4.3 and 6.7% for $H/B = 1.0$ and negligible for $H/B = 1.5$.

Figure 4 shows the variations of $N_{\text{cf}}$ with $R_E$ plotted on a logarithmic scale (Fig. 4) for different normalized thicknesses and settlement ratios of 0.5, 1.0, 2.5, 5.0, and 10%. $N_{cf}$ increases linearly to marginally bilinearly with stiffness ratio. For $H/D$ = 0.25, $N_{\text{cf}}$ increases from 1.33 at $R_E = 2.5$ at SR = 0.5% to 1.54 at $R_E = 10$; similarly for a settlement ratio of 10%, $N_{cf}$ increases from 3.84 for $R_E = 2.5$ to 4.0 for $R_E$ = 10. For a normalized thickness of 1.5 and at a settlement ratio of 0.5% and 5%, the

**Table 2** Comparison of bearing capacity factors with Merifield and Nguyen [5]

| $H/D$ | Merifield and Nguyen [5] | Present analysis $N_{cf}$ at SR 10% |
|---|---|---|
| 0.25 | 4.02 | 4.02 |
| 0.5 | 5.23 | 5.23 |
| 1.0 | 6.06 | 6.11 |
| 1.5 | 6.04 | 6.12 |

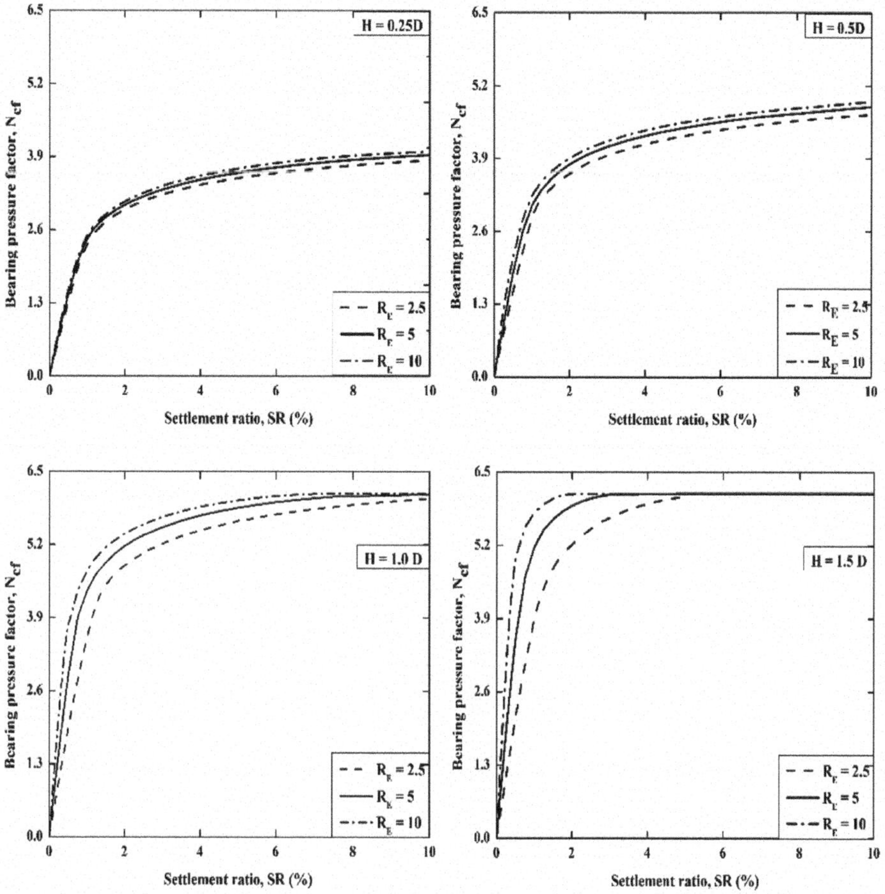


**Fig. 3** $N_{cf}$ versus SR: effect of $R_E$ for **a** $H/D = 0.25$, **b** $H/D = 0.5$, **c** $H/D = 1$, and **d** $H/D = 1.5$

values of $N_{cf}$ are 2.15 and 5.0 and 6.07 and 6.12 for $R_E = 2.5$ and 10, respectively (Fig. 4d) implying the effect of compressibility on bearing pressure to be minimal or negligible at higher settlements.

Bearing pressure coefficient, $R_q$ is defined as the ratio of bearing pressure factor, $N_{cf}$, to that at 10% settlement $N_f$. Variation of $R_q$ with stiffness ratio, $R_E$, for SR of 0.5% and 1% is shown in Fig. 5. Bearing pressure coefficient, *Rq*, at SR = 0.5% and 1% increases from 0.35 to 0.82 and from 0.64 to 0.95 for $H/D = 1.5$ for relative stiffness of top layer increasing 4 folds from 2.5 to 10. The effect of rigidity is significant at smaller displacements and reduces with displacement becoming negligible at larger displacements. For $H/D = 0.25$, bearing pressure coefficients, $R_q$, at SR = 0.5% are 0.59 and 0.63 and 0.64 and 0.95 for $H/D = 0.25$ and 1.5, respectively, for $R_E = 2.5$ and 10. The effect of compressibility is significant for the larger thickness of top strong layer.

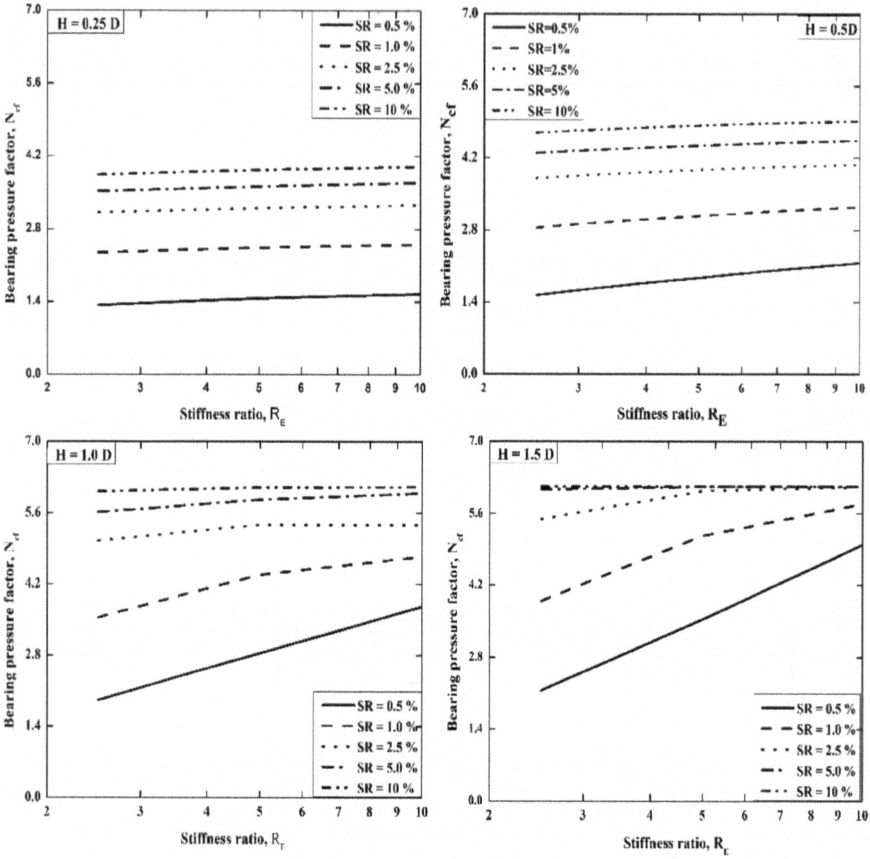


Fig. 4 Variation of $N_{cf}$ with $R_E$ at different SR for **a** $H/D = 0.25$ **b** $H/D = 0.5$ **c** $H/D = 1$ **d** $H/D = 1.5$

# 5 Conclusions

Circular footing resting on strong stiff clay layer overlying weak soft soil is analyzed for different thicknesses of the top layer and different modular ratios. Bearing pressure–settlement responses are obtained for different compressibilities of the upper layer. Bearing pressure coefficients at specific settlement increase with increase in stiffness ratio of clay layers. For $H/D = 1.5$, $R_q$ at SR $= 0.5\%$ increases from 0.35 to 0.82 as $R_E$ increased from 2.5 to 10. Bearing pressure coefficients also increase with increase in thickness of top strong and stiff layer. For $R_E = 5$, $R_q$ at SR $= 0.5\%$ increases from 0.37 to 0.58 as H/D increased from 0.25 and 1.5. Effect of compressibility is significant at smaller settlements and larger thicknesses of the layer beneath the rigid footing.

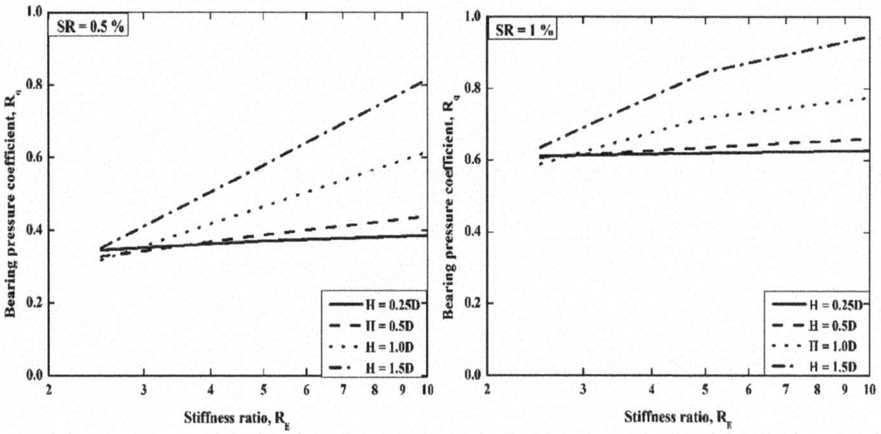


**Fig. 5** Bearing pressure coefficient, $R_q$ versus stiffness ratio, $R_E$ **a** SR = 0.5% **b** SR = 1%

# References


1. Button SJ (1953) The bearing capacity of footings on a two-layer cohesive subsoil. In: 3rd international conference in soil mechanics foundation engineering, pp 332–335
2. Siva Reddy A, Srinivasan RJ (1971) Bearing capacity of footings on clays. Soils Found 11(3):51–64. https://doi.org/10.3208/sandf1960.11.3_51
3. Brown JD, Meyerhof GG (1969) Experimental study on bearing capacity in layered clays. In: 7th international conference on soil mechanics and foundation engineering (Mexico). Pp, 45–51
4. Merifield RS, Sloan SW, Yu HS (1999) Rigorous plasticity solutions for the bearing capacity of two-layered clays. Geotechnique 49:471–490. https://doi.org/10.1680/geot.1999.49.4.471
5. Merifield RS, Nguyen VQ (2006) Two- and three-dimensional bearing-capacity solutions for footings on two-layered clays. Geomech Geoengin 1:151–162. https://doi.org/10.1080/17486020600632637
6. Papadopoulou K, Gazetas G (2020) Shape effects on bearing capacity of footings on two-layered clay. Geotech Geol Eng 38:1347–1370. https://doi.org/10.1007/s10706-019-01095-6
7. Pusadkar SS, Baral SM (2015) Behaviour of square footing resting on two layered clay deposits. In: Proceedings of 50th Indian Geotechnical Conference, Pune, India, 17-19th December, 2015
8. Benmebarek S, Benmoussa S, Belounar L, Benmebarek N (2012) Bearing capacity of shallow foundation on two clay layers by numerical approach. Geotech Geol Eng 30:907–923. https://doi.org/10.1007/s10706-012-9513-6
9. Benmebarek S, Saifi I, Benmebarek N (2017) Undrained vertical bearing capacity factors for ring shallow footings. Geotech Geol Eng 35:355–364. https://doi.org/10.1007/s10706-016-0110-y
10. Benmebarek S, Saifi I, Benmebarek N (2017) Depth factors for undrained bearing capacity of circular footing by numerical approach. J Rock Mech Geotech Eng 9:761–766. https://doi.org/10.1016/j.jrmge.2017.01.003
11. McMahon BT, Haigh SK, Bolton MD (2013) Cavity expansion model for the bearing capacity and settlement of circular shallow foundations on clay. Geotechnique 63:746–752. https://doi.org/10.1680/geot.12.P.061
12. Shiva Bhushan JYV, Madhav M, Narasimha Reddy GV (2022) Effect of compressibility on bearing pressure of soft ground. In: Muthukkumaran K, Satyanarayana Reddy CNV, Joseph A, Senthamilkumar S (eds) Foundation and forensic geotechnical engineering. Springer, Singapore. https://doi.org/10.1007/978-981-19-6359-9_8

# Interaction Effects for Large Piled Rafts in Clay Soil

**Rajib Modak and Baleshwar Singh**

## 1 Introduction

In general, the foundations types that are commonly in use are shallow (raft) and deep (pile) foundations. However, in recent decades a composite structure called piled raft foundation (PRF) consisting of both raft and piles has gained a significant recognition. The PRF has proven to be a cost-effective foundation option compared to the traditional foundations to carry heavy superstructure loads due to the contribution of both raft and piles that are taken into account in the PRF design. Several practical examples of PRFs have been reported in the literature [1, 2]. The PRFs can be classified into two types, *i.e.*, small and large piled rafts [3]. In the case of small piled rafts, the width of raft ($B_r$) is smaller than the pile length ($L_p$) ($B_r/L_p < 1$). However, in the case of large piled rafts, the width of raft is greater than the length of piles ($B_r/L_p > 1$). As both the raft and piles in a PRF contribute to the load sharing, a major design concept is the proportion of load carried by each structural element [4].

The overall load response of a PRF is governed by the complex interaction effects involved in the load-sharing mechanism, as shown in Fig. 1a. The interactions are caused due to the overlapping of stress and displacement fields of individual structural elements (raft and piles). The different interaction effects of a PRF are: (1) piles to piles (P-P); (2) raft to piles (R-P); and piles to raft (P-R) [5]. The P-P interaction effect is the changes in the load response of pile group (PG) and single pile (SP). The R-P interaction effect represents the changes in the load response of PG due to the raft in PRF. The P-R interaction effect signifies the changes in the raft load response

R. Modak (✉) · B. Singh
Department of Civil Engineering, Indian Institute of Technology Guwahati, Guwahati 781039, India
e-mail: rmodak@iitg.ac.in

B. Singh
e-mail: baleshwar@iitg.ac.in

 
B. T. Jose et al. (eds.), *Proceedings of the Indian Geotechnical Conference 2022 Volume 2*, Lecture Notes in Civil Engineering 477,
https://doi.org/10.1007/978-981-97-1741-5_2

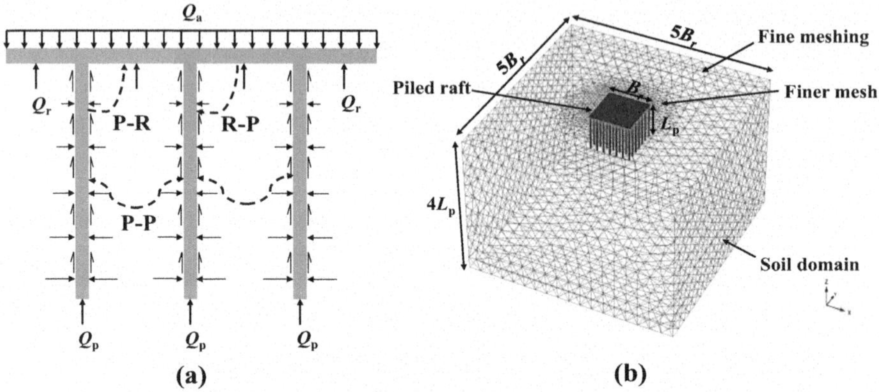


**Fig. 1** **a** Load-sharing mechanism and PRF interactions; **b** typical FE mesh of the soil domain and PRF

due to the piles in PRF. Taking into account the PRF interactions (Fig. 1a), Park and Lee [6] expressed the load-carrying capacity of a PRF as follows:

$$Q_{\text{pr}} = \alpha_r Q_{\text{ur}} + \alpha_p Q_{\text{pg}} = \alpha_r Q_{\text{ur}} + \alpha_p \eta_p Q_{\text{sp}}, \tag{1}$$

where $\eta_p$, $\alpha_p$, and $\alpha_r$ are the P-P, R-P, and P-R interaction effects; $Q_{\text{ur}}$, $Q_{\text{pg}}$, and $Q_{\text{sp}}$ are the load-carrying capacities of unpiled raft (UR), PG where the raft is not in contact with the ground, and SP, respectively.

Several investigations on evaluating the PRF interactions have been reported in the literature [6–11]. Furthermore, studies on the load-sharing behavior of piled rafts have also been reported [12–15]. For PRFs in sand, centrifuge tests were carried out Park and Lee [6], and they found that both P-R and R-P interaction effects initially decreased and then increased with the increase in settlement. Halder and Manna [8] conducted experimental investigations on PRF in sand and suggested equations for calculating various PRF interactions. Kumar and Choudhury [10] carried out 3D FE analysis for PRFs in sand and proposed expressions for determining the bearing capacity using PRF interactions. Reul [11] carried out three-dimensional (3D) finite element (FE) analysis of PRF in overconsolidated clay and reported that at a higher load level due to PRF interactions, pile skin friction increases. Bhaduri and Choudhury [12] presented a displacement-based FE approach to evaluate the load-sharing behavior of PRF. Lee et al. [15] studied the load-settlement response of PRFs in clay experimentally and suggested a load-sharing model based on PRF settlement. However, most investigations on PRF interactions and load-sharing behavior were expressed as a function of foundation geometry and property conditions. A few literature have reported the effect of settlement on the interactions and load-sharing behavior of large PRF in stiff clay through numerical modeling. Thus in the present paper, the settlement-dependent variation of PRF interactions and the load-sharing behavior are evaluated for large PRFs in stiff clay soil with varying pile numbers, pile spacings, pile lengths, and pile diameters. For this purpose, 3D FE analyses have been performed using Plaxis 3D software [16].

# 2 3D Numerical Modeling

## 2.1 *Boundary Conditions, Meshing, and Soil Modeling*

The drained soil behavior is considered and water table is assumed to be at ground surface. Based on the domain analysis study, the soil-domain boundaries were considered large enough to avoid boundary effects (Fig. 1b). The domain of soil in the lateral directions is extended up to a distance equivalent to twice of the width of raft ($B_r$) from the edges of raft, with displacements allowed in vertical direction and restrained in horizontal direction. The bottom of the soil domain is extended up to thrice the length of pile ($L_p$) from the tip of pile, with displacements restricted in both directions (vertical and horizontal). The ground surface displacements are free in all directions. To determine the optimum mesh size, mesh convergence study has been performed, and "Fine" mesh is selected. A finer mesh has been generated nearby the raft and piles by applying local mesh refinement. Figure 1b illustrates the soil-domain finite element mesh.

The Hardening soil model that is based on stress-dependent stiffness of soil is used to model the elastoplastic soil behavior. This is an advanced soil model which includes both compression and shear hardening. Three input parameters are required to represent the elastoplastic soil deformation, *i.e.*, ref. secant stiffness modulus ($E_{50}^{\text{ref}}$), ref. odometer stiffness modulus ($E_{\text{oed}}^{\text{ref}}$), and ref. unloading–reloading stiffness modulus ($E_{\text{ur}}^{\text{ref}}$) (Table 2). The different stress-dependent stiffnesses ($E$) are estimated using Eq. 2.

$$E = E^{\text{ref}}\left(\frac{c' \cot \emptyset' + \sigma'}{c' \cot \emptyset' + p^{\text{ref}}}\right)^{m}, \tag{2}$$

where $E^{\text{ref}}$ is the reference stiffness value corresponding to the reference stress ($p^{\text{ref}}$) level of 100 kPa, $\sigma'$ is the stress level, $c'$ and $\phi'$ are the shear strength parameters, and $m$ is the stress dependency factor.

## 2.2 *Validation*

The validation of the present study is performed by comparing the load-settlement variations of PRFs as indicated by Cho et al. [17] using the finite element based software Abaqus. A 10 × 10 m and 1 m thick raft with nine piles arranged in 3 × 3 square pattern of 0.5 m diameter, with spacing equivalent to thrice the diameter of

**Table 1** Material properties [17]

| | Clay (stiff) | Bearing layer | Raft | Piles |
|---|---|---|---|---|
| Elasticity modulus, $E'$(MPa) | 45 | 500 | 30,000 | 12,500 |
| Cohesive strength, $c'$(kPa) | 20 | 0.1 | – | – |
| Angle of friction, $\phi'$(°) | 20 | 45 | – | – |
| Poisson's ratio, $v'$ | 0.3 | 0.3 | 0.2 | 0.25 |
| Total unit weight, $\gamma_t$ (kN/m$^3$) | 19 | 20 | 25 | 25 |

**Table 2** Material properties of soil, raft, and piles

| Materials | Properties | Values |
|---|---|---|
| Soil | Unit weights $\gamma_{\text{unsat}}/\gamma_{\text{sat}}$ (kN/m$^3$) | 20/20 |
| | Poisson's ratio, $\nu_s$ | 0.2 |
| | Cohesion, $c´$ (kN/m$^2$) | 20 |
| | Angle of friction $\phi´$ (°) | 20 |
| | Coefficient of lateral earth pressure ($K_o$) | 0.8 |
| | Ref. secant stiffness modulus, $E_{50}^{\text{ref}}$ (kN/m$^2$) | $3.50 \times 10^4$ |
| | Ref. odometer stiffness modulus, $E_{\text{oed}}^{\text{ref}}$ (kN/m$^2$) | $4.28 \times 10^4$ |
| | Ref. unloading–reloading stiffness modulus, $E_{\text{ur}}^{\text{ref}}$ (kN/m$^2$) | $1.05 \times 10^5$ |
| | Stress dependency power, $m$ | 1 |
| Raft | Unit weight (kN/m$^3$) | 25 |
| | Elastic modulus, $E_r$ (MN/m$^2$) | 34,000 |
| | Poisson's ratio, $\nu_r$ | 0.2 |
| Piles | Unit weight (kN/m$^3$) | 25 |
| | Elastic modulus, $E_p$ (MN/m$^2$) | 30,000 |
| | Poisson's ratio, $\nu_p$ | 0.2 |

pile and length of piles of 8 and 16 m, is selected. The properties of the materials are given in Table 1. As shown in Fig. 2, the results obtained from the present study are in good agreement with the reported results.

# 3 Numerical Analyses

## *3.1 Piled-Raft Interactions and Load Sharing*

A squared raft of 38 × 38 m size [18] and 2 m thickness has been considered for the numerical analyses. The particular raft thickness corresponds to that of an intermediate flexible raft based on the raft–soil stiffness ratio [19]. Table 2 presents the soil [20], raft [18], and pile [18] properties used for the study. Three different PRF interaction effects, as mentioned in Sect. 1, are considered, *i.e.*, piles-to-piles (P-P), raft-to-piles (R-P), and piles-to-raft (P-R) effect. The settlement-dependent variation

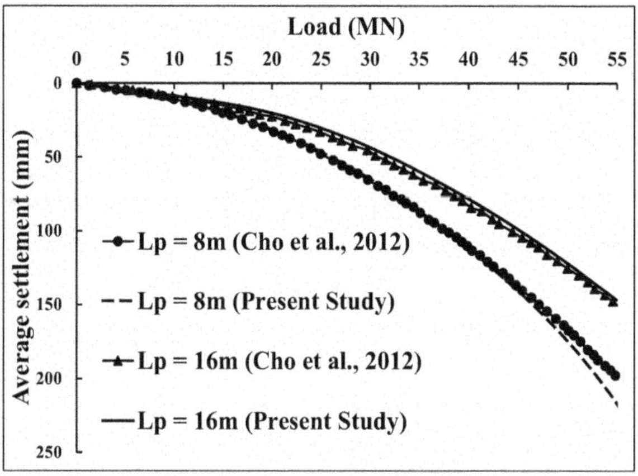


**Fig. 2** Load-settlement behavior of Plaxis 3D compared with Cho et al. [17]

of these interactions has been evaluated for the variation in the number of piles, spacing between piles, length of piles, and diameter of piles (Table 3). The number of piles ($N_p$) varies from 25 to 81 piles such that the area enclosed by the pile group ($B_g$), as shown in Fig. 3, is the same for all the PRFs. For 49 piles, the pile spacing ($s$) is varied from minimum to maximum possible pile spacing. The minimum pile spacing as per IS 2911 Part 1 [21] for group piles in clay soil is three times the pile diameter ($D_p$). For the same 49 piles, the pile length ($L_p$) is now varied between 20 and 30 times the diameter ($D_p$) of the pile. Again for 49 piles, the pile diameter is varied from 0.5 m to 1.5 m. In the study, while varying one parameter, the standard values (Table 3) of all the other parameters are considered. In addition to the PRF models, corresponding load-settlement responses of UR, PG, and SP are also analyzed to evaluate the different PRF interactions. The settlement-dependent variation of the P-P interaction effect is obtained by evaluating the ratio of load capacity of PG (with '$N_{\text{p}}$' number of piles) to the load capacity of '$N_{\text{p}}$' SP at the same settlement level. For R-P interaction effect, the ratio of the load capacity of piles in PRF to the PG load capacity with a similar number of piles as PRF is calculated. Furthermore, to determine the variation of P-R interaction effect with PRF settlement, the ratio of the load capacities of the raft in a PRF to that of the UR is evaluated.

The study considers settlement criteria of 0.35% $B_{\text{r}}$ [14]. The settlement-dependent variation of the PRF interactions is observed up to the maximum average settlement ($w_{\text{avg}}$) of 0.35% $B_r$. Thus, all the foundations are subjected to a load corresponding to which $w_{\text{avg}}$ is equivalent to 0.35% $B_r$. A uniformly distributed load (*UDL*) over the raft area is applied in the case of UR, PG, and PRF, whereas for SP, point load at the pile head is applied to determine the load-settlement response. The $w_{\text{avg}}$ is evaluated using Eq. 3 [18], where $w_{\text{center}}$ and $w_{\text{corner}}$ are the raft's center and corner settlements, respectively. To understand the variation in the load-sharing behavior of PRF with settlement, the load shared by raft and piles are evaluated. The pile load share is evaluated as the sum of axial load at all pile heads, and the raft load

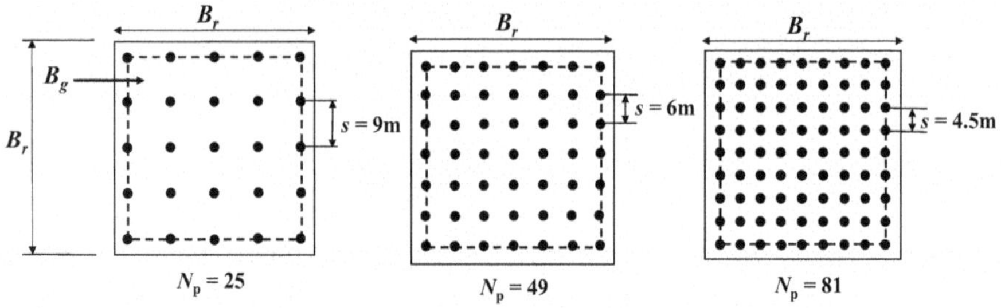


**Fig. 3** Various pile configurations used in the study

**Table 3** Piled raft geometric configurations

| Parameters | Values |
|---|---|
| Width of raft, $B_r$ (m) | 38 |
| Thickness of raft, $t_r$ (m) | 2 |
| Pile number, $N_p$ | 25, 49*, 81 |
| Pile spacing, $s$ (m) | 3, 4, 5, 6*, |
| Pile length, $L_p$ (m) | 20, 25, 30* |
| Pile diameter, $D_p$ (m) | 0.5, 1*, 1.5 |

*Standard value if not varied

share is then obtained by deducting the pile load share from the total applied load.

$$w_{avg} = \frac{1}{3}(2w_{center} + w_{corner}). \tag{3}$$

## 3.2 Results and Discussion

**Influence of pile number.** The influence of pile number on the settlement-dependent variation of the PRF interactions and the load-sharing behavior is presented in Fig. 4. Figure 4a shows that for all the pile configurations, *i.e.*, 5 × 5, 7 × 7, and 9 × 9, the P-P interaction effect ($\eta_p$) is lower within the initial settlement range and it approaches a constant value of unity at a higher settlement value. This implies piles in PG having a lower load capacity than '$N_p$' SP within the initial settlement range and this is due to greater overlapping of stresses and displacement fields within the piles in PG. Also, within the initial settlement range, the $\eta_p$ value is lower for a higher number of piles. However, as PRF settlement increase, once the capacity of piles in PG is mobilized fully, $\eta_p$ approaches unity for all the pile configurations. The settlement value corresponding to which the $\eta_p \approx 1$ is larger for a higher pile number. de Sanctis and Mandolini [7] suggested adopting a $\eta_p$ of unity based on FE analysis results for PRF in soft clay deposits. The R-P interaction effect ($\alpha_p$) influences the load response of PRF in two ways, *i.e.*, positive and negative. The positive aspect corresponds to the fact that due to the raft's presence, the confining stress of soil underneath the raft increases, which further increases the pile skin friction resistance. The negative aspect is related to the lesser relative pile–soil displacement due to the downward movement of the soil underneath the raft upon loading. Figure 4b shows the settlement-dependent variation of $\alpha_p$ for different pile numbers. For all the pile configurations, almost a similar trend of $\alpha_p$ variation with PRF settlement is observed. The negative effect is encountered initially, but with the increase in settlement, the positive effect is more pronounced. At the maximum settlement of 0.35% $B_r$ ($\approx$ 135 mm), the $\alpha_p$ values for piles varying from 25 to 81 lie within a close range of 1.0–1.04.

The P-R interaction effect ($\alpha_r$) indicates the changes in the load-settlement response of the raft due to the presence of piles underneath in PRF. The variation of $\alpha_r$ with respect to PRF settlement for varying pile numbers is shown in Fig. 4c. The figure illustrates that for all the pile configurations, $\alpha_r$ decreases initially and then gradually increases with the increase in the settlement. The initial decrease is because of the reduction of raft–soil contact pressure due to the downward soil movement. However, with the increase in settlement, once the pile capacity of piles in PRF is fully mobilized, the raft contribution in PRF increases, and thus $\alpha_r$ increases. Park and Lee [6] have also reported similar results by means of centrifuge tests conducted for PRF in sands. Moreover, the $\alpha_r$ values are lower for a higher number of piles due to lower raft–soil contact pressure. The $\alpha_r$ value at the maximum settlement of 0.35% $B_r$ ranges from 0.48 to 0.92 for piles varying from 81 to 25, respectively. The load-sharing behavior in terms of percentage of load carried by each structural element at different settlement levels for varying pile numbers is shown in Fig. 4d. It can be seen that with the increasing pile number, the pile load share increases and that of the raft decreases. For any pile number, the pile load share increases to a certain settlement value and then decreases with the increasing PRF settlement. The settlement values at which the dip in the pile load share is observed for all pile configurations are similar to the settlements at which $\eta_p$ approaches unity and $\alpha_r$ increases after showing a decreasing trend. The settlement values for which a decrease in pile load share is encountered for 25, 49, and 81 piles are 30, 55, and 75 mm, respectively. This shows that at these settlements, the pile capacity of piles in PRF is fully mobilized, and further increment in load is carried by the raft.

**Influence of pile spacing.** Figure 5 shows the influence of pile spacing on PRF interactions and load-sharing behavior variation with settlement. The standard 7 × 7 pile configuration with different $s/D_p$ (pile spacing to pile diameter) ratios of 3, 4, 5, and 6 is considered. For the different $s/D_p$ ratios considered, a lower $\eta_p$ value is observed initially, and then it approaches unity at a higher settlement level, as shown in Fig. 5a. Also, within the initial settlement range, the $\eta_p$ value is higher for a larger $s/D_p$ ratio. The reason is that, at a larger $s/D_p$ ratio the overlapping of stress and displacement fields are lesser, and the pile behavior in PRF is similar to that of SP. The figure also shows that the PRF settlement corresponding to which $\eta_p \approx 1$ is lower

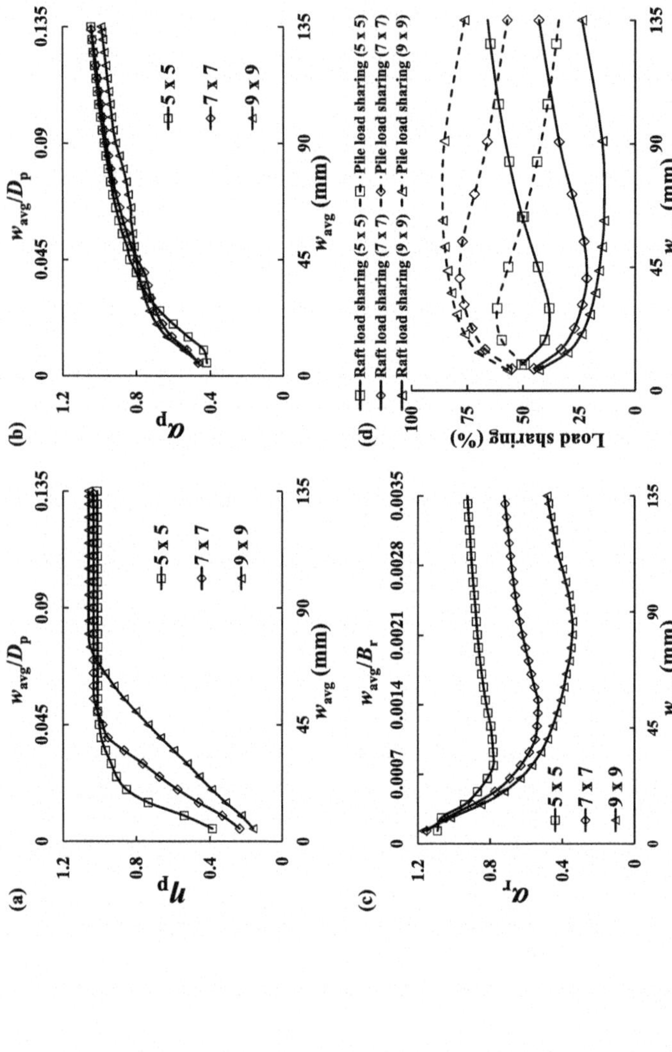


**Fig. 4** Influence of pile number on PRF interactions and load-sharing behavior

for a larger pile spacing. The variation of $\alpha_p$ with PRF settlement for varying pile spacing is presented in Fig. 5b. It can be observed that for all $s/D_p$ ratios, the negative effect is significant initially, followed by the positive effect at a higher settlement value. In addition, the $\alpha_p$ variation is almost similar initially at the lower settlement range, and at a higher settlement level, the $\alpha_{\mathrm{p}}$ value is slightly greater for a larger $s/D_p$ ratio. The $\alpha_p$ value lies within the range of 0.9–1.04 at the settlement of 0.35% $B_{\mathrm{r}}$ for $s/D_p$ ratio varying from 3 to 6, respectively.

The settlement-dependent variation of $\alpha_{\mathrm{r}}$ for varying $s/D_p$ ratios is shown in Fig. 5c. The figure shows that $\alpha_r$ decreases continuously for smaller $s/D_p$ ratios (3 and 4) up to the maximum settlement of 0.35% $B_{\mathrm{r}}$. However, for larger $s/D_p$ ratios (5 and 6), the $\alpha_r$ value initially decreases and then increases with the increase in PRF settlement. This is because at a smaller $s/D_p$ ratio due to greater superimposition of stresses and displacement fields, the pile capacity is not fully mobilized within the considered settlement range. The $\alpha_{\mathrm{r}}$ value at 0.35% $B_r$ settlement lies within the close range of 0.68–0.75 for the different $s/D_p$ ratios considered. Figure 5d illustrates the load shared by the piles and raft in PRF at different $s/D_p$ ratios. With the increase in the pile spacing, the pile load share increases and that of the raft decreases. This is because, with the increase in pile spacing, the pile behavior in PRF is comparable to SP as no neighboring piles' influence is experienced. However, it can be observed

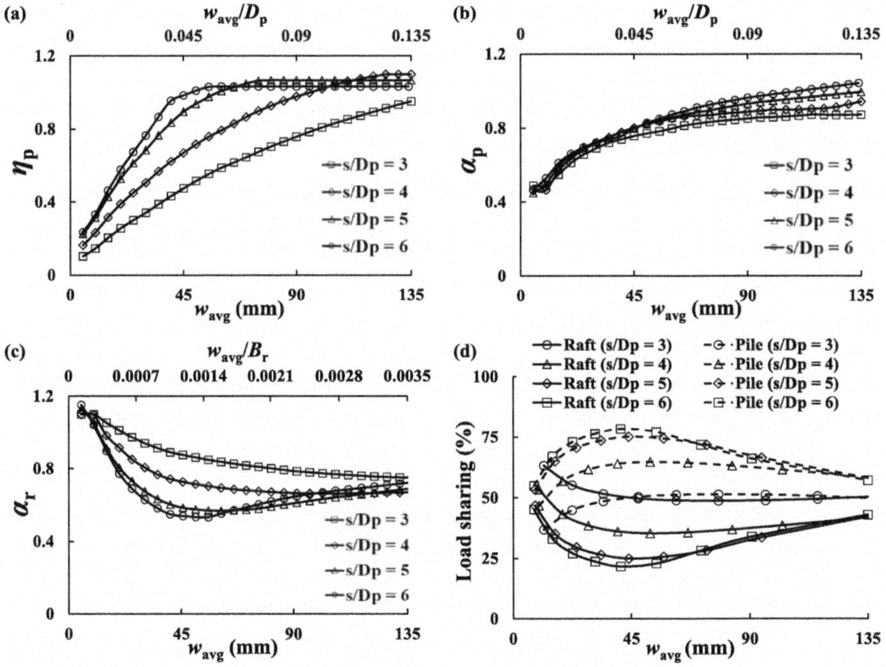


**Fig. 5** Influence of pile spacing on PRF interactions and load-sharing behavior

that the increase in the pile load share is not that significant for the increase in $s/D_p$ ratio from 5 to 6.

**Influence of pile length**. To understand the effect of pile length on the PRF interactions and load-sharing behavior, the 7 × 7 pile configuration with three pile length-to-pile diameter ratios ($L_p/D_p$) has been considered, which are 20, 25, and 30. The other standard parameters have been selected from Table 3. Figure 6a illustrates the variation of $\eta_p$ with respect to settlement for different $L_p/D_p$ ratios. The settlement value corresponding to which $\eta_p$ approaches unity increases as the length of pile increases. This is because a higher settlement is required for the pile capacity to be fully mobilized for longer length piles. The settlement-dependent variation of $\alpha_p$ for PRFs with varying $L_p/D_p$ ratios is shown in Fig. 6b. From the figure, it can be seen that a variation in the length of pile has insignificant effect on $\alpha_p$ variation with PRF settlement. The $\alpha_p$ value varies from 1.01 to 1.04 at the maximum settlement of 0.35% $B_r$ for the $L_p/D_p$ ratio varying from 20 to 30, respectively. The $\alpha_r$ value, as shown in Fig. 6c, initially decreases and then increases with the increase in PRF settlement for all $L_p/D_p$ ratios considered. The decrease in $\alpha_r$ value is observed to be more significant for longer length piles. The minimum value of $\alpha_r$ (0.53) is encountered for the $L_p/D_p$ ratio of 30 at the settlement of 50 mm. At the settlement of 0.35% $B_{r,}$ the $\alpha_r$ value ranges from 0.72 to 0.85 for $L_p/D_p$ ratios varying from 30 to 20, respectively. From the load shared by the individual structural elements, as shown in Fig. 6d, it can be seen that the pile load share increases with the increase in the pile length. This is due to the increase in the pile load capacity with the increasing length of piles in PRF. Furthermore, it can also be observed that the settlement value corresponding to which the raft load share rises and that of pile decreases increases with the increase in the pile length.

**Influence of pile diameter**. The standard 7 × 7 pile configuration is selected with pile diameter varying from 0.5 to 1.5 m to understand the effect of pile diameter on the settlement-dependent variation of PRF interactions and load sharing. For all the other parameters, the standard values are considered (Table 3). Figure 7a presents the $\eta_p$ variation with PRF settlement for different pile diameters. The $\eta_p$ value is lower for larger diameter piles at the initial settlement range. The settlement value corresponding to which $\eta_p \approx 1$ increases with the increase in pile diameter. The $\alpha_p$ variation with settlement as shown in Fig. 7b shows no significant difference for the different pile diameters considered. However, within this range of settlement, up to 0.35% $B_r$, the $\alpha_p$ value increases gradually with PRF settlement. The $\alpha_p$ value at 0.35% $B_r$ settlement ranges from 0.98 to 1.04 for the pile diameter varying from 1.5 to 0.5 m, respectively. The $\alpha_r$ variation with PRF settlement for varying pile diameters is shown in Fig. 7c. It can be observed that the $\alpha_r$ values significantly decrease as the pile diameter increases. This is because with increasing diameter, the pile capacity increases, resulting in the reduction of raft contribution in PRF, and thus, $\alpha_r$ decreases. The minimum value of $\alpha_r$ (0.39) is observed in the case of PRF with piles of 1.5 m diameter and corresponding to the settlement of 65 mm. At the maximum settlement of 0.35% $B_{r,}$ the $\alpha_r$ value ranges from 0.5 to 0.91 for pile diameter varying from 1.5 to 0.5 m, respectively. de Sanctis and Mandolini

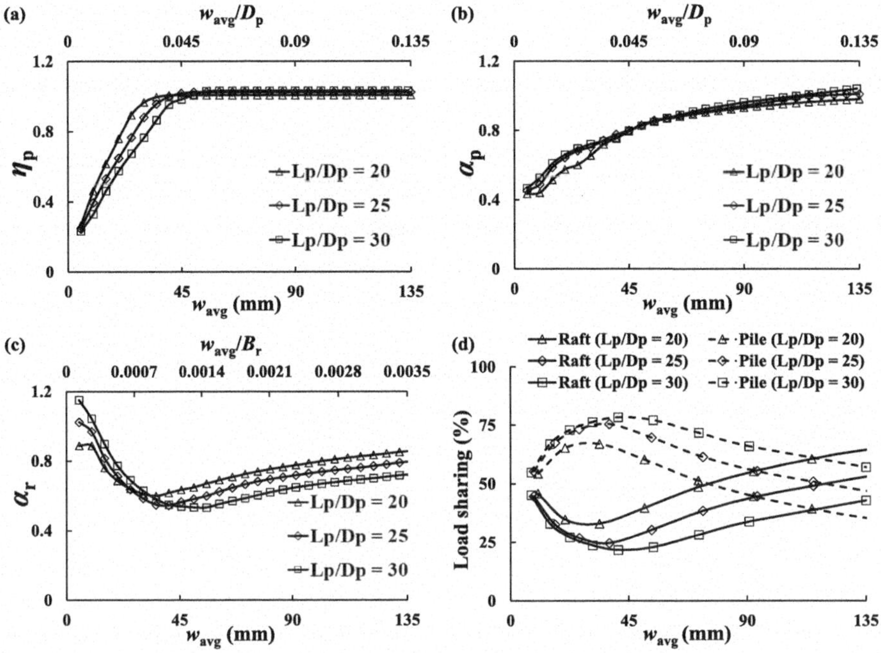


**Fig. 6** Influence of pile length on PRF interactions and load-sharing behavior

[7] have reported the range of $\alpha_r$ from 0.4 to 1.0 for PRF in soft clays. From the percentage load share, as shown in Fig. 7d, it can be seen that with the increase in the pile diameter, the pile load share increases. Also, within the range of settlement considered, the piles carry the majority of the load except in the case of PRF with 0.5 m diameter piles. This can be attributed to full pile load capacity mobilization for smaller diameter piles at a lower settlement level.

## 4 Conclusions

In the study, 3D numerical analyses were performed on large PRFs in stiff clay soil with varying pile numbers, pile spacings, pile lengths, and pile diameters to determine the settlement-dependent variation of PRF interactions and load-sharing behavior. The load-settlement responses of UR, PG, and SP were analyzed to evaluate the different PRF interactions. The P-P interaction effect approached unity at a higher settlement level for all the PRFs considered in the study. The settlement corresponding to which the P-P interaction effect reaches unity increased with the increase in pile number, pile length, and pile diameter, and it decreased with the increase in the pile spacing. In the scope of the study, the negative R-P interaction

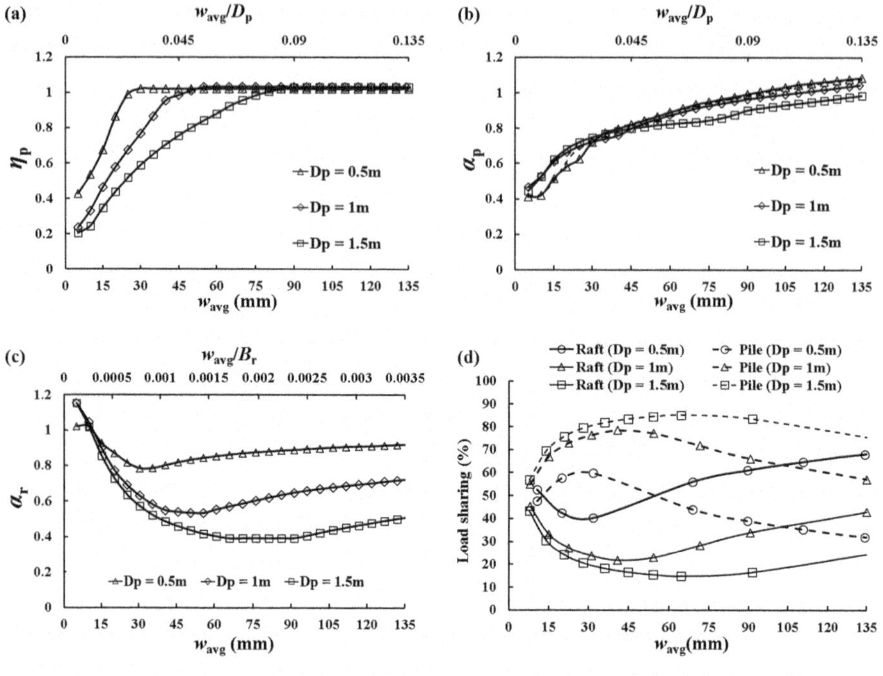


**Fig. 7** Influence of pile diameter on PRF interactions and load-sharing behavior

effect was observed initially, and with the increase in the PRF settlement, the positive R-P interaction effect was more prevalent. No significant difference in the R-P interaction effect variation with respect to PRF settlement was observed for varying pile numbers, pile lengths, and pile diameters. However, at a settlement of 0.35% $B_{r,}$ the R-P interaction effect is marginally higher for a larger pile spacing. The R-P interaction effect showed an increasing trend for all PRFs, and it ranged from 0.9 to 1.04 with an average of about 0.97 at the maximum average settlement of 0.35% $B_r$.

The P-R interaction effect initially decreased due to lower raft-soil contact stress and then increased as the pile capacity of the piles in PRF was fully mobilized. The settlement value up to which a decreasing trend in the P-R interaction effect was observed increased for a higher pile number, pile length, and pile diameter. However, for PRFs with a smaller pile spacing, the P-R interaction effect decreases continuously within the settlement range considered in the study. Overall, the P-R interaction effect varied in the range from 0.48 to 0.92, with an average value of 0.7. From the load-sharing behavior, it was observed that the pile load share increased and that of raft decreased as pile number, spacing, length, and diameter increase for the entire settlement range considered. The pile load share initially increased with the PRF settlement and then decreased once the pile load capacity was fully mobilized. The settlement value at which the pile load share reaches a peak is greater for a higher pile number, length, and diameter.

## References


1. El-Mossallamy Y (2002) Innovative application of piled raft foundation in stiff and soft subsoil. In: Deep foundations 2002: an international perspective on theory, design, construction, and performance, pp 426–440
2. Poulos HG (2001) Piled raft foundations: design and applications. Geotechnique 51(2):95–113
3. Russo G, Viggiani C (1998) Factors controlling soil-structure interaction for piled rafts. Darmstadt Geotech 4:297–321
4. Clancy P, Randolph MF (1996) Simple design tools for piled raft foundations. Geotechnique 46(2):313–328
5. Katzenbach R, Arslan U, Moormann C (2000) Piled raft foundation projects in Germany. In: Hemsley JA (ed) Design applications of raft foundations. Thomas Telford, London, pp 323–391
6. Park D, Lee J (2015) Interaction effects on load-carrying behavior of piled rafts embedded in clay from centrifuge tests. Can Geotech J 52(10):1550–1561
7. de Sanctis L, Mandolini A (2006) Bearing capacity of piled rafts on soft clay soils. J Geotech Geoenviron Eng 132(12):1600–1610
8. Halder P, Manna B (2021) A new model for the prediction of load sharing in piled raft system-an experimental investigation. Arab J Sci Eng 46(11):10667–10680
9. Horikoshi K, Randolph MF (1998) A contribution to optimum design of piled rafts. Geotechnique 48(3):301–317
10. Kumar A, Choudhury D (2018) Development of new prediction model for capacity of combined pile-raft foundations. Comput Geotech 97:62–68
11. Reul O (2004) Numerical study of the bearing behavior of piled rafts. Int J Geomech 4(2):59–68
12. Bhaduri A, Choudhury D (2020) Serviceability-based finite-element approach on analyzing combined pile–raft foundation. Int J Geomech 20(2):04019178
13. Conte G, Mandolini A, Randolph MF (2003) Centrifuge modelling to investigate the performance of piled rafts. In: Van Impe WF (ed) Proceedings of 4th international geotechnical seminar on deep foundations on bored and auger piles. Millpress, Rotterdam, Netherlands, pp 359–366
14. Lee J, Kim Y, Jeong S (2010) Three-dimensional analysis of bearing behavior of piled raft on soft clay. Comput Geotech 37(1–2):103–114
15. Lee J, Park D, Choi K (2014) Analysis of load sharing behavior for piled rafts using normalized load response model. Comput Geotech 57:65–74
16. Brinkgreve R, Swolfs W, Engin E (2015) PLAXIS user's manual, version 6.1. Balkema, Rotterdam, The Netherlands
17. Cho J, Lee JH, Jeong S, Lee J (2012) The settlement behavior of piled raft in clay soils. Ocean Eng 53:153–163
18. Reul O, Randolph MF (2004) Design strategies for piled rafts subjected to nonuniform vertical loading. J Geotech Geoenviron Eng 130(1):1–13
19. Fraser RA, Wardle LJ (1976) Numerical analysis of rectangular rafts on layered foundations. Geotechnique 26(4):613–630
20. Engin HK, Brinkgreve RBJ Investigation of pile behaviour using embedded piles. In: Proceedings of the 17th international conf. on soil mechanics and geotechnical engineering, Alexandria, pp 1189–1192
21. IS 2911–Part 1, Section 4: Design and construction of pile foundations–Code of practice, Bureau of Indian Standards (2010)

# Analysis of Vertically Loaded Pile-Raft Foundation on Cohesionless Soil Using ABAQUS

**Kabiraj Mahata, Md Sadiqur Rahaman, and Ashis Kumar Bera**

## 1 Introduction

The foundation system is one of the most significant aspects of any Geotechnical Engineering project. Due to increase of population and land scarcity problem, complexity of geotechnical problem also increases, and to overcome the new challenges, geotechnical engineers try to established these two separate foundation systems, namely shallow and deep foundations. The foundation engineer will be able to supply the appropriate values for the design in order to provide the requisite safety while also providing a more cost-effective solution by integrating these two systems. As a result, a hybrid system known as "piled-raft foundation" has been developed. A number of researchers already documented study on behaviour of pile-raft foundation on different types of soils. By employing a model test and numerical analysis, in soft clay, Wu et al. [1] were able to estimate the load-settlement response of a combined piled-raft system. They discovered that sum of individual resistance capacity of pile-raft system is more than combined pile-raft system. Mrouehand Shahrour [2] investigated the impact of urban tunnel development on surrounding piling foundations. It was performed by utilising 3Delastoplastic FE modelling. Based on a numerical investigation employing a 3D finite element approach, Lee et al. [3] established three-dimensional behaviour of a piled-raft on soft clay. Using an elasto-plastic soil model, Peng et al. [4] developed a nonlinear 3D analysis. To model the pile interaction, they used an elastic pile material and interface element. Hyuang et al. [5] proposed a simplified nonlinear analysis for combined piled-raft system in stratified soil medium under vertical loading. They found that the suggested technique can forecast the load distribution of single piles, pile groups, and pile-rafts in stratified soils

K. Mahata · M. S. Rahaman · A. K. Bera (✉)
Research Scholar, Department of Civil Engineering, Indian Institute of Engineering Science and Technology, Howrah 711103, India
e-mail: ashis@civil.iiests.ac.in

 
B. T. Jose et al. (eds.), *Proceedings of the Indian Geotechnical Conference 2022 Volume 2*, Lecture Notes in Civil Engineering 477,
https://doi.org/10.1007/978-981-97-1741-5_3

reasonably well. Zhan et al. [6] developed a numerical modelling of vertical loading pile such as self-weight stress field, bearing resistance by ABAQUS which one is finite element-based software. They predicted that bearing capacity of foundation increases and differential settlement is decreased with combined effect of pile-raft. Nguyen et al. [7] suggested a design method of piled-raft foundation under larger external load considering interaction effects for pile-raft foundation for obtaining the ultimate load-carrying capacity of pile-raft and settlement of the foundation. Vertically loaded floated large pile group is explored using a multiphase approach by Bourgeois et al. [8], with special attention on soil–pile interactions, which carry an essential influence in foundation performance. To explore the behaviour of rafts on settlement decreasing piles, Garhy et al. [9] conducted an experimental programme on small-scale model stacked rafts on sandy soil. Lee et al. [10] investigated on the load-sharing behaviour of piled-rafts. They proposed a load-sharing model on the basis of settlement which behaves nonlinear response and depends on piled-raft interaction effect. Cho et al. [11] proposed a nonlinear 3D analytical method, numerical method, and case studies to examine the load deformation characteristics pile-raft foundations subjected to vertical and lateral load. They discovered that this method provides a significantly bigger piled-raft settlement than the linear elastic analysis results. Lee et al. [12] analysed the load-sharing behaviour of piled-raft embedded in granular backfill material using finite element method with the help of PLAXIS-3D software. Wulandari et al. [13] estimated the settlements unpiled raft foundation then piled foundation and finally piled-raft foundation under the same loading using PLAXIS-2D considering pile number as a variable. From the outcomes' result, they found that the settlement decreases in case of piled-raft foundation. Based on the review of the literature, objective of present work has been taken to carry out an interaction factor-based load-sharing study and non-linearload settlement characteristics of pile-raft foundation with the help of the finite element method.

## 2 Numerical Modelling

In general, a theoretical solution found by finite element method must satisfy equilibrium, compatibility, constitutive material and boundary conditions. To find a solution for a foundation problem with finite element method, certain idealisation and some assumptions have to be made. In particular, the soil behaviour must be defined as a material constitutive relationship, the geometry and boundary requirement for pile-raft analysis. The assumptions are considered in the present study are as a homogeneous, isotopic, linear elastic behaviour, dry condition of soil, to develop an FE model in ABAQUS environment for pile-raft embedded in homogenous cohesion less soil. Friction contact is used in interaction module. An eight-noded linear brick element is considered for this study.

## 3 Geometric Modelling

Numerical modelling and analysis have been carried out for pile-raft embedded in cohesion less soil with different pile diameters and raft size. Figure 1 shows the representational diagram of soil pile-raft system. For the purpose of analysis, it is required to choose a finite domain. The combined piled-raft foundation (CPRF) system geometry has been modelled as a 3D model in this study. Due to symmetry of the soil pile-raft, only quarter for FE simulation of the domain has been chosen, and this will also speed up the analysis approach. In this present numerical modelling and analysis, the diameter of pile has been taken as '$B_p$' and length of pile has taken as '*l*'. In FE simulation, the border must be shifted such that the boundary effect is minimal. In this study, boundary of the domain for the soil block has been shifted to a distance of 3*l* in vertical direction and *l in horizontal* direction from centre of pile-raft.

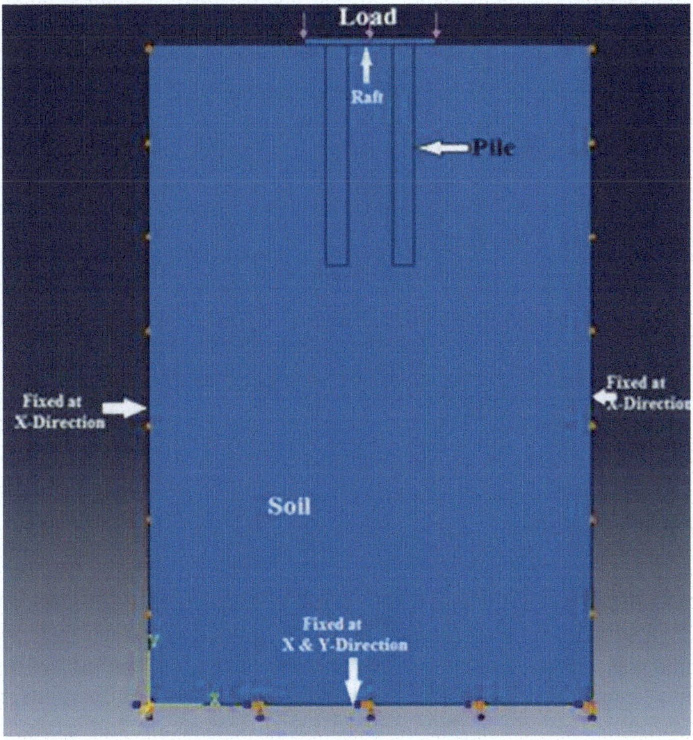


**Fig. 1** Representational diagram of soil pile-raft foundation system

**Table 1** Properties of sand

| Density index (%) | Void ratio (%) | Density (kg/m$^3$) | Young modulus (kPa) | Angle of internal friction (degree) |
|---|---|---|---|---|
| 30 | 69 | 1550 | 32,508 | 32 |
| 50 | 63 | 1610 | 37,108 | 35 |
| 70 | 57 | 1670 | 42,261 | 38 |

# 4 Material Properties

## *4.1 Material Properties of Sand*

The soils employed in the FE analyses were sands with three distinct relative densities, $D_R = 30$, 50, and 70%. The value of relative density has been chosen based on the earlier researcher [12]. The Mohr-Coulomb constitutive model is used for finite element analysis of these soils. In accordance with Bowles [14], the Poisson's ratio of sand is kept as 0.32 for entire analysis. The values of Young modulus, angle of internal friction, and density of the sand for respective relative density are presented in Table1.

## *4.2 Properties of Pile-Raft*

In the present investigation, the values of pile-raft materials, viz., unit weight (kN/m$^3$), E (kPa), and Poisson's ratio, have been considered as 25, $30 \times 10^6$, and 0.15, respectively. The value of properties of the pile-raft has been chosen based on the earlier researcher Lee et al. [12].

# 5 Mesh Generation

The geometry has been partitioned into a huge into a number of small elements in order to do finite element computation. Eight nodded brick elements in soil and pile-raft represent a three-dimensional finite element mesh. One hundred and fifteen numbers of elements have been generated on pile-raft and 8803 elements have been generated on soil. C3D8R is the standard for this element; hourglass control is employed with eight-node linear brick with reduced integration. An element size of 600 mm × 600 mm has been created in the case of pile-raft; an element size of 600 mm × 600 mm is adopted in loading area of soil, and minimum size of 600 mm and maximum size of 900 mm have been used in the case of rest of the soil. The deformed shape of model is shown in Fig. 2.

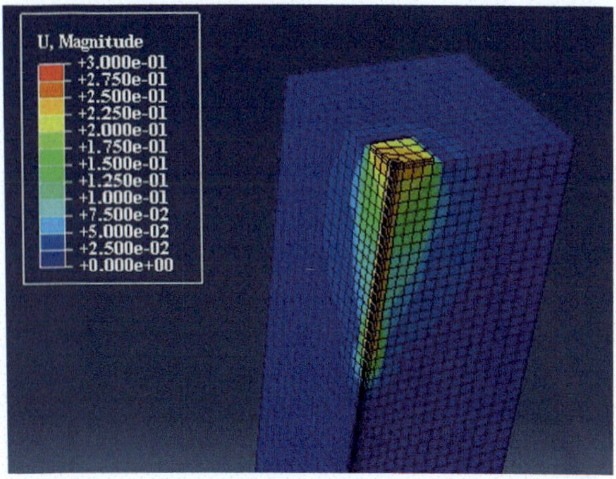


**Fig. 2** Deformed shape of pile-raft model

# 6 Result and Discussion

## *6.1 Load Variation of Raft, Pile, and Piled-Raft Foundation*

Figures 3, 4, and 5 show the variation of load versus displacement curve for raft, pile, and piled-raft system for different relative densities ($D_R$) of 30%, 50%, and 70%, respectively, with pile diameter 0.5 m and raft size 2 m × 2 m. Figures 6, 7, and 8 show the variation of load versus displacement of raft, pile, and combined piled-raft foundation (CPRF) for different relative densities ($D_R$) of 30%, 50%, and 70% with pile diameter 0.75 m and raft size 3 × 3 m, respectively. Figures 9, 10, and 11 show the variation of load versus displacement of raft, pile, and combined piled-raft foundation (CPRF) for different relative densities ($D_R$) of 30%, 50%, and 70%, respectively, for pile diameter 1.0 m and raft size 4 × 4 m. From Figs. 3, 4, and 5, it is found that with increase in relative density loose sand ($D_R = 30\%$) to medium dense sand ($D_R = 70\%$), the load-carrying capacity of pile group, raft, pile-raft increases irrespective of settlement under study. Similar trend is also observed in case of other figures (Figs. 6, 7, 8, 9, 10, and 11). Due to that, with increase in relative density as the result of an increase in shear strength.

## *6.2 Combined Pile-Raft Load Sharing*

The load receiving capabilities of pile-raft components are determined by the achieved settlement level and hence play a very crucial part in determining the structure serviceability requirement. It denotes symbolically as a $\alpha_{CPRF}$, defined as the proportion of pile load in pile-raft system to the total applied load, as follows:

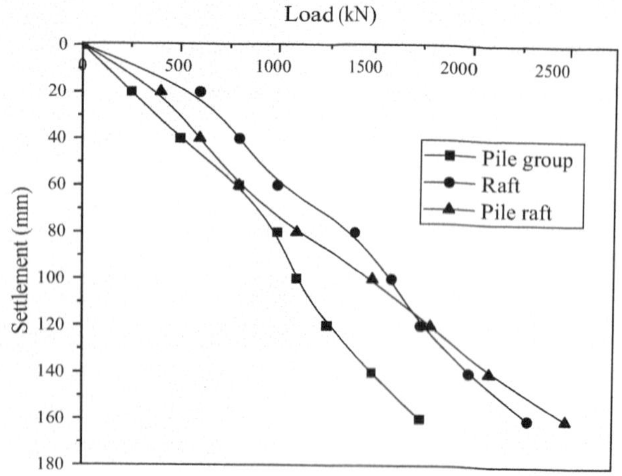


**Fig. 3** Load versus settlement curve for pile dia. 0.5 m and raft 2 × 2 m for $D_R = 30\%$

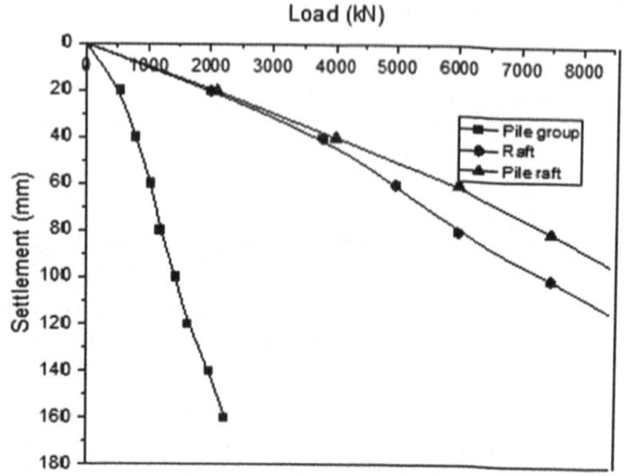


**Fig. 4** Load versus settlement curve for pile dia. 0.5 m and raft 2 × 2 m for $D_R = 50\%$

$$\alpha_{\mathrm{CPRF}} = \frac{\alpha_{\mathrm{pr}}}{\eta\left(1 + \frac{Q_{\mathrm{UR}}}{Q_{\mathrm{PG}}}\right)}.$$

Tables 2, 3, and 4 present the load-carrying capacity of pile/piles in combined, single, and group pile-raft for relative density 30%, 50%, and 70%, respectively, for measured value (ABAQUS value) as well as calculated value (calculation has been done from equation presented by Kumar and Choudhury [15]), where $\acute{\eta}$ and $\alpha_{\mathrm{pr}}$ are pile-raft efficiency and pile-to-raft interaction factor, respectively. From the tables (Tables 2, 3, and 4), it is observed that with increase of pile diameter from 0.5 to 1.0 m, the value of $\alpha$ CPRF increases with pile diameter and number of piles in group in group pile-raft system increases. It may be due to that with increase in

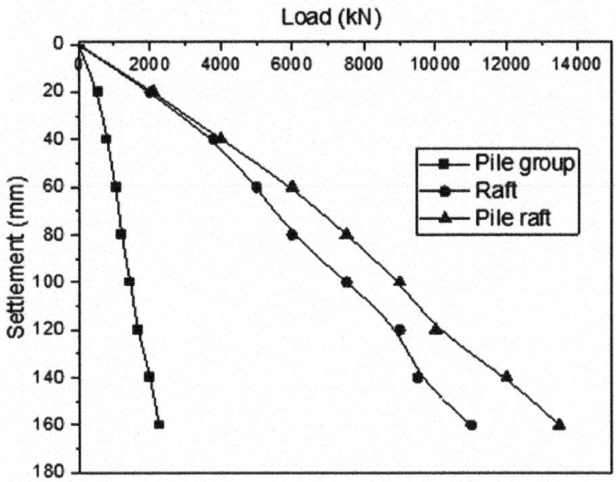


**Fig. 5** Load versus settlement curve for pile dia. 0.5 m and raft 2 × 2 m for $D_R = 70\%$

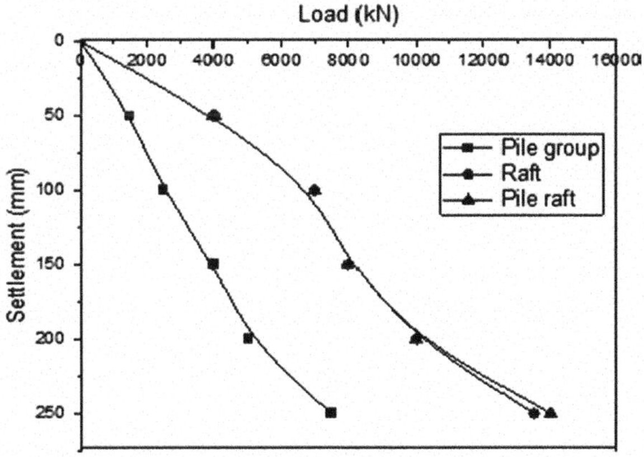


**Fig. 6** Load versus settlement curve for pile dia. 0.75 m and raft 3 × 3 m for $D_R = 30\%$

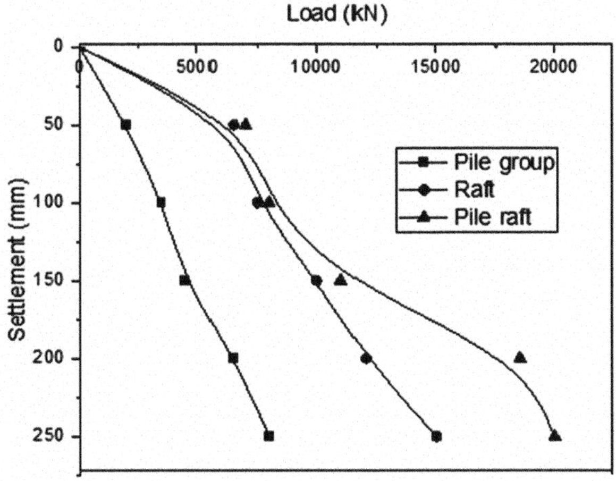


**Fig. 7** Load versus settlement curve for pile dia. 0.75 m and raft 3 × 3 m for $D_R = 50\%$

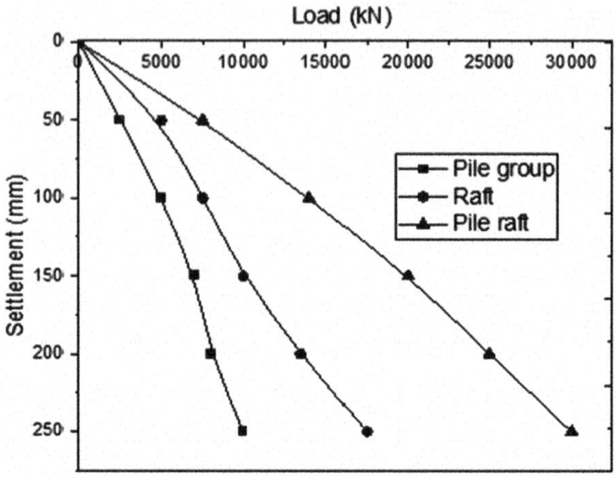


**Fig. 8** Load versus settlement curve for pile dia. 0.75 m and raft 3 × 3 m for $D_R = 70\%$

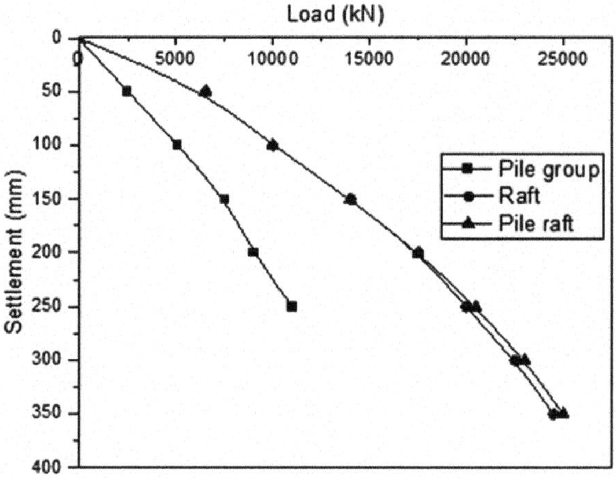


**Fig. 9** Load versus settlement curve for pile dia. 1.0 m and raft 4 × 4 m for $D_R = 30\%$

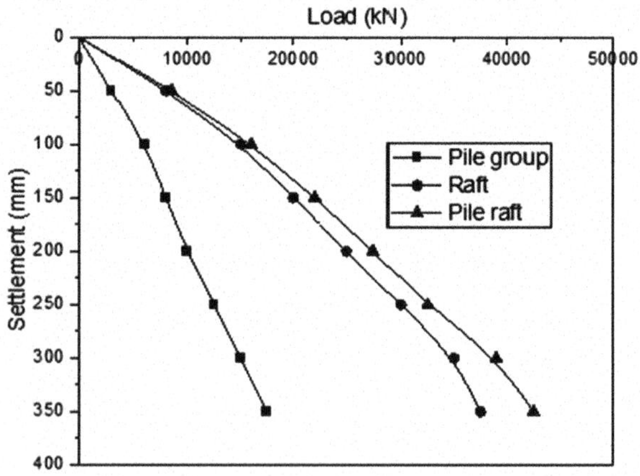


**Fig. 10** Load versus settlement curve for pile dia. 1.0 m and raft 4 × 4 m for $D_R = 50\%$

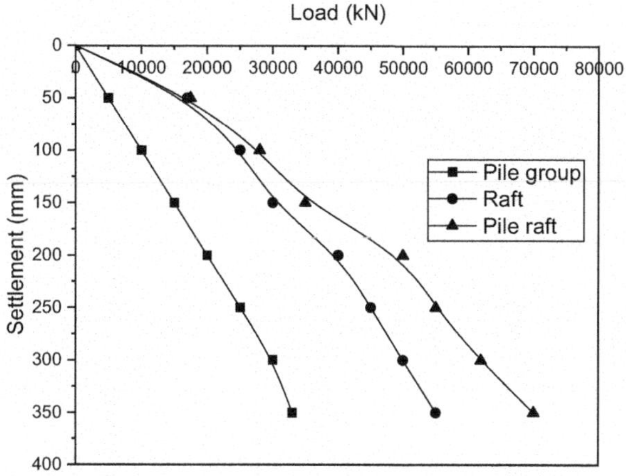


**Fig. 11** Load versus settlement curve for pile dia. 1.0 m and raft 4 × 4 m for $D_R = 70\%$

pile diameter and also number of pile in pile-raft system, the surface area of the pile increases.

## 7 Conclusions

From the above study, a finite element-based analysis has been worked out to explore the load-sharing behaviour of combined piled-raft system in homogeneous sand with different relative densities. This study focuses on the impact of pile-raft interaction factor on the load-sharing behaviour.

The following conclusion can be derived from above studies, findings.

(1) The analysis shows that with increase of density index, the ultimate load of pile, raft, and piled-raft also increases.
(2) The ultimate load-bearing capability is increased with increasing size of raft.
(3) More considerable load capacity results in a higher load-sharing ratio in piles in a combination piled-raft foundation.
(4) The values of $\alpha_{CPRF}$ increases by increasing number of piles and increasing diameter of pile as well.

**Table 2** Load-carrying capacity of pile/piles in combined single and group pile-raft for relative density 30%

$D_R = 30\%$

| Model | Dia. (m) | $\alpha_{pr}$ | $\acute{\eta}$ | $Q_{ult}$ (kN) (measured raft) | $Q_{theo,u}$ (kN) (calculated raft) | $Q_{PG}$ (kN) (measured) | $Q_{PG}$ (kN) (calculated) | $\alpha_{CPRF}$ (measured) | $\alpha_{CPRF}$ (calculated) |
|---|---|---|---|---|---|---|---|---|---|
| Single pile-raft (2 × 2) | 0.5 | 0.99 | 1.245 | 2040 | 1619.44 | 640 | 565.29 | 0.190 | 0.206 |
| Single pile-raft (3 × 3) | 0.75 | 0.99 | 1.12 | 7360 | 5465.61 | 2640 | 2310.15 | 0.233 | 0.263 |
| Single pile-raft (4 × 4) | 1 | 0.99 | 0.96 | 15,440 | 12,955.52 | 4640 | 4274.31 | 0.238 | 0.256 |
| Group pile-raft (2 × 2) | 0.5 | 0.99 | 1.245 | 2040 | 1619.44 | 4400 | 4610.74 | 0.543 | 0.588 |
| Group pile-raft (3 × 3) | 0.75 | 0.99 | 1.12 | 7360 | 5465.61 | 20,600 | 20,837.94 | 0.651 | 0.700 |
| Group pile-raft (4 × 4) | 1 | 0.99 | 0.96 | 15,440 | 12,955.52 | 44,000 | 48,705.79 | 0.763 | 0.815 |

**Table 3** Load-carrying capacity of pile/piles in combined single and group pile-raft for relative density 50%

| $D_R = 50\%$ | | | | | | | | | |
|---|---|---|---|---|---|---|---|---|---|
| Model | Dia. (m) | $\alpha$pr | $\acute{\eta}$ | $Q_{ult}$ (kN) (measured raft) | $Q_{theo,u}$ (kN) (calculated raft) | $Q_{PG}$ (kN) (measured) | $Q_{PG}$ (kN) (calculated) | $\alpha_{CPRF}$ (measured) | $\alpha_{CPRF}$ (calculated) |
| Single pile-raft (2 × 2) | 0.5 | 0.99 | 1.245 | 2680 | 2474.51 | 1240 | 1251.54 | 0.252 | 0.267 |
| Single pile-raft (3 × 3) | 0.75 | 0.99 | 1.12 | 8760 | 8351.46 | 4360 | 3280.29 | 0.294 | 0.249 |
| Single pile-raft (4 × 4) | 1 | 0.99 | 0.96 | 19,780 | 19,796.04 | 6640 | 5377.33 | 0.259 | 0.220 |
| Group pile-raft (2 × 2) | 0.5 | 0.99 | 1.245 | 2680 | 2474.51 | 8800 | 10,438.12 | 0.610 | 0.643 |
| Group pile-raft (3 × 3) | 0.75 | 0.99 | 1.12 | 8760 | 8351.46 | 28,000 | 29,791.6 | 0.673 | 0.690 |
| Group pile-raft (4 × 4) | 1 | 0.99 | 0.96 | 19,780 | 19,796.04 | 64,000 | 68,995.02 | 0.788 | 0.801 |

**Table 4** Load-carrying capacity of pile/piles in combined single and group pile-raft for relative density 70%

$D_R = 70\%$

| Model | Dia. (m) | $\alpha$pr | $\acute{\eta}$ | $Q_{ult}$ (kN) (measured raft) | $Q_{theo,u}$ (kN) (calculated raft) | $Q_{PG}$ (kN) (measured) | $Q_{PG}$ (kN) (calculated) | $\alpha_{CPRF}$ (measured) | $\alpha_{CPRF}$ (calculated) |
|---|---|---|---|---|---|---|---|---|---|
| Single pile-raft (2 × 2) | 0.5 | 0.99 | 1.245 | 3960 | 4534.92 | 1620 | 2520.07 | 0.231 | 0.284 |
| Single pile-raft (3 × 3) | 0.75 | 0.99 | 1.12 | 13,160 | 15,305.35 | 5960 | 5018.18 | 0.276 | 0.218 |
| Single pile-raft (4 × 4) | 1 | 0.99 | 0.96 | 33,600 | 36,279.35 | 11,160 | 8340.79 | 0.257 | 0.193 |
| Group pile-raft (2 × 2) | 0.5 | 0.99 | 1.245 | 3960 | 4534.92 | 14,800 | 15,260.9 | 0.672 | 0.613 |
| Group pile-raft (3 × 3) | 0.75 | 0.99 | 1.12 | 13,160 | 15,305.35 | 44,000 | 47,521.4 | 0.680 | 0.669 |
| Group pile-raft (4 × 4) | 1 | 0.99 | 0.96 | 33,600 | 36,279.35 | 96,000 | 111,557.4 | 0.764 | 0.778 |

## References

1. Wu WJ, Chai JC, Huang JZ (2002) Interaction between pile and raft in piled raft foundation. Adv Build Technol I:603–610
2. Mroueh H, Shahrour I (2002) Three-dimensional finite element analysis of the interaction between tunnelling and pile foundations. Int J Numer Anal Meth Geomech 26:217–230
3. Lee JH, Kim Y, Jeong S (2010) Three-dimensional analysis of bearing behavior of piled raft on soft clay. Comput Geotech 37(1–2):103–114
4. Peng JR, Rouainia Mand Clarke BG (2010) Finite element analysis of laterally loaded fin piles. Comput Struct88:1239–1247
5. Hyung JL, Young K, Sangseom J (2010) Three-dimensional analysis of bearing behaviour of piled raft on soft clay. Comput Geotech 37:103–114
6. Zhan Y, Hong W, Fu-Chen L (2012) Modelling vertical bearing capacity of pile foundation by using ABAQUS.EJGE, Bund L 17
7. Nguyen DDC, Jo SB, Kim DS (2013) Design method of piled-raft foundations under vertical load considering interaction effects. Comput Geotech 47:16–27
8. Bourgeois E, Buhan P, Hassen G (2012) Settlement analysis of piled-raft foundations by means of a multiphase model accounting for soil-pile interactions. Comput Geotech 46:26–38
9. Garhy EB, Galil A, Youssef FA, Mohamed AR (2013) Behaviour of raft on settlement reducing piles: experimental model stud. J Rock Mech Geotech Eng 5:389–399
10. Lee J, Donggyu P, Kyujin C (2014) Analysis of load sharing behaviour for piled rafts using normalized load response model. Comput Geotech 57:65–74
11. Jeong S, Cho J (2014) Proposed nonlinear 3-D analytical method for piled raft foundations. Comput Geotech 59:112–126
12. Lee J, Daesung P, Donggyu P, Keunbo P (2015) Estimation of load-sharing ratios for piled rafts in sands that includes interaction effects. Comput Geotech 63:306–314
13. Wulandari P, Tjandra D (2015) Analysis of piled raft foundation on soft soil using PLAXIS 2D. In: The 5th international conference of euro Asia civil engineering forum, pp 363–367
14. Bowles EJ (2005) Foundation analysis and design, 5th edn. The McGraw-Hill, Singapore
15. Kumar A, Choudhury D (2018) Development of new prediction model for capacity of combined pile-raft foundation. Comput Geotech 97:62–68

# Undrained Bearing Pressure of Desiccated Non-homogeneous Ground Accounting for Compressibility

**J. Y. V. Shiva Bhushan, G. V. Narasimha Reddy, and Madhira R. Madhav**

## 1 Introduction

Normally consolidated alluvial or marine soils, when exposed to atmosphere, are subjected to seasonal changes and form a crust or desiccated layer because of diurnal heating and cooling, lowering of water table, etc. [1]. Desiccated normally consolidated soils exhibit complex variations of strength with depth. Undrained shear strength of normally consolidated soil varies linearly with depth for a young normally consolidated deposit. With time, the deposit may gain in shear strength due to aging. Solutions for bearing capacity of strip footing resting on non-homogeneous aged clays with shear strength profile linearly increasing with depth given by Davis and Booker [2] are

$$q_{uf} = F_R\left[c_0 N_c + \frac{\rho B}{4}\right], \tag{1}$$

where $N_c = 5.14$; $\rho$—the rate of increase of shear strength with depth, $c_o$—undrained shear strength at ground level, $B$—width of the footing. $F_R = f(\rho B/c_o)$—correction factor for roughness of the footing [2]. Shiva et al. [3] derived an equation for estimating the bearing capacity of footing resting on desiccated NC soil considering strength variation with depth. Desiccated normally consolidated soil (DNC) properties from few different locations are listed in Table 1 [4–6]. The thickness of the desiccated layer ranges from 1.5 m to 3.7 m, the rate of increase of undrained strength

J. Y. V. Shiva Bhushan (✉) · G. V. Narasimha Reddy · M. R. Madhav
Department of Civil Engineering, JNT University, Hyderabad 500085, India
e-mail: jyvshivabhushan@gmail.com

G. V. Narasimha Reddy
e-mail: gvnreddy@jntuh.ac.in

 
B. T. Jose et al. (eds.), *Proceedings of the Indian Geotechnical Conference 2022 Volume 2*, Lecture Notes in Civil Engineering 477,
https://doi.org/10.1007/978-981-97-1741-5_4

**Table 1** Properties of desiccated normally consolidated ground

| Parameter | Vankleek silo | Richmond silo | Bergado 2002 | Bergado 2012 |
|---|---|---|---|---|
| Unit weight of soil, $\gamma$ (kN/m$^3$) | 15 | 15 | 15 | 17 |
| Thickness of desiccated zone, $h_0$ (m) | 3.7 | 2.4 | 2.3 | 1.5 |
| Rate of increase of shear strength, $\rho$ (kPa/m) | 1.35 | 1 | 4 | 4.5 |
| Rate of decrease of shear strength, $\kappa$ (kPa/m) | 22.7 | 12.5 | 13.08 | 18.03 |
| Undrained strength at the interface of desiccated and NC soil, $c_{u0}$ (kPa) | 16 | 30 | 16 | 16 |

in NC layer from 1.0 to 4.5, the rate of decrease of undrained strength in the desiccated zone from 23 to 12, and the undrained strength at the top on NC layer from 16 to 30 kPa, indicating large ranges in nature.

Most of the bearing capacity theories consider the soil to be homogeneous and incompressible. Shear modulus of clays also has a significant effect on the undrained bearing pressure [7–9]. Rowe and Booker [10] proposed influence charts for estimation of settlement of circular footings resting on non-homogeneous elastic soil with a crust. The effect of crust depth and stiffness, as well as underlying non-homogenous soil stiffness on settlement of rigid footing, was reported. Based on the range of values from actual shear strength variations, a parametric study was carried out on the effect of desiccation on undrained bearing pressure of soft ground.

## 2 Problem Description

Rigid circular footing (diameter, D) located on desiccated layer of thickness, $h_O$, overlying normally consolidated soil was considered (Fig. 1). Desiccated normally consolidated (DNC) ground was considered as a two-layer soil whose undrained strengths vary linearly with depth, strength decreasing and increasing with depth in the two layers, respectively. The rates of decrease in the top desiccated layer and increase in the lower NC soil layer of shear strengths with depth are denoted by $\kappa$ and $\rho$, respectively. The shear strength of the soil at the interface of the desiccated layer and NC soil was $c_{u0}$.

Desiccated layer of thickness, $h_0$, with undrained strength $c_{uT}$, at the base of footing deceasing to $c_{u0}$, at the interface with the NC soil was considered (Fig. 1). Poisson's ratio of desiccated normally consolidated ground was taken as 0.495 to simulate the undrained behavior. Rigidity index, $G/c_u$, was assumed constant for both the layers by increasing the shear stiffness, $G$, at the same rate as that of undrained strength with depth. This study considers desiccation layer over NC soil as a two-layered system with the non-homogeneity and compressibility of the ground, in addition to the effect of desiccation on the strength profile.

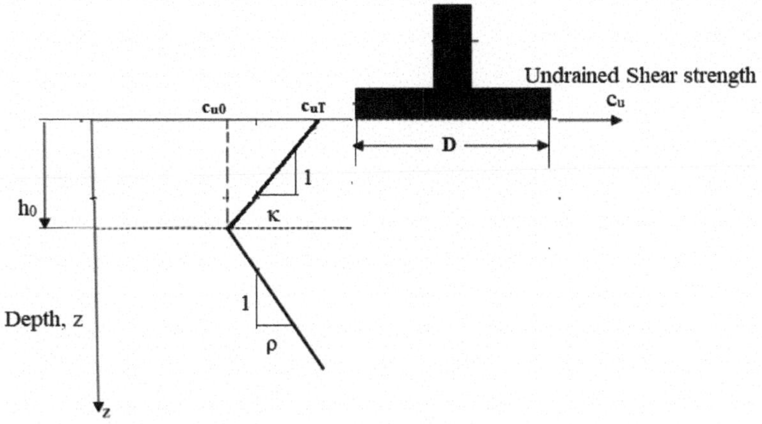


**Fig. 1** Sketch describing the problem statement with parameters

## 3 Finite Element Model

Finite element software—PLAXIS 2D, 2020—was used to analyze the problem. A circular footing of 5 m diameter was considered. A vertical displacement was prescribed and the displacement–stress response was obtained. Fifteen-noded triangular elements were used to discretize the geometry. The bottom boundary was fixed in both radial and vertical directions, while the side boundaries were fixed in the radial direction and the top boundary was free. The boundaries were fixed at 6 times the diameter in vertical and horizontal directions. Mesh convergence was checked and the mesh size was fixed as very fine.

The analysis was carried out in two stages—initially, the effect of desiccation on NC soil was studied considering the compressibility and later the heterogeneity of the ground is accounted for. The study was limited to a vertical displacement of 25 cm, i.e., 5% of diameter of the footing. Numerical model was validated with available literature for incompressible homogeneous ground. Parametric study was carried out on the results obtained from numerical modeling.

## 4 Results

The non-homogeneity (strength variation with depth) of DNC soil was considered alongside the desiccation effect in the top portion of the clay (Fig. 1). Obtained bearing pressure for a prescribed settlement was normalized with undrained strength at the interface and was defined as bearing pressure factor, $N_{cf}$ $(= q/c_{u0})$.

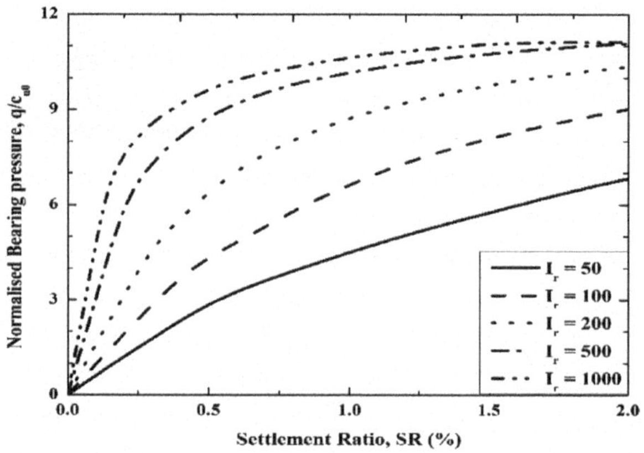


**Fig. 2** $N_{cf}$ versus SR curves for varying rigidity indices, $I_r$

## Effect of Compressibility on Bearing Pressure of Desiccated NC Ground

Rigidity index, $G/c_u$, was assumed constant for the two layers by increasing the shear stiffness, $G$, at the same rate as that of undrained strength. Rate of decrease in strength with depth, due to desiccation within top few meters, $\kappa$, was taken as 10 kPa/m. Degree of non-homogeneity in desiccated portion, $\kappa D/c_{u0}$, was obtained as 5. Rigidity index of desiccated layer, $I_r'$, was assumed as 200.

Rate of increase in strength with depth in NC soil, $\rho$, was taken as 5 kPa/m. The degree of non-homogeneity, $\rho D/c_{u0}$, becomes 2.5. Rigidity indices, $I_r$, of NC soil considered are 50 (soft) and 100, 200, 500 and 1000 (rigid). Figure 2 presents the variation of bearing pressure factor with settlement ratio for different rigidity indices. $N_{cf}$ increases with the rigidity of NC ground. $N_{cf}$ for $I_r = 500$ soil increased 2.25 times that for $I_r = 50$ at SR $= 1\%$.

## Effect of Non-homogeneity of NC Soil

The rate of increase in strength with depth, $\rho$, of NC soil was increased in steps from 0 through 5, keeping the undrained strength of 10 kPa at the top. The degree of non-homogeneity values is 0, 1, and 2.5. Assuming all the other parameters as constant, the effect of $\rho D/c_{u0}$ was evaluated in this section.

Figure 3 shows the variation of bearing pressure factor with settlement ratio for different $\rho D/c_{u0}$. $N_{cf}$ is nearly the same for settlement ratios, *SR*, less than about 0.6% for different $\rho D/c_{u0.}$ $N_{cf}$ increases with $\rho D/c_{u0}$ for settlement ratios greater than about 0.6%. $N_{cf}$ increased by 17% for $\rho$ increasing from 0 to 5 kPa/m at SR $=$ 2%.

## Effect of Non-homogeneity of Desiccation Zone

Degree of non-homogeneity of NC ground, $\rho D/c_{u0}$, was kept constant at 5. Thickness of desiccation layer, $h_0$, was taken as 2 m. The rate of decrease in strength, $k$, in the desiccated layer is varied from 0 through 10 kPa/m. Here $\kappa = 0$ represents a crust with strength constant with depth in the desiccation zone. Rigidity indices of both

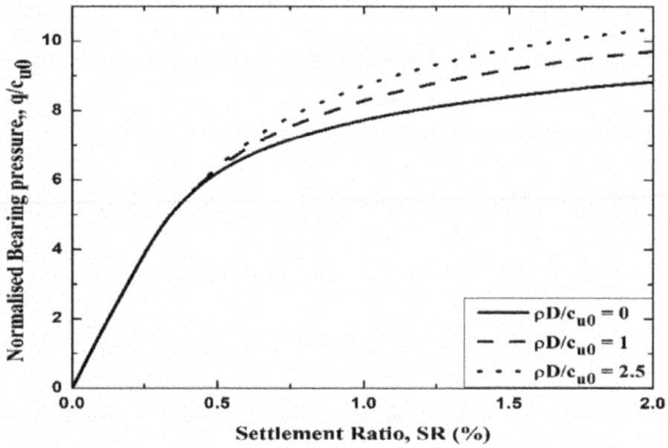


**Fig. 3** $N_{cf}$ versus SR curves for varying degrees of non-homogeneity, $\rho D/c_{u0}$

the layers are 200. Strength and stiffness decrease gradually with a rate of $\kappa$ till the interface of desiccated and NC layers.

For $\rho D/c_{u0} = 2.5$, normalized bearing pressure–settlement ratio curves obtained for various $\kappa$ are depicted in Fig. 4. Figure 4a–c show the curves obtained for DNC ground with thickness of desiccation layer, $h_0 = 1.0$ m, 1.5 m, and 2.0 m, respectively. For a given thickness of desiccation layer, it is observed that DNC ground with higher rates of decrease in strength in desiccation zone resulted in higher bearing capacity. This is because of higher average undrained strength in the top desiccated zone, with a constant undrained strength of 10 kPa maintained at the interface of desiccation and NC layers.

## Effect of Thickness on NBC

Normalized bearing capacity, NBC ($= q_u/c_{u0}$), of circular footing resting on DNC ground is plotted against the degree of non-homogeneity parameter in desiccation layer, $\kappa D/c_{u0}$, for different thicknesses, $h_0$, and presented in Fig. 5. It is seen that NBC increases linearly with increase in $\kappa D/c_{u0}$ for a given thickness of desiccation layer.

NBC increases 1.82 times for $\kappa$ increasing by 5 times (strong and stiff desiccation layer) compared to an increase of 1.45 times for $\kappa D/c_{u0}$ of 2.5. The percentage of incremental variation in NBC is higher for higher $\kappa D/c_{u0}$. It is evident that the desiccated layer's degree of non-homogeneity and thickness correspond to the layer's strength in magnitude and depth of extent, respectively. As a result, the influence of desiccated layer thickness on undrained bearing capacity increases as the degree of non-homogeneity in the desiccation zone increases.

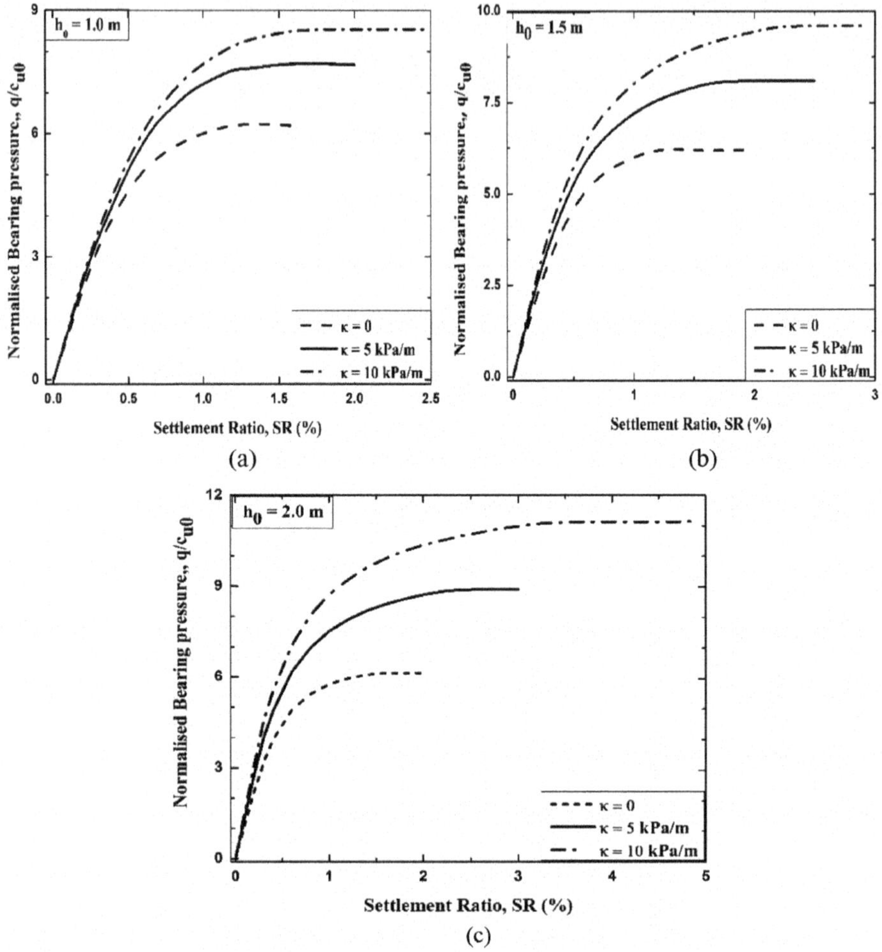


**Fig. 4** $N_{cf}$ versus SR (%): effect of $\kappa$ for **a** $h_0 = 1.0$ m, **b** $h_0 = 1.5$ m, **c** $h_0 = 2.0$ m

## 5 Conclusions

Effect of desiccation on bearing pressure of footing resting on DNC soil considering the compressibility and non-homogeneity of the ground is studied. The present study quantifies the individual effects of compressibility, non-homogeneity, and thickness of desiccation layer on bearing pressure factors. Bearing pressure factor increases with increase in rigidity index and degree of non-homogeneity for a specific settlement. Undrained bearing capacity increases with increase in degree of non-homogeneity for a given thickness of desiccation layer and its effect is more at higher values of non-homogeneity parameter in desiccation.

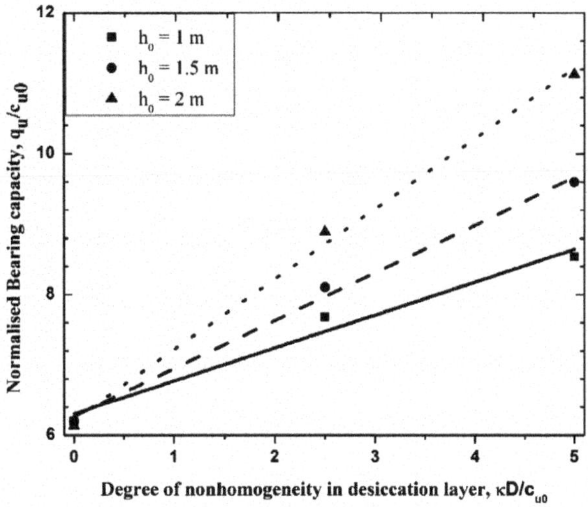


**Fig. 5** NBC versus $\kappa D/c_{u0}$: effect of $h_0$

# References


1. Day RW (1996) Desiccation theory for soft cohesive soils. J Geotech Eng 122:943–947. https://doi.org/10.1017/CBO9781107415324.004
2. Davis EH, Booker JR (1973) The effect of increasing strength with depth on the bearing capacity of clays. Geotechnique 23:551–563. https://doi.org/10.1680/geot.1973.23.4.551
3. Shiva Bhushan JYV, Mohammed AA, Madhav M (2021) Re-evaluation of failure of silo tower foundations. In: Satyanarayana Reddy CNV, Muthukkumaran K, Satyam N, Vaidya R (eds) Ground characterization and foundations, lecture notes in civil engineering, vol 167. Springer, Singapore, pp 831–843. https://doi.org/10.1007/978-981-16-3383-6_72
4. Bozozuk M (1972) Foundation failure of the vanleek hill tower silo. In: Proceedings of the ASCE specialty conference on performance of earth and earth-supported structures, vol 1(2). Purdue University, pp 885–902
5. Bozozuk M (1974) Bearing capacity of clays for tower silos. Can Agric Eng 16:13–17
6. Bergado DT, Lorenzo GA, Long PV (2002) Limit equilibrium method back analyses of geotextile-reinforced embankments on soft bangkok clay—a case study. Geosynth Int 9:217–245. https://doi.org/10.1680/gein.9.0216
7. Shiva Bhushan JYV, Madhav M, Narasimha Reddy GV (2022) Effect of compressibility on bearing pressure of soft ground. In: Muthukummaran K, Reddy CNV, Joseph A, Senthamilkumar S (eds) Foundation and forensic geotechnical engineering, lecture notes in civil engineering, vol 295. Springer, Singapore, pp 75–80. https://doi.org/10.1007/978-981-19-6359-9_8
8. McMahon BT, Haigh SK, Bolton MD (2013) Cavity expansion model for the bearing capacity and settlement of circular shallow foundations on clay. Geotechnique 63:746–752. https://doi.org/10.1680/geot.12.P.061
9. McMahon BT, Haigh SK, Bolton MD (2014) Bearing capacity and settlement of circular shallow foundations using a nonlinear constitutive relationship. Can Geotech J 51:995–1003. https://doi.org/10.1139/cgj-2013-0275
10. Rowe RK, Booker JR (1981) The behaviour of footings resting on a non-homogeneous soil mass with a crust. Part II. Circular footings. Can Geotech J 18:265–279. https://doi.org/10.1139/t81-029

# Pile Termination Criteria for Rock Socketed Piles in Chennai Metro Project

**R. Rajaraman**

## 1 Introduction

In all metro construction projects of India, large diameter bored piles are employed, with sizes varying from 1000 and 1500 mm. These piles must be designed for heavy loads. Hence, socketing into the rock is essential and unavoidable. The termination of piles into the rock is not an easy exercise at the field due to degree of uncertainties involved in determination of grade of rock and suitable parameters at the field. This paper tries to develop a methodology for determining the pile termination criteria for socketed piles with due consideration to the safety of the structure.

## 2 Geotechnical Parameters

Determination of parameters of soil and rock are still a tough engineering platform where the value of judgement is highly precious. It has also been observed quite widely in metro projects that grade of rock estimated during pre-construction stages is inconsistent with grade of rock evaluated during construction stage. Hence, a flexible and site friendly tool is required to deal with these uncertainties prevailing at estimation of geotechnical parameters.

R. Rajaraman (✉)
SMEC Rail India Pvt. Ltd., Chennai, India
e-mail: rajaraman.ramanathan@gmail.com



B. T. Jose et al. (eds.), *Proceedings of the Indian Geotechnical Conference 2022 Volume 2*, Lecture Notes in Civil Engineering 477,
https://doi.org/10.1007/978-981-97-1741-5_5

## 3 Pile Termination Criteria

Hydraulic rotary rigs make penetration of piles faster in all types of rocks including hard rocks. At field, decision on pile termination is a challenging task as various inputs required for deciding the pile termination are qualitative and subjective in nature. There is a limiting value of resistance of chisel, based on pile termination decision cannot be arrived at. IS 2911 Part 1-Section 2 can be utilized for determining pile termination criteria at field based on some simple and easy tests at field. Pile termination criteria has been used as a quality control tool in IRC 78. There are various approaches to determine the pile termination. They are elaborated in the following sections.

### *3.1 IS 2911 Part 1-Section 2 Method*

$$Q_a = c_{u1} N_c \frac{\pi B^2}{4F_s} + \alpha c_{u2} \frac{\pi BL}{F_s}$$

where

$c_{u1}$ Shear strength of rock at pile base, in kN/m$^2$.
$N_c$ Bearing capacity factor = 9.
$F_s$ Factor of safety = 3.
$\alpha$ 0.9.
$c_{u2}$ Average shear strength of rock in the socketed length, in kN/m$^2$.
$B$ Minimum width/diameter of pile shaft, in m.
$L$ Socket length of pile, in m.

This method is applicable when SPT value is > 60.

This method is quite empirical in nature. Estimation of shear strength plays a crucial role in determining pile capacity. Shear strength has been recommended in accordance with SPT values, which could not be conducted to good quality rock mass and would not be performed at field during piling. Ideally, best way to get shear strength of rock mass at site is by doing breakability and scratch tests (Table 1).

### *3.2 IRC 78 Method*

There are two approaches given in IRC 78. Method 1 is suggested when (CR + RQD)/2 < 30 and Method 2 is recommended when (CR + RQD)/2 > 30, where CR is Core Recovery in % and RQD is Rock Quality Designation in %.

**Table 1** Shear strength of rock for various SPT values as per IS 2911 Part 1 Section 2 [1]

| Value of *N* | Shear strength kN/m² | Strength | Grade | Breakability | Penetration | Scratch |
|---|---|---|---|---|---|---|
| 600 | 400 | Strong | A | Difficult to break against solid object with hammer | | Cannot be scratched with knife |
| | 200 | | | | | |
| | 100 | Moderately Strong | B | Broken against solid object with hammer | | Can be just scratched by thumb nail |
| 400 | 80 | | | | | |
| | 60 | | C | Broken in hand by hitting with hammer | | Can be just scratched by thumb nail |
| | 40 | Moderately weak | D | Broken by leaning on sample with hammer | No penetration with knife | |
| 200 | 20 | | | | | |
| | 10 | Weak | E | Broken by hand | 2 mm with knife | |
| | 8 | Hard or very weak | F | Easily broken by hand | 5 mm with knife | |
| 100<br>80 | 6 | | | | | |
| 60 | 4 | | | | | |

*Method 1*

The ultimate load carrying capacity shall be estimated by using the following equation:

$$Q_u = R_e + R_{af} = K_{\text{sp}} q_c d_f A_b + A_s C_{\text{us}}$$

where

$Q_u$ Ultimate capacity of pile socketed into rock.
$R_e$ Ultimate end bearing capacity of pile.
$R_f$ Ultimate side socket shear resistance.
$K_{\text{sp}}$ An empirical co-efficient, to be referred from the Table 2.
$q_c$ Average unconfined compressive strength of rock core below base of pile for the depth twice the diameter/least lateral dimension of pile in MPa.
$d_f$ Depth Factor $= 1 + 0.4 \frac{\text{Length of socket}}{\text{Diameter of socket}}$.

Maximum value of $d_f$ should be 1.2.

$A_b$ Cross-sectional area of base of pile.
$A_s$ Surface area of socket.
$c_{\text{us}}$ Ultimate shear strength of rock along socket length $= 0.225\sqrt{q_c}$.

Maximum value of $c_{\text{us}}$ shall be taken as 3.0 MPa for M 35 concrete. For other grades of concrete, this shall be modified by a factor $\sqrt{(fck/35)}$ (Table 2).

*Method 2*

This method is applicable when cores and/or core testing results are not available, or when geo-material is highly fragmented. The shear strength of geo-material is obtained from its correlation with extrapolated SPT values for 300 mm of penetration as presented in Table 3.

$$Q_u = R_e + R_{af} = c_{ub}N_cA_b + A_sC_{us}$$

$c_{\text{ub}} =$ Average shear strength below base of pile, for the depth equal to twice the diameter/least lateral dimension of pile, based on average '$N$' value of this region $N_c = 9$.

In method 2, $c_{\text{us}}$ is ultimate shear strength along socket length, to be obtained based on average '$N$' value of socket portion. This also shall be restricted to shear capacity of concrete of the pile, to be taken as 3.0 MPa for M 35 concrete in confined condition, which for other strengths of concrete can be modified by a factor $\sqrt{(fck/35)}$.

Allowable capacity on the pile in both methods is given below:

$$Q_a = R_e/3 + R_{af}/6$$

**Table 2** Values of empirical co-efficient, $K_{\text{sp}}$ [1]

| S. No. | (CR + RQD)/2 (%) | $K_{\text{sp}}$ |
|---|---|---|
| 1 | 30 | 0.3 |
| 2 | 100 | 1.2 |

**Table 3** Shear strength of rock for various SPT values [1]

| Shear strength/Consistency | Moderately weak | Weak | Very weak |
|---|---|---|---|
| $N$ value | 300–200 | 200–100 | 100–60 |
| Shear strength/Cohesion in MPa | 3.3–1.9 | 1.9–1.7 | 0.7–0.4 |

## 3.3 IS 14593 Method

Depending upon the type of rock, any one of the following methods shall be used for computing compression capacity:

a. Based on uniaxial compressive strength of rock.
b. Based on limit pressure of rock.
c. Based on shear strength of rock.
d. Structural strength of pile.

*Based on uniaxial compressive strength of rock*

Where the rock is sound, the strength of the foundation rock is generally much in excess of the design requirements, provided the walls of the discontinuity are closed and they are favourably oriented.

$$Q_s = q_c N_j N_d A_p + q_c \pi D l_s A_s \alpha \beta$$

$Q_s$ Safe Load Capacity of Pile
$N_j$ Empirical co-efficient depending on the spacing of discontinues $= \frac{3+S/B}{10\sqrt{(1+300\frac{\delta}{S})}}$
$S$ Spacing of discontinuities.
$B$ Pile diameter.
$\delta$ Thickness of discontinuities.
$N_d$ Depth Factor $= 0.8 + 0.2\ l_s/D, \leq 2$.
$l_s$ Socket length into rock.
$\alpha$ Rock socket slide resistance reduction factor.
$\beta$ Rock socket correction factor.

*Based on limit pressure of rock*

This method is suited for weathered or closely jointed rocks and for soft rocks.

$$Q_u = [P_O + K_b(P_1 - P_0)]A_p + \pi f_1 D l_s$$

$P_0$ Overburden pressure at the elevation of pile tip.
$K_b$ Bearing capacity co-efficient depending on socket length.
$P_1$ Limit pressure determined from pressure metre test in the zone extending two times the pile diameter above and below pile toe (top).
$f_1$ Frictional resistance component.

*Based on shear strength of rock*

$$Q_a = c_{u1} N_c \frac{\pi B^2}{4F_s} + \alpha c_{u2} \frac{\pi B L}{F_s}$$

This equation is same as 3.1. In this code, factor of safety recommended for getting allowable capacity is 6. This method is applicable for siltstone, mudstone, and weathered sandstone. The partially mobilized frictional resistance offered by the soil may be neglected.

*Structural capacity*

The piles shall have necessary structural strength to transmit loads imposed on it, ultimately to the rock as per IS 2911 (Part l/Set 2).

### *3.4 Pile Penetration Rate Method*

The concept of pile penetration ratio (PPR) can be utilized in such case. The pile penetration ratio (PPR) reflects the energy in ton-metre required to advance the pile bore of one sq. metre cross-sectional area by 1 cm.

Energy E spent for N blows = 63.5 kg × 75 cm × N blows (in kg—cm units) = E × $10^{-5}$ tonne metre.

Area of samples is 0.758 × $(5.2)^2$ $cm^2$ = 21.24 $cm^2$, penetrating 30 cm.

Hence PPR = 63.5 × 75 × $N$ × $10^{-5}$/(21.24 × 1 0^ × 30) = 0.747 $N$.

PPR for $N = 50 = 37.35$ tm/$m^2$/cm and for $N = 200 = 149.4$ tm/$m^2$/cm.

From $N$ values, pile capacity can be back calculated using IS 2911 Part Section 2 method/IRC Method 2.

## 4 Comparison

In this section, a typical case study from a stretch of Chennai metro project is used for comparing the results obtained from various methods (Table 4).

Strength/Consistency-Weak-Moderately Weak

Pile Diameter = 1.2 m

UCS of intact rock = 50 MPa

It is clear from the above table that the pile capacity calculated from various methods is not matching with each other. Total pile capacity calculated using IRC method is on the lower side and shall be conservatively considered for design.

But during pile drilling, getting the values like RQD, CR, UCS is practically difficult considering huge numbers of piles involved for a metro project.

It is very difficult to penetrate through rock stratum and even penetration through 10 cm may take around 3 days sometimes. In such cases, a method of pile termination criteria based on penetration rate is recommended for speedy and safe execution of piles in such a large scale.

The input parameters as required for estimating the pile capacity are very difficult to be attained. The conservative estimation always causes delay in construction and the implication of these delays is very severe especially in metro projects. If we advise large termination lengths to socketed piles conservatively, penetration requirement

**Table 4** Comparison of pile strength parameters by various methods

| Strength/Consistency | Weak-moderately weak |
|---|---|
| Shear strength of rock below the base of the pile | 1900 kN/m$^2$ |
| Pile diameter | 1.2 m |
| UCS of intact rock | 50 MPa |
| Shear strength of rock below the base of the pile | 2000 kN/m$^2$ |
| Socket length | 3.6 m |
| End bearing component as per IS 2911 Part 1 Section 2 | 678 Tonne |
| Socket resistance as per IS 2911 Part 1 Section 2 | 814 Tonne |
| (CR + RQD)/2 | 50% |
| End bearing component as per IRC 78 | 565 Tonne |
| Socket resistance as per IRC 78 | 275 Tonne |
| End bearing component as per IS 14593 | 380 Tonne |
| Socket resistance as per IS 14593 | 633 Tonne |

will also be high. Hence, an optimization of pile termination lengths is an important aspect in pile foundation design. In conventional methods using Indian standards and IRC codes, parameters like cohesion of rock mass, uniaxial compressive strength of intact rock play a key role. It is not logical to apply limited values of uniaxial compressive strength of intact rock everywhere in all piles considering the extent of pile foundations available in a metro project.

Shear strength of rock mass is an input for 2 methods IS 2911 Part 1/Section 2 method and IRC 78 Method 2. The value of shear strength assigned as per these standards is only based on empirical approaches.

Hence, a methodology based on pile penetration ratio (PPR) is more economical, safe, and time saving.

The correlation between PPR rate and N value can be easily derived using the formula given in IRC 78 as given below.

Allowable capacity of pile has been calculated using the formula given in Appendix 5 of IRC 78-Method 2 (Table 5).

We need a reliable correlation between PPR and pile capacity so that we can guide the contractor during construction based on PPR values to determine the pile termination. Pile capacity required for the structure is around 6000 kN.

**Table 5** Correlation between PPR and *N* value

| *N* | Pile penetration rate (PPR) | Cohesion in kPa |
|---|---|---|
| 60 | 44.82 | 400 |
| 100 | 74.7 | 700 |
| 200 | 149.4 | 1900 |
| 300 | 224.1 | 3300 |
| 400 | 298.8 | 5000 |

$$Q_a = c_{\mathrm{ub}} N_c A_b/3 + A_s c_{\mathrm{us}}/6$$

Formula reduces to

$$\frac{Q_a - c_{\mathrm{ub}} N_c A_b/3}{c_{\mathrm{us}}/6} = A_s$$

$$\frac{Q_a - c_{\mathrm{ub}} N_c A_b/3}{c_{\mathrm{us}}/6} = \pi D L_s$$

For the hinged piles resting on rock, proper seating has to be ensured. The minimum socket length should be 300 mm in hard rock, and 0.5 times the diameter of the pile. The allowable end bearing component after dividing by factor of safety shall be restricted to 5 MPa. For calculation of socket friction capacity, the top rock 300 mm depth of rock shall be neglected. The friction capacity shall be further limited to depth of six times diameter of pile. Additionally, 50% of end bearing capacity has been considered for design due to chances of further reduction of end bearing capacity during construction.

Accordingly, socket length for various PPR values is derived as given below.

| $N$ | Pile penetration rate (PPR) | Socket length in m | Remarks |
|---|---|---|---|
| 60 | 44.82 | 31.4 | Frictional capacity limited to six times diameter of pile and hence socket length not feasible |
| 100 | 74.7 | 16.9 | |
| 200 | 149.4 | 4.7 | |
| 300 | 224.1 | 1.8 | |
| 400 | 298.8 | 0.30 | |

If PPR is < 149.4, socketed piles would not be effective and very less socket length is achieved if PPR > 298.8. This would give proper awareness to site personnel regarding the penetration length and construction shall be conveniently adjusted accordingly.

## 5 Conclusion

Pile capacity calculated from various methods given in Indian standards is not matching with each other since the input parameters used in these methods are different. All these parameters are not easily available at field during pile drilling. A site friendly method is a method which will aid in determining pile termination criteria based on pile penetration ratio. This method has been used in Chennai metro and was successfully executed in hundreds of piles.

## References

1. Cole KW, Stroud MA (1977) Rock socketed piles at Coventry point, Marketway, Coventry. In: Proceedings of piles in weak rock, 47–62. Institution of Civil Engineers, London
2. IS 2911 Part 1 Section 2-Design and construction of pile foundations—code of practice Part 1-concrete piles-Section 2 -Bored Cast In-Situ Concrete Piles
3. IRC 78-Standard specifications and code of practice for road bridges Section: VII-Foundations and Substructure
4. IS 14593-Design and construction of bored cast-in-situ piles founded on rocks-guidelines

# Assessment of Predictive Equations of Extents of Failure Zone in Sand Beneath the Cutting Edge of Open Caisson Using Image Analysis

**Jitesh T. Chavda and G. R. Dodagoudar**

## 1 Introduction

The soil in contact with the cutting edge fails in bearing when the cutting edge of the caisson is loaded with the subsequent raising of the steining. The bearing failure leads to formation of influence zone termed as failure zone beneath the cutting edge. This bearing failure of soil initiates the sinking of the open caisson. The estimation of the vertical and radial extents of the failure zone and bearing capacity of the cutting edge will help in planning of proper excavation strategies to achieve the controlled sinking of the caisson.

The open caisson of larger diameter (i.e., $Do > 10$ m) is generally used for launch and reception pits for tunnel boring machines and underground units like pumping stations, storm water tanks, storage and attenuation tanks (e.g., [2, 13, 14, 18, 24]). Figure 1 shows the excavation of soil at the cutting edge of the large diameter caisson by machine excavators. The sinking of such larger diameter caisson needs proper excavation strategies to allow the uniform sinking. The uneven sinking of the caisson may result in stress concentration and structural failure of the steining.

In such cases, the sequential excavation is adopted for the uniform and control sinking of the caisson (see Fig. 2). A rough estimation of the extent of the influence zone in the soil based on the configuration of the cutting edge and the soil type helps in proper planning of sequential excavation strategies [13]. The soil is excavated near the cutting edge in sequence as shown in Fig. 2. This allows the failure of

J. T. Chavda (✉)
Department of Civil Engineering, Sardar Vallabhbhai National Institute of Technology Surat, Surat 395007, India
e-mail: jiteshchavda03@yahoo.in

G. R. Dodagoudar
Department of Civil Engineering, Indian Institute Technology Madras, Chennai 600036, India
e-mail: goudar@iitm.ac.in



B. T. Jose et al. (eds.), *Proceedings of the Indian Geotechnical Conference 2022 Volume 2*, Lecture Notes in Civil Engineering 477,
https://doi.org/10.1007/978-981-97-1741-5_6

**Fig. 1** Large diameter caisson for launch and reception pits for tunnel boring machines and underground units like pumping stations, storm water tanks, storage and attenuation tanks [2, 13, 14, 18, 24]

soil in contact with cutting edge altogether and permits uniform sinking. However, the guidelines on how much deeper and up to what radial extent the soil needs to be excavated is limited. For estimation of the extent of influence zone in the soil beneath the cutting edge of the open caisson, [6] have carried out sensitivity analysis and the extent of the failure zone corresponding to the various parameters is evaluated. The parameters considered in the sensitivity analysis are: radius ratio ($r_i/r_o$) and tapered angle ($\beta$) of the cutting edge, unit weight ($\gamma$), friction angle ($\phi$), and cohesion of the soil ($c$), and magnitude of surcharge ($q$) representing depth of sinking of the caisson. The predictive equations are developed to estimate the extent of the failure zone for the given range of parameters and units as given in Table 1. The developed predictive equations are:

$$X/B = -0.9817 + 5.6522(r_i/r_o) - 0.0276(\beta) + 0.0374(c) + 0.0325(\phi) + 0.0028(q) - 0.0307(\gamma), \quad (1)$$

$$Y/\beta = 1.0596 + 1.5214(r_i/r_o) - 0.0167(\beta) + 0.0120(c) + 0.0101(\phi) + 0.0013(q) - 0.0093(\gamma). \quad (2)$$

The measurement of soil displacements in geotechnical experimentations using particle image velocimetry (PIV) technique is reported in the literature [7, 9, 15, 16, 19, 22, 23]. The soil flow corresponding to the varying cutting angles of the cutting edge is evaluated using PIV technique by Royston et al. [14]. They performed the

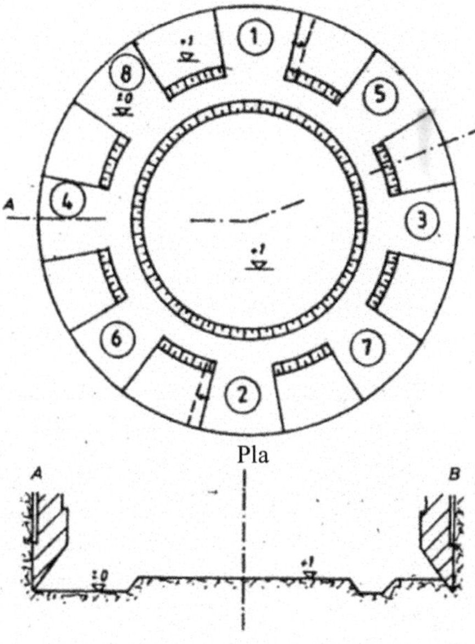


**Fig. 2** Excavation sequences, 1–8, for uniform sinking of caisson [13]

**Table 1** Range of parameters used in the regression [8]

| Component | Parameter | Unit | Range |
|---|---|---|---|
| Soil | $\gamma$ | kN/m$^3$ | 14–22 |
| | $c$ | kPa | 0–30 |
| | $\phi$ | Degree | 5–30 |
| Geometry of cutting edge | $r_i/r_o$ | – | 0.5–0.9 |
| | $\beta$ | Degree | 30–90 |
| Surcharge | $q$ | kPa | 0–150 |

experiments corresponding to the plane strain condition, applicable for the circular open caisson with higher radius ratio, approximately equal to 1. The soil flow corresponding to the varying cutting angles, radii ratio, and depth of sinking of the cutting edge is evaluated using open-source MATLAB module, and GeoPIV-RG has been reported by Chavda et al. [7]. The module GeoPIV-RG follows the principles of PIV, originally developed in the field of experimental fluid mechanics [1]. Certain modifications to the PIV technique have been proposed by White et al. [21] and Stanier et al. [16] to enhance its applicability in geotechnical engineering.

In the present study, the soil displacements around the cutting edge of the circular open caissons with different cutting angles have been investigated by performing the image- based deformation analysis using GeoPIV-RG. The failure zones in the sand at the cutting edge of the circular open caisson are evaluated using 1$g$ model test. The extent of failure zone in sand corresponding to the cutting edge with different radii ratios, varying cutting angles, and different sinking depths is evaluated using image-based deformation measurement technique. The results obtained from the 1$g$ model half-open caisson tests are used to assess the predictive equations (Eqs. 1 and 2). The extent of the failure zone obtained from the experiments is compared with the extent of the failure zone obtained from the predictive equations. The results of the present study provide a clear picture of the extent of the failure zone around the cutting edge.

## 2 Experimental Program

### 2.1 *Test Setup and Sand Used in the Study*

The loading frame used for CBR test is used for the evaluation of soil flow mechanism (see Fig. 3). The tank is fabricated with one side transparent, also termed as a viewing window, to capture the soil movement around the cutting edge of the half-open caisson during the penetration. The Indian standard (IS) sand, properties reported in Table 2, is used in the model tests [10]. All tests are performed at relative density of the sand at $D_r = 50\%$ to limit the scope of the study to one density.

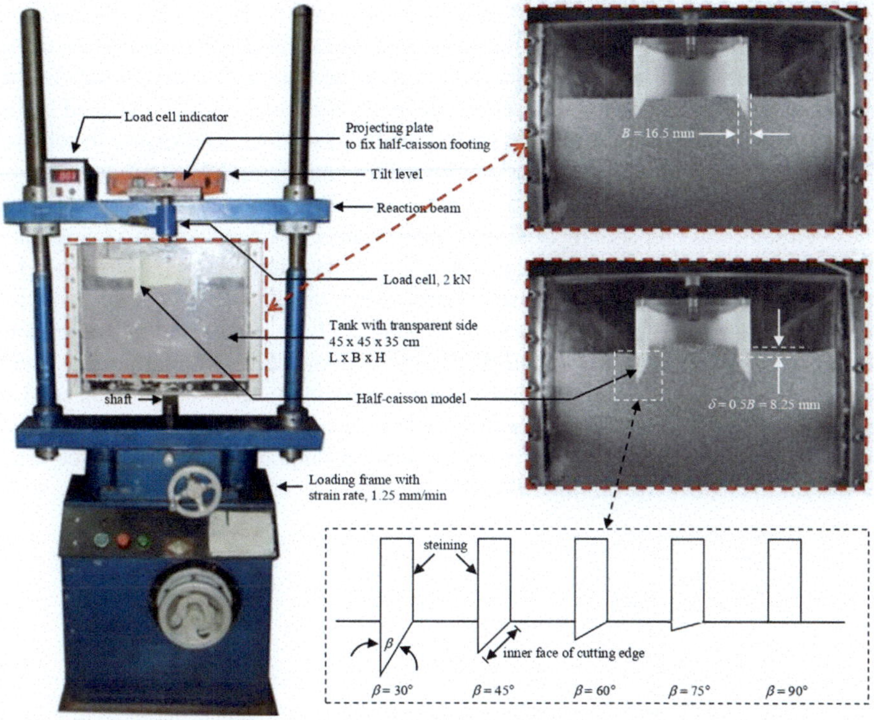


**Fig. 3** Experimental setup of half-open caisson model showing loading frame, closer view of viewing window of the tank, and configuration of the cutting edge of the half-open caisson

**Table 2** Index properties of Indian standard sand (Grade II)

| Parameter | Unit | Value | References |
|---|---|---|---|
| Specific gravity | – | 2.64 | ASTM [3] |
| Max. unit weight | kN/m$^3$ | 16.81 | ASTM [3] |
| Min. unit weight | kN/m$^3$ | 14.50 | ASTM [4]: Method B |
| Relative density | % | 50 | ASTM [4] |
| Unit weight of sand | kN/m$^3$ | 15.57 | ASTM [4] |
| $\phi$ ($D_r = 50\%$) | degree | 35° | IS: 2720-13 [11] |
| $\psi$ ($D_r = 50\%$) | degree | 11° | IS: 2720-13 [11] |
| Young's modulus | MPa | 5.768 | Plate load test |
| Poisson's ratio | – | 0.3 | $K_0 = 1 - \sin(\phi)$ |
| Height of fall ($D_r = 50\%$) | mm | 270 | ASTM [4] |
| Unified soil classification | – | SP[a] | [5] |

[a]SP—poorly graded sand

For the image-based deformation measurement analysis, the sand texture has been optimized to maximize the precision in the measurement of the soil displacements [17, 20, 21, 25]. White et al. [21] reported that the natural texture present in the river sand is sufficient to carry out the image analysis. However, the sand used in the present study has an off-white color. Hence, artificial seeding is necessary to achieve the required texture in the sample. The procedure and methodology reported in [7] are used for the preparation of the dyed sand and the optimum texture, respectively. The dyed sand was prepared by mixing 60 ml of black color pen ink with 5 kg of sand. Then, the sand was kept in an oven at 105 °C for 24 h so that the color is absorbed by the sand particles. In order to remove the excess ink, the dyed sand was washed with the fresh water and kept in the oven for another 24 h to remove the moisture. The black colored dyed sand thus obtained is termed as the seeding sand in the present study. The texture in the sand is generated by mixing the IS sand and seeding sand at 65 and 35% by weight. The sand used in all the experiments is seeded with 65% of the dyed sand, i.e., it comprises 35% of the natural sand and 65% of the dyed sand by weight to achieve the required texture.

## 2.2 *Caisson Models*

For the image analysis, the caisson models are half cut to represent an axisymmetric problem. The configuration of the half-cut open caisson models is depicted in Fig. 4. The half-open caisson models are fabricated from the Polytetrafluoroethylene (PTFE) Teflon tubes. The transparent side of the tank is made of Perspex acrylic sheet. For maximum repetitive use of the transparent acrylic glass, the half-open caisson models are fabricated with PTFE for the investigation of the soil flow mechanism. The properties of the PTFE caisson models are given in Table 3. The Young's modulus of the PTFE is almost 80 times higher than the sand, and, hence the half-cut caisson models are relatively rigid and hence they experience negligible deformations.

## 2.3 *Data Processing*

The images obtained from the tests consist of series of images from the initial to the final penetration depth same as the width of the cutting edge. In any PIV analysis, the deformations at any point of time are calculated using a particular image by comparing it to the reference image (initial undeformed image). The reference image represents the image which corresponds to zero penetration of the caisson and the final image represents the image which corresponds to the particular penetration at which the displacement field of the sand medium is required to be evaluated. In the study, the depth of penetration is kept same as half the width of the cutting edge. At this penetration, the fully developed failure envelops are obtained.

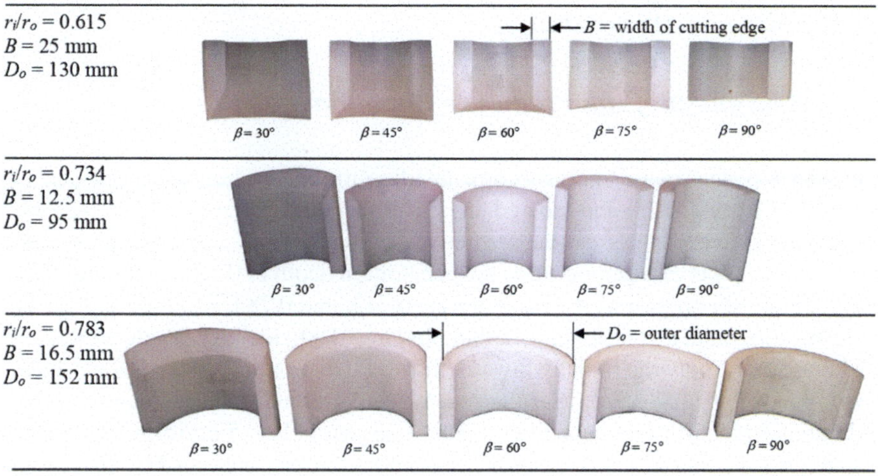


**Fig. 4** Configuration of the half-open caisson models

**Table 3** Properties of caisson model

| Material | Parameter | Unit | Value |
|---|---|---|---|
| Caisson model: Polytetrafluoroethylene | Unit weight | kN/m$^3$ | 11 |
| | Young's modulus | MPa | 500 |
| | Poisson's ratio | – | 0.46 |

## 2.4 *Camera and Test Procedure*

In the present study, a Nikon D5300 digital single-lens reflex (DSLR) camera, with AF-S NIKKOR lens, *f*/3.5–5.6, 18–55 mm focal length, is used which employs a complementary metal–oxide–semiconductor (CMOS) sensor (23.5 mm × 15.6 mm) with a rolling shutter arrangement. The use of such camera in large deformation analysis has already been reported by Mishra et al. [12], in measuring localized strains in geotextiles. In order to compensate for the low optical sensitivity of the CMOS sensor and to reduce the shutter speed, an external lighting is provided by the two 16W LED light panels. The test is performed by moving the tank in the upward direction with a constant strain rate of 1.25 mm/min and the photographs are captured at an interval of 3 s (0.33 Hz). This frequency of image acquisition is decided based on the strain rate and displacement of the soil medium between the two successive images.

# 3 Results and Discussion

## 3.1 Assessment of the Predictive Equations

The results of the image analysis corresponding to the varying radii ratio ($r_i/r_o =$ 0.613, 0.743, and 0.783) and tapered angles of the cutting edge ($\beta = 30°, 45°, 60°,$ $75°$ and $90°$) are used to assess the predictive equations. The vectorial and horizontal displacement plots corresponding to the penetration depth of $0.5B$ are obtained from the GeoPIV_RG program for $r_i/r_o = 0.613, 0.743$, and 0.783.

In the horizontal displacement contour plot, in order to identify the clear picture of the failure zone, only the radially inward horizontal displacement vectors are shown. The typical plot of the vectorial displacement, horizontal displacement contours, and radially inward displacement contours is shown in Fig. 5 corresponding to the cutting edge having a radius ratio of 0.783 and a cutting angle of 60°. The extent of the failure zone is evaluated from these plots for varying radii ratio and cutting angles. Equations (1) and (2) are used to predict the extent of the failure zone for the same configuration of the cutting edge and material properties of the sand as used in the tests. The experimental values and those predicted from the equations are compared and shown in Fig. 6. It is observed from the figure that Eqs. (1) and (2) estimated the extent of the failure zone within 20% of the error band. The error band of 20% is with respect to the value of $X/B = Y/B = 4$. These predictive equations can be used to identify the extent of the failure zone in the field and accordingly the suitable excavation strategies can be adopted.

# 4 Conclusions

In the study, the predictive equation to estimate size of failure zone in soil beneath the cutting edge of open caisson is examined. The equation accounts the effect of variation of geometry of the cutting edge, soil properties, and depth of sinking on the extents of the failure zone. The experimental results from the $1g$ caisson tests are used to assess the predictive equations. The predictive equations can be used in the preliminary estimation of the extent of the failure zone in the radial and vertical directions for the selected configuration of the open caisson. Following conclusions are drawn from the present study:

- The easy-to-use predictive equations have been assessed to estimate the radial and vertical extent of the failure zone in the soil beneath the cutting edge of the open caisson. The extent of the failure zone is determined from the $1g$ caisson tests and is compared with the estimated extent of the failure zone from the predictive equations. The predictions made by the equations are within 20% error band of the experimental results.

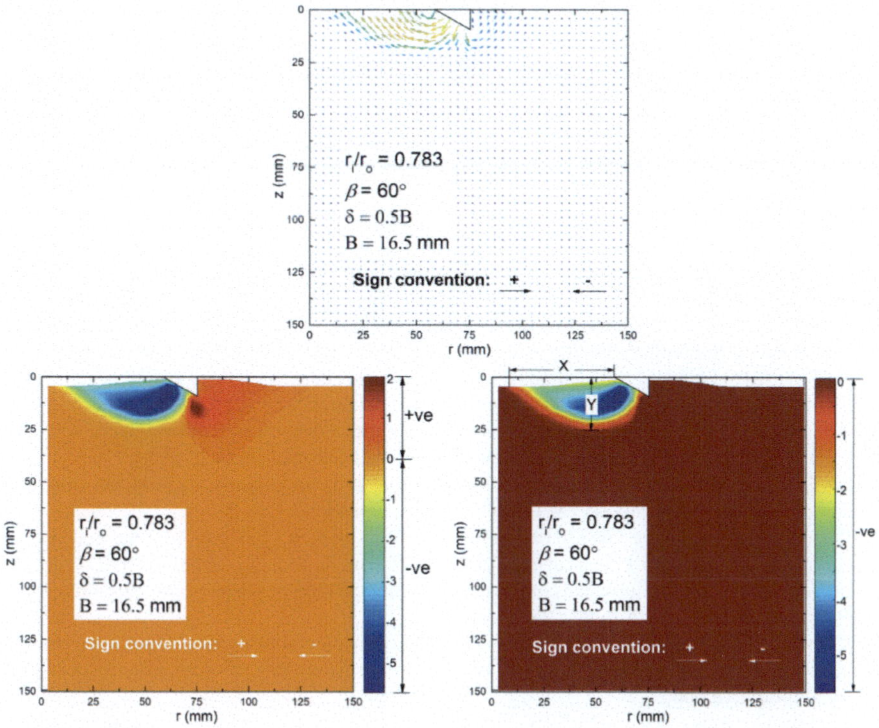


**Fig. 5** Typical vectorial displacements, horizontal displacement contours, and radially inward displacement contours [8]

**Fig. 6** Comparison of experimental results and predicted values of the extent of failure zone at the cutting edge

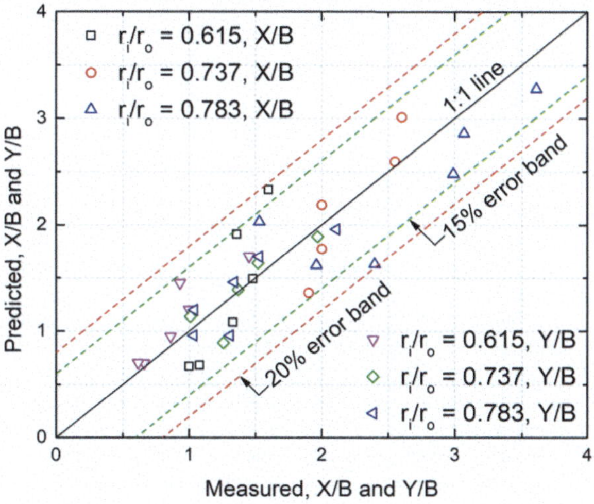

- Based on the assessment, the predictive equation can be used as the guidelines on how much deeper and up to what radial extent the soil needs to be excavated for controlled and uniform sinking of large diameter caisson with confidence.

## References

1. Adrian RJ (1991) Particle imaging techniques for experimental fluid mechanics. Ann Rev Fluid Mech 23:261–304
2. Allenby D, Waley G, Kilburn D (2009) Examples of open caisson sinking in Scotland. Proce ICE Geotech Eng 162(1):59–70
3. ASTM (2016) Standard test methods for maximum index density and unit weight of soils using a vibratory table. ASTM D4253-16, West Conshohocken, PA
4. ASTM (2016) Standard test methods for minimum index density and unit weight of soils and calculation of relative density. ASTM D4254-16, West Conshohocken, PA
5. ASTM (2011) Standard practice for classification of soils for engineering purposes (uni- fied soil classification system) ASTM D2487-11, West Conshohocken, PA
6. Chavda JT, Dodagoudar GR (2018) Finite element modelling of extent of failure zone in $c$-$\varphi$ soil at the cutting edge of open caisson. In: Numerical methods in geotechnical engineering IX: proceedings of the 9th European conference on numerical methods in geotechnical engineering, Porto, Portugal, 25–27th June 2018
7. Chavda JT, Mishra SR, Dodagoudar GR (2019) Experimental evaluation of ultimate bearing capacity of the cutting edge of open caisson. Int J Phys Modelling Geotech. https://doi.org/10.1680/jphmg.18.00052
8. Chavda JT, Dodagoudar GR (2019) Experimental evaluation of failure zone in sand beneath the ring footing and cutting edge of open caisson using image analysis. In: Indian geotechnical conference 2019, SVNIT Surat, Gujarat, India, Dec 2019
9. Iskander, M. (2010) *Modelling with Transparent Soils Visualizing Soil Structure Inter- action and Multi Phase Flow, Non-Intrusively.* Springer, New York.
10. IS: 650 (1991) Standard sand for testing cement—specification. Bureau of Indian Standards, New Delhi
11. IS: 2720-13 (1996) Methods of test for soils, Part 13: direct shear test. Bureau of Indian Standards, New Delhi
12. Mishra SR, Mohapatra SR, Sudarsanan N, Rajagopal K, Robinson RG (2017) A simple image-based deformation measurement technique in tensile testing of geotextiles. Geosynth Int 24(3):306–320
13. Nonveiller E (1987) Open caissons for deep foundations. J Geotech Eng, ASCE 113(5):424–439
14. Royston R, Phillips BM, Sheil BB, Byrne BW (2016) Bearing capacity beneath tapered edges of open dug caissons in sand. In: Proceedings of civil engineering research in Ireland, Galway, Ireland
15. Rechenmacher AL, Finno RJ (2004) Digital image correlation to evaluate shear banding in dilative sands. Geotech Test J 27(1):13–22
16. Stanier SA, Blaber J, Take WA, White DJ (2016) Improved image-based deformation measurement for geotechnical applications. Can Geotech J 53(5):727–739
17. Stanier SA, White DJ (2013) Improved image-based deformation measurement for the centrifuge environment. Geotech Test J 36(6):915–927
18. Sheil B, Royston R, Byrne B (2018) Real-time monitoring of large-diametercaissons. In: Proceedings of China-Europe conference on geotechnical engineering, pp 725–729. https://doi.org/10.1007/978-3-319-97112-4_162
19. Take WA (2015) Thirty-Sixth Canadian geotechnical colloquium: advances in visualization of geotechnical processes through digital image correlation. Can Geotech J 52(9):1199–1220

20. Take W (2003) The influence of seasonal moisture cycles on clay slopes, Ph.D. Thesis, University of Cambridge, Cambridge, U.K.
21. White DJ, Take WA, Bolton MD (2003) Soil deformation measurement using particle image velocimetry (PIV) and photogrammetry. Géotechnique 53(7):619–631
22. Yuan B, Xu K, Wang Y, Chen R, Luo Q (2017) Investigation of deflection of a laterally loaded pile and soil deformation using the PIV technique. Int J Geomech ASCE 17(6):04016138, 1–9
23. Xiao Y, Yin F, Liu H, Chu J, Zhang W (2017) Model tests on soil movement during the installation of piles in transparent granular soil. Int J Geomech, ASCE 17(4):06016027, 1–7
24. Yao Q, Yang X, Li H (2014) Construction technology of open caisson for oversize surge shaft in drift gravel stratum. Electron J Geotech Eng 19:5725–5738
25. White DJ, Take WA (2005) Discussion on: application of particle image veloc imetry (PIV) in centrifuge testing of uniform clay. In: Zhang YD, Tan TS, Leung CF (eds) Int J Phys Modell Geotech 5(4):27–31

# Analysis of Settlement Profiles of Shallow Strip Footings Resting on Geosynthetic-Reinforced Sands

**Quamar Tabish, Saptarshi Kundu, and Ambarish Ghosh**

## 1 Introduction

The vertical and horizontal stresses from the superstructure on the footings may cause foundation soil failures and excessive settlements detrimental to the overlying structure. Considering the preceding, numerous ground improvement procedures have been developed, and soil reinforcement has been extensively praised for its versatility in technical, economic, and environmental feasibility and ease of use in many incarnations. Soil reinforcements, starting with metallic strip reinforcements [1, 2], broadened their domain with sheet type reinforcements, and finally, revolutionary geosynthetics in various forms have superseded them all. Geosynthetic has been successfully adopted as a cost-effective solution to improve the bearing capacity and prevent soil settlement beneath the foundation. Researchers have conducted an increasing number of studies on the subject in the last few years [3–5]. The role of geosynthetic material are diverse, as it can serve multiple purposes of reinforcement, separation, filtration, drainage, containment, and confinement of soil.

One such viable utility involves placement of high tensile strength reinforcing members under the footing for increasing the ultimate bearing capacity of sandy deposits [1, 6, 7]. A ground-breaking study was conducted to this effect [8] on the influence of reinforcing layers placed under a foundation on the bearing capacity of sandy soil underlain by soft clay. Placement of a low-stiffness polymer net under the footing was observed to increase the bearing capacity of the soft soil significantly. Addition of geosynthetic reinforcement to the soil provides the soil with more lateral constraint, which may enhance the load-spread angle and decrease the bearing pressures. Generally, the load transfer in soil occurs through grain-to-grain contact. For

Q. Tabish · S. Kundu (✉) · A. Ghosh
Department of Civil Engineering, Indian Institute of Engineering Science and Technology, Shibpur, Howrah 711103, India
e-mail: saptarshi@civil.iiests.ac.in



B. T. Jose et al. (eds.), *Proceedings of the Indian Geotechnical Conference 2022 Volume 2*, Lecture Notes in Civil Engineering 477,
https://doi.org/10.1007/978-981-97-1741-5_7

the sand grains below the footing, each grain transfers the load to the grain below it. To spread the load over a larger area, the entire soil layer in the reinforced soil layer behaves like a semi-rigid slab. There is a lot of friction generated at the interface of soil and geosynthetic layer to confine the soil. A geosynthetic layer, as compared to an unreinforced soil, aids in spreading the applied surface load across a wider area by providing a restraint for the granular fill. As per traditional method, a load dispersion ratio of 2:1 (vertical: horizontal) is generally assumed beyond the edges of the surface-loaded zone, and this assumption is used to estimate the vertical stresses in the foundation soil empirically.

The pressure is transferred from below the footing base of width "$B$" to the underlying soil layer and redistributed over a wide width of "$B + \Delta B$" by the geosynthetic, thereby reducing the intensity of the pressure and increasing the bearing capacity. Due to the semi-rigid nature of the geosynthetic-reinforced foundation, the additional load dispersion width "$\Delta B$" in a reinforced section is greater than that in an unreinforced section. In order to gain further insight into this phenomena, a series of small-scale physical model tests were carried out under displacement-controlled conditions to investigate the settlement reduction and dispersion of a load under a strip footing resting on geosynthetic-reinforced sand deposits. In order to comprehend and interpret accurately the load dispersion mechanism of geosynthetic-reinforced foundation, Particle Image Velocimetry (PIV) technique [9] has been employed, as explained in subsequent sections.

## 2 Materials

### 2.1 Sand and Geosynthetic

The air-dried fine sand used below the model strip footing in the present analysis has an average particle size ($D_{50}$) of 0.48 mm. Figure 1 represents the grain size distribution of the sand used in the experiment. Geotechnical investigations were conducted on the sand as per BIS specifications, including sieve analyses, specific gravity tests, maximum and minimum density tests, and direct shear testing. Table 1 summarizes the engineering properties of dry sand. The maximum and minimum dry densities were found to be 17.27 kN/m$^3$ and 15.23 kN/m$^3$, respectively. The sand was densely packed into the model tank during subsequent investigations, resulting in a density of 17.04 kN/m$^3$ ($D_r = 90\%$). The sand pluviation method was used to obtain specific relative densities, which is reportedly an excellent way to make uniform and homogeneous sand beds with a given relative density [10, 11]. In this investigation, a sand pluviation hopper was employed to release sand into the container from a predetermined height to achieve a defined average relative density. The desired height was determined from the calibrated graph shown in Fig. 2. Direct shear tests conducted on a typical shear box (60 × 60 × 20 mm) by pluvaiting the sand sample

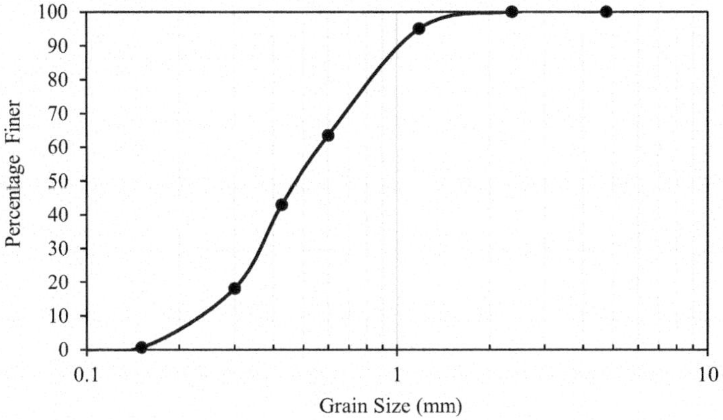


**Fig. 1** Grain size distribution curve of sand

**Table 1** Properties of dry sand

| Property | Value |
|---|---|
| Specific gravity | 2.67 |
| Maximum dry density (kN/m$^3$) | 17.27 |
| Minimum dry density (kN/m$^3$) | 15.23 |
| $D_{50}$ (mm) | 0.48 |
| Friction angle ($\phi°$) | 32.5 |
| Maximum void ratio ($e_{max}$) | 0.753 |
| Minimum void ratio ($e_{min}$) | 0.546 |

in layers revealed the angle of internal friction of the sand as about 32.5°. Additionally, the friction coefficient between sand and reinforcement was determined from a conventional direct shear test apparatus and observed to be 0.997 (Table 2).

During testing of model footing, a woven polypropylene geotextile was used as the reinforcing geosynthetic. The tensile strength of the geotextile is 0.025 kN/m and 0.0625 kN/m at 2% strain and 5% strain, respectively. While the strength of this geosynthetics would be regarded as low in a full-scale application, considering appropriate scaling laws linking the full-scale prototype and the corresponding model, it replicates a highly tensile model reinforcement. The interface friction angle for woven polypropylene geotextile in a direct shear box is evaluated as 32.4°. Table 2 summarizes the basic properties of the geosynthetics used in the experiment. Figure 3 depicts the load-elongation behavior of the model geosynthetic as evaluated from standard wide-width tensile strength tests, as per ASTM D4595-17.

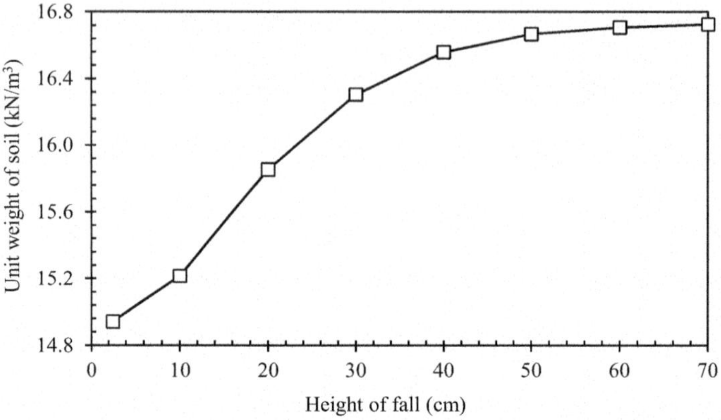


**Fig. 2** Calibration curve used in sand pluviation method

**Table 2** Properties of model geosynthetics

| Property | Non-woven geotextile (NGT) |
|---|---|
| Material | PP |
| Tensile strength at 2% strain (kN/m) | 0.025 |
| Tensile strength at 5% strain (kN/m) | 0.075 |
| $C_i$ | 0.997 |

$C_i$ = Sand-reinforcement interaction coefficient at R.D of 90%

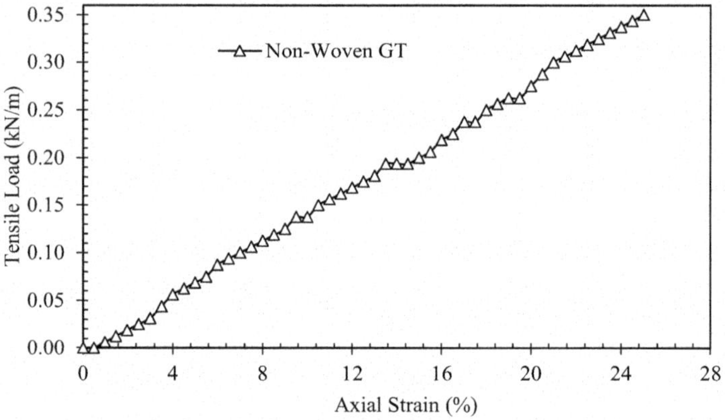


**Fig. 3** Load-elongation behavior of model geosynthetic used in the study

## 3 Experimental Set-Up

### 3.1 Model Parameters and Instrumentation Details

The model footing tests were conducted within a container of length 0.76 m, a height of 0.46 m, and a width of 0.1 m. The schematic view and details of the container are shown in Fig. 4. The front face is made of perspex sheet to observe deformations in the soil mass during experiments. The container was sufficiently rigid to maintain plane strain conditions for the tests. The model strip footing had a width of 60 mm, and was centered in the model container with its length running parallel to the container's width. Several uniformly spaced permanent markers were placed on the perspex sheet to aid deformation measurements below the footing using the particle image velocimetry (PIV) technique. Particle image velocimetry (PIV) is a novel geotechnical testing technique that uses digital cameras to monitor soil deformation within the soil mass [9]. PIV technique is a unique subset of the Digital Image Correlation approach. This approach combines numerical computing with digital image processing. A digital camera was thus used in the study to capture the images of the front side of the model container at regular intervals of 10 s during the progress of the experiment. The camera can monitor the entire area below the footing for subsequent utilization in the PIV technique, to determine the horizontal and vertical displacements of the soil face. The ratio of container length to footing width was kept at 13 in this experiment, which was enough to eliminate boundary effects. This is based on a similar experimental observation [12], where a ratio of the length of container to width of model footing exceeding 12 was reported to have no effect on load-settlement behavior of the model footing.

### 3.2 Sand Bed Preparation and Testing Program

At the outset, the sand was filled in the tank at the required density to achieve general shear failure in course of experimentation. The relative density (Dr) was maintained at 90% ($\rho_{dry} = 1.704$ Mg/m$^3$) throughout the testing. The sand was pluviated in layers and was lightly compacted in the tank with a wooden block to achieve a consistent layer thickness of roughly 50 mm. Horizontal lines at 50 mm intervals were drawn on the outer face of the container to establish a thickness of 50 mm. This process was repeated for unreinforced tests until the height of sand deposit attained 0.40 m. For tests involving reinforcement, the sand surface was leveled to place a geotextile layer on the foundation, once the sand reached the height of a reinforcement layer. The sand was then again pluviated and densified to the desired relative density. The procedure was continued until the overall thickness of sand deposit reached the desired level.

Several sand depositions were done in the tank before the actual experiments. In these trials, the relative density error was less than 3%. The container was put underneath the loading system and adjusted to ensure that the strip footing was in

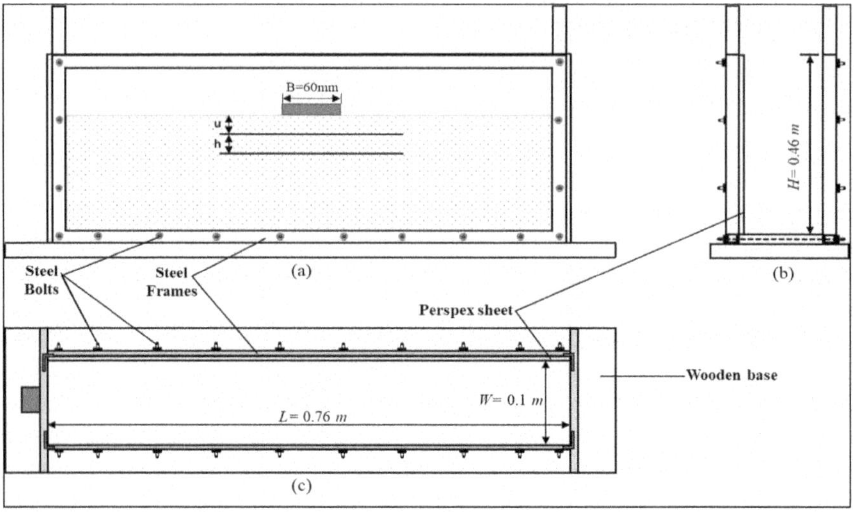


**Fig. 4** Schematic views of model container **a** Front view **b** Side view **c** Plan

the middle of the foundation model after the reinforced sand foundation model was prepared. Following that, a load cell was placed to measure the load, and a linear variable displacement transducer (LVDT) was used to measure the footing settlement induced by loading. The front side of the model container was made of a perspex sheet, on which several uniformly spaced permanent markers were placed to aid in deformation measurements below the footing using the particle image velocimetry (PIV) technique. To remove the effect of surrounding lights on the photographs, a strip of lights was positioned in front of the container in a rectangular pattern. The complete set-up was covered with a large piece of cloth at the time of testing to guarantee the same intensity of lighting during the experiment. The displacement vectors and deformation patterns of sand were subsequently computed using GeoPIV software.

The following studies on model strip footing were carried out using the experimental set-up and loading method stated before to explore the behavior of the strip footing on the geosynthetic-reinforced sand under displacement-controlled conditions. Table 3 shows the outline of laboratory model tests conducted during the experiment. As can be observed from Table 3, tests were conducted on sand without reinforcement, and on sand with multi-layer geotextile reinforcement embedded at different depths.

**Table 3** Description of laboratory model test

| Reinforcing material | Number of layers | *l/B* | *u/B* | *h/B* |
|---|---|---|---|---|
| Constant parameters: relative density = 90%, B = 60 mm, $\gamma_d$ = 17.04 kN/m$^3$ | | | | |
| Unreinforced | – | – | – | – |
| Reinforced with NGT | 1 | 12 | 0.5 | 0.25 |
| Reinforced with NGT | 2 | 12 | 0.5 | 0.25 |
| Reinforced with NGT | 3 | 12 | 0.5 | 0.25 |

# 4 Results and Discussion

The main objective of the current research is to investigate the advantages of utilizing geosynthetic reinforcement to reduce the settlement of foundations by employing digital image correlation technique. For this purpose, extensive laboratory model tests were conducted on unreinforced and geosynthetic-reinforced sand, as stated previously. The parameters investigated in the model tests include the number of reinforcement layers (*N*) and the vertical spacing between reinforcement layers (*h*). Figure 5 represents the load-settlement curves measured for model footing tests with different layers of reinforcement at different h/B values and for the same *l*/*B* values, where *l* is length of reinforcement and *h* is the spacing between reinforcements.

The benefit of applying reinforcement is assessed in terms of bearing capacity ratio (BCRs) at a particular settlement based on the observed load-settlement behavior of the model footing. The bearing capacity ratio (BCRs) is the ratio of the bearing capacity of the reinforced soil foundation to that of the corresponding unreinforced at a specified settlement. Table 4 presents the BCRs obtained at settlement ratios (*s/B*) of 10, 20, and 30%, wherein the settlement ratio (*s*/*B*) is defined as the ratio of footing settlement (*s*) to footing width (*B*).

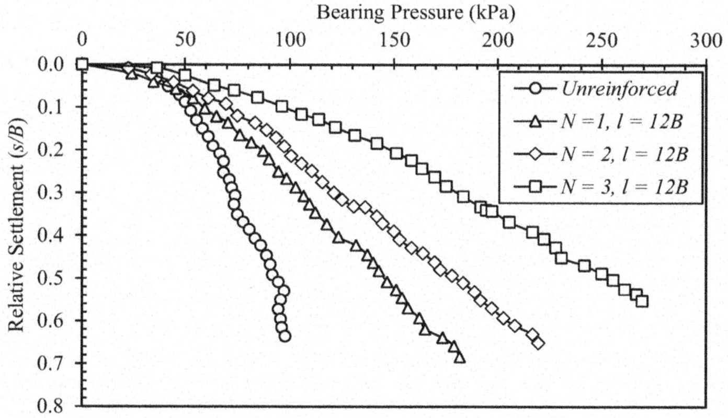


**Fig. 5** Load-settlement curve obtained from model test

**Table 4** Summary of laboratory model test

| Reinforcement configuration | $u$ (mm) | $h$ (mm) | $s/B = 10\%$ | | $s/B = 20\%$ | | $s/B = 30\%$ | |
|---|---|---|---|---|---|---|---|---|
| | | | $q$ (kPa) | BCR | $q$ (kPa) | BCR | $q$ (kPa) | BCR |
| Unreinforced | – | – | 52.78 | – | 66.67 | – | 73.89 | – |
| $N = 1$ | 30 | 15 | 59.44 | 1.13 | 87.22 | 1.31 | 106.67 | 1.44 |
| $N = 2$ | 30 | 15 | 69.44 | 1.32 | 100.56 | 1.51 | 121.11 | 1.64 |
| $N = 3$ | 30 | 15 | 96.11 | 1.82 | 151.11 | 2.27 | 183.33 | 2.48 |

In the model footing test, when the top reinforcement layer was placed at 0.5 times the width of the footing, and the reinforcement length was six times the width of the footing, it was observed that upon load application, the reinforcement pulled out the sand present above it. Since tests were performed at 1 g, the confinement capacity was less, which could have contributed to the pull-out of the geosynthetic prior to failure. However in the field, failure occurs reportedly by the elongation of geosynthetics. Hence, for simulating field conditions, the length of reinforcement was subsequently increased, and for the remaining experiments, the top reinforcement spacing was considered as 0.5 times the width of the footing, and the length of reinforcement as 12 times the width of the footing. By examining the load-settlement curves presented in Fig. 5, it can be observed that the bearing pressure increases as the settlement increases for both reinforced and unreinforced sand deposit. The observed settlement pattern is comparable to that of general shear failure, as anticipated.

## *4.1 Effect of Number of Reinforcement Layers on the Load-Spread Angle*

The load-spread angle ($\beta$) is the angle at which the load is distributed. It is reported that the reinforced bed has a higher load-spread angle value than the unreinforced bed, and usually ranges from 25 to 30 degrees [13]. Further, the L/B ratio of rectangular footings and the number of reinforcement layers can affect the load-spread angle [14]. The displacement vectors obtained from PIV analysis in the present study indicated that increase in number of reinforcements cause a decrease of increment rate.

Using the PIV approach, the load distribution mechanism of the geosynthetic-reinforced section and unreinforced section is shown in Fig. 6. The vector movements was used to calculate the width of the load distribution area at the base of the geosynthetic layer. The following process was used to calculate the load-spread angle from the picture analysis: (i) The image was digitized and output data generated, (ii) Vertical displacement vectors were measured at the horizontal line indicating the position of each reinforcement layer, (iii) The displacement of the soil particles from their orientation at the reinforcement line was used to determine the width of the load distributed area, as beyond this line, the displacement vectors became almost horizontal, (iv) From the horizontal distance contour ($X$ mm), the base width of the load dispersion region was calculated as ($B + \Delta B$), and the height of the soil above this layer was calculated as ($h + u$), which is equal to the height of the reinforcing layer. The dispersion angles were computed for each test configuration using the above procedure, as summarized in Table 5.

## 5 Conclusions

The study involves the development of a small-scale physical model for analyzing the settlement profiles under shallow strip footing resting on geosynthetic-reinforced sands. A series of experiments were performed by pluviating sand at a relative density of 90%, and by varying the number and spacing between the subsequent reinforcing layers. The tests were performed under displacement-controlled conditions, and instrumented using a load cell and linearly variable displacement transducer (LVDT). The results were interpreted by plotting the load–displacement curve and the displacement vectors obtained through PIV analysis. Based on the analysis and interpretation of the results, the following major conclusions can be drawn:

(1) The load-settlement curves registered higher values for the reinforced foundation in comparison to an unreinforced foundation under displacement-controlled conditions.
(2) The reinforcement was effective in controlling heaving of the foundation sand.
(3) It was found that adding a third layer of reinforcement significantly improves the bearing capacity ratio (BCR) of the footing, as compared to adding the first two layers.
(4) The load-spread angle was found to increase with the number of reinforcement layers, and the rate of this increase reduced with increasing reinforcement layers.

In conclusion, it may be stated that the present research is thus helpful in situations where it is required to construct heavy structures supported on robust foundation systems on sandy soils, by providing geosynthetic reinforcements at different layers within the sandy soil. Geosynthetics reinforce the soil and are cost-effective, and additionally reduce the foundation settlement and increase the bearing capacity.

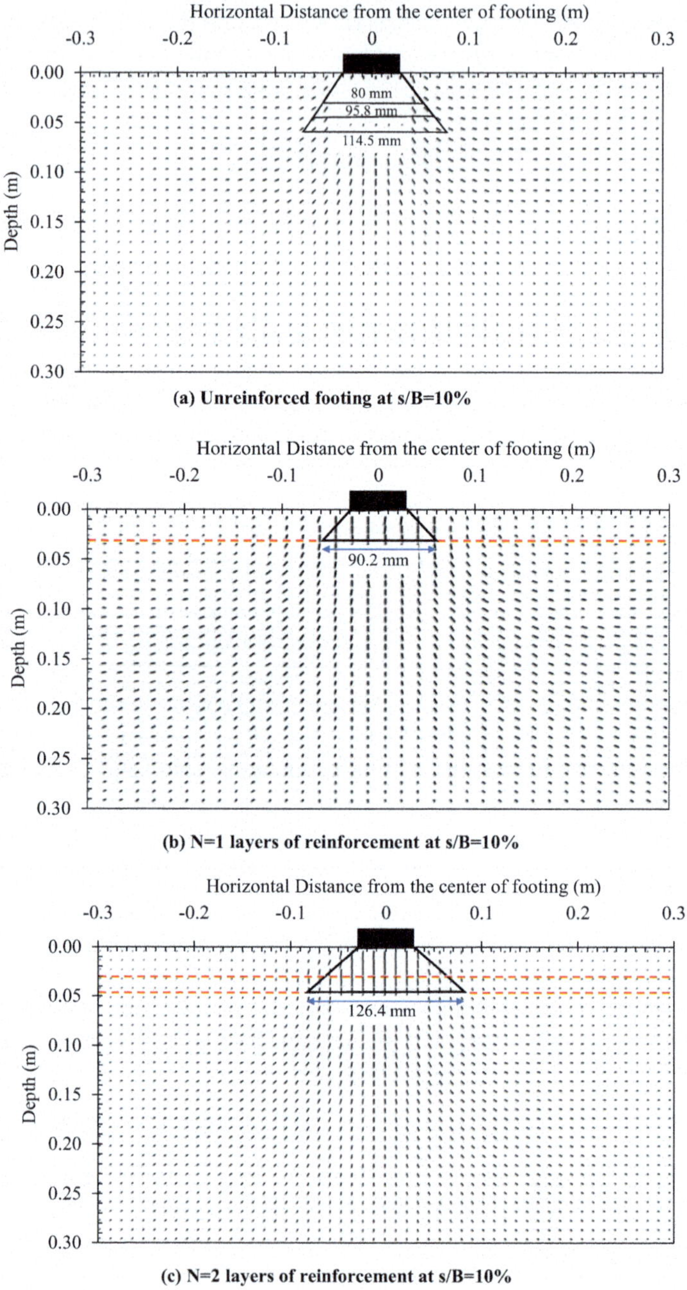


**Fig. 6** Load distribution mechanism of unreinforced and geosynthetic-reinforced sand obtained using PIV technique

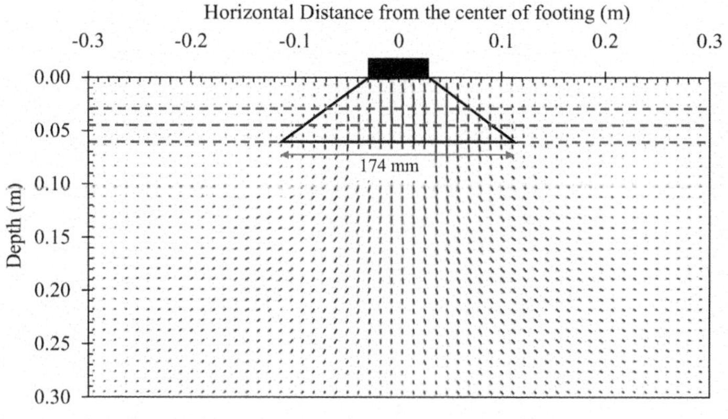


**(d) N=3 layers of reinforcement at s/B=10%**

**Fig. 6** (continued)

**Table 5** Summary of spread angle observed in model test

| Reinforcement configuration | $u$ (mm) | $h$ (mm) | $s/B = 10\%$ | | | |
|---|---|---|---|---|---|---|
| | | | $B + \Delta B$ (mm) | $h + u$ (mm) | ΔB/2 (mm) | $\alpha$ (°) |
| Unreinforced | | | 80 | 30 | 10.0 | 18.4 |
| $N = 1$ | 30 | 15 | 90.2 | 30 | 15.1 | 26.7 |
| $N = 2$ | 30 | 15 | 126.4 | 45 | 33.2 | 36.4 |
| $N = 3$ | 30 | 15 | 174 | 60 | 57.0 | 43.5 |

# References


1. Binquet J, Lee KL (1975) Bearing capacity tests on reinforced earth slabs. J Geotech Eng Division ASCE 101(12):1241–1255
2. Binquet J, Lee KL (1975) Bearing capacity analysis of reinforced earth slabs. J Geotech Eng Division ASCE 101(12):1257–1276
3. Pinto MIM (2003) Applications of geosynthetics for soil reinforcement. Ground Improvem 7(2):61–72
4. Ghosh A, Ghosh A, Bera AK (2005) Bearing capacity of square footing on pond ash reinforced with jute geotextile. Geotext Geomembr 23(2):144–173
5. Abu-Farsakh M, Chen QM, Sharma R (2013) An experimental evaluation of the behavior of footings on geosynthetic-reinforced sand. Soils and Found 53(2):335–348
6. Guido VA, Chang DK, Sweene MA (1986) Comparison of geogrid and geotextile reinforced earth slabs. Can Geotech J 23:435–440
7. Manjunath VR, Dewaikar DM (1996) Bearing capacity of inclined loaded footing on geotextile reinforced two-layer soil system. In: Proceedings of the international symposium on earth reinforcement, Fukuoka, Japan, pp 619–622
8. Yamanouchi T (1970) Experimental study on the improvement of the bearing capacity of soft ground by laying a resinous net. In: Proceedings of the symposium on foundations on interbedded sands, Perth, Australia, pp 102–108

9. White DJ, Take WA, Bolton MD, Munachen SE (2001) A deformation measurement system for geotechnical testing based on digital imaging, close-range photogrammetry, and PIV image analysis. In: Proceedings of 15th international conference on soil mechanics and geotechnical engineering, Istanbul, Turkey, pp 539–542
10. Rad NS, Tumay MT (1987) Factors affecting sand specimen preparation by raining. Geotech Test J 10(1):31–37
11. LoPresti DCF, Pedroni S, Crippa V (1992) Maximum dry density of cohesion-less soils by pluviation and by ASTM D4253–83: a comparative study. Geotech Test J 15(2):180–189
12. Kumar J, Bhoi MK (2008) Interference of multiple strip footings on sand using small scale model tests. Geotech Geol Eng 26(4):469–477
13. Love JP, Burd HJ, Milligan GWE, Houlsby GT (1987) Analytical and model studies of reinforcement of a layer of granular fill on a soft clay subgrade. Can Geotech J 24(4):611–622
14. Saha Roy S, Deb K (2017) Bearing capacity of rectangular footings on multilayer geosynthetic-reinforced granular fill over soft soil. Int J Geomech 17(9):04017069–1–17

# Group Efficiency Ratio of Pile Groups Subjected to Harmonic Vibration

**Deepthi Sudhi, Sanjit Biswas, and Shiva Shankar Choudhary**

## 1 Introduction

Recent advancement in industries have upgraded the machine efficiency and thereby its vibration, which will ultimately demand an ideally suited foundation to ensure better stability. In-order to cope with heavy dynamic loads by machines, piles are generally arranged in groups, with varying spacing and configurations. This spacing and arrangement of pile chosen is having a significant role in its dynamic response. Designers generally culminate in dilemma during designing pile groups subjected to dynamic loading, as pile-soil-pile interactions have huge impact on its dynamic response. Therefore, safe and precise estimation of dynamic response of group pile is possible only by considering all contributing factors related to pile-soil-pile interaction in pile group.

Numerous theoretical approaches have been conducted to analyse piles subjected to dynamic loading, out of which boundary method, Winkler's method, continuum approach, finite element method, lumped soil mass model are some of the prominent ones. Continuum approach being a versatile method, several significant studies have been performed [1, 2, 11, 13, 15] for understanding the nonlinear behaviour of soil-pile system. Novak and Mitwally [14] has proposed a method for analysing group piles utilizing dynamic interaction factors. Later, [9], put forward a method

D. Sudhi (✉) · S. Biswas
Department of Civil Engineering, National Institute of Technology, Warangal, Telangana 506004, India
e-mail: ds720024@student.nitw.ac.in

S. Biswas
e-mail: sbiswas@nitw.ac.in

S. S. Choudhary
Department of Civil Engineering, National Institute of Technology Patna, Patna 800005, India
e-mail: shiva@nitp.ac.in

 
B. T. Jose et al. (eds.), *Proceedings of the Indian Geotechnical Conference 2022 Volume 2*, Lecture Notes in Civil Engineering 477,
https://doi.org/10.1007/978-981-97-1741-5_8

for estimating dynamic interaction factors using static interaction factors. Experimental studies are mandatory for checking the reliability of analytical response and for attaining the actual response. Several small scale tests had been performed as it is inexpensive. Recently manifold studies [1, 3, 4, 7, 10] have been broadened over different aspects of pile groups to get a better intuition on the dynamic behaviour of group piles.

This paper focusses on the estimation of stiffness and damping of pile group, based on the studies by Choudhary et al. [3] and Choudhary et al. [4] for 3-pile, 2 × 2, 2 × 3 and 3 × 3, under coupled vibration. Subsequently, Group Efficiency Ratio (GER) which is referred as ratio of stiffness of pile group to sum of stiffness of individual piles is determined, to acquire an extensive idea about the dynamic pile-soil-pile interaction.

## 2 Background

### *2.1 Experimental Study*

Field tests were done on pile groups (3-pile, 2 × 2, 2 × 3 and 3 × 3), with a spacing as, 3 times the diameter of pile, under coupled vibration, to estimate the response of soil-pile system. Forced vibration tests were done on hollow steel piles of outer diameter 114 mm and length 3 m with a thickness of 3 mm, under static loads of 12 kN. Pile and pile cap were connected using specially fabricated connectors and Fig. 1 illustrates the same. As per the SPT test (IS:2131, 1981) performed, the soil around the pile was mostly characterized as clayey and silty soil and detailed soil properties are given in [3]. Figure 2, illustrates the schematic diagram along with its original field set up.

**Fig. 1** Illustration of specially fabricated connectors for connecting pile cap with pile

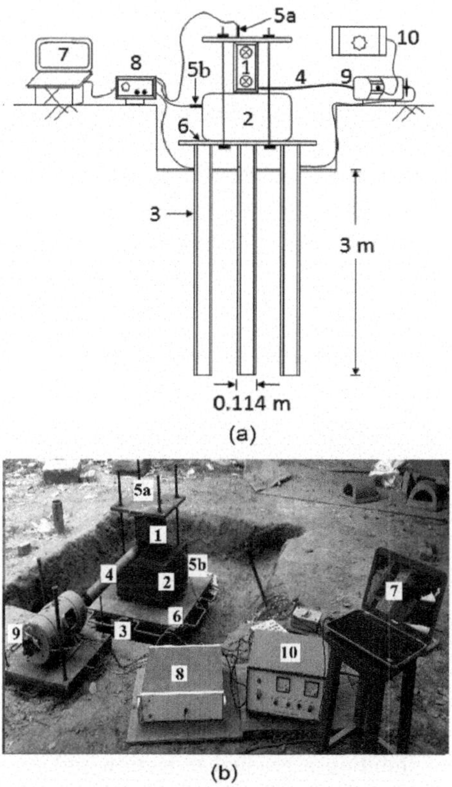


(1) Mechanical oscillator (2) Static load (3) Hollow steel pile (4) Flexible shaft (5) Accelerometer (6) Pile Cap (7) Laptop with software package (8) Data logger (9) DC motor (10) Speed Control Unit

**Fig. 2** **a** Schematic diagram of test set up. **b** Original field test set up

From time versus acceleration response, frequency-amplitude responses were acquired for all pile groups under horizontal and rocking modes separately at various eccentric moments of 0.868, 1.269, 1.631, 1.944 Nm. The plots were exhibiting nonlinear behaviour with an increased resonant amplitude and decreased resonant frequency for an increasing eccentric moments. Details of the experiments and results are expounded in Choudhary et al. [3] and Choudhary et al. [4].

## *2.2 Continuum Approach*

Numerical analysis had been performed on group piles using software DYNA 5, which works based on continuum approach and superposition method. Novak et al. (1974), [14] had estimated stiffness of a single pile and group pile, respectively, in layered medium by assuming linear elastic behaviour of soil and a perfect bond between soil-pile. For indulging nonlinearity of soil into analysis, model proposed by [11] has been used in this particular study.

With a motive on replicating the original field condition, soil-pile separation length was considered along with boundary zone parameters like shear modulus, thickness ratio and damping. Frequency-amplitude responses were obtained for each eccentric moments and boundary zone parameters were adjusted for acquiring a well harmonious behaviour between analytical and experimental response. All the analytical assumptions and responses are given in concerned literature.

## 3 Results and Discussion

### 3.1 Stiffness and Damping of Soil-Pile System

Along with frequency-amplitude response, impedance functions of soil-pile system, which comprises of stiffness and damping were also being estimated for both modes of vibrations for all groupings and single pile. Figures 3 and 4 illustrates the plot of impedance functions of single pile in horizontal mode and rocking mode respectively, for different eccentric moments along with frequency.

From Fig. 3, it is noted that, with an increase in eccentric moments, both stiffness and damping is reducing. This reduction in stiffness can be inferred due to loosening of soil-pile bond at higher vibration. However, for an increased frequency, stiffness is exhibiting a gradual reduction and damping has showcased a rapid decrement, with maximum value at lower frequency. In case of rocking mode of vibration Fig. 4,

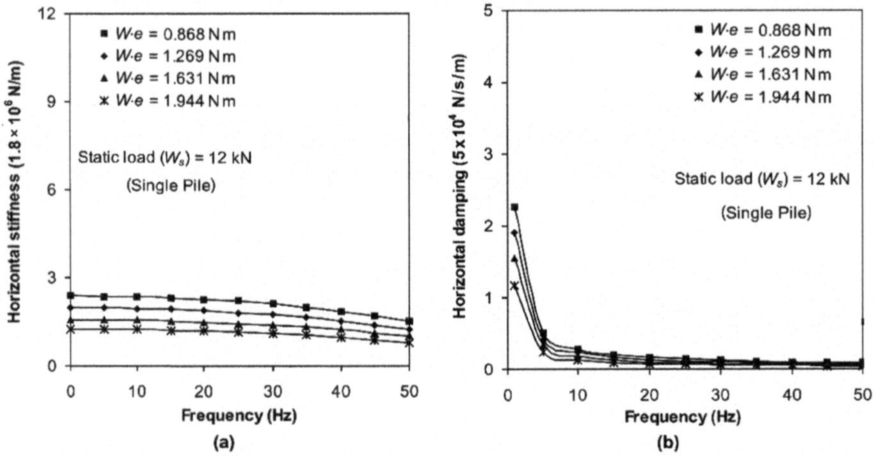


**Fig. 3** Variation of impedance functions of single pile with frequency for various eccentric moments in horizontal mode of vibration **a** stiffness **b** damping

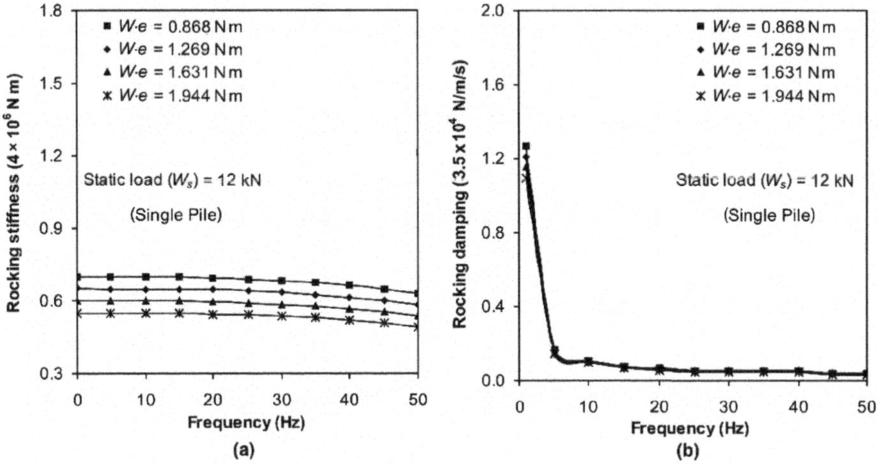


**Fig. 4** Variation of impedance functions of single pile with frequency for various eccentric moments in rocking mode of vibration **a** stiffness **b** damping

both stiffness and damping are decreasing with increasing eccentric moments and responses are similar to horizontal mode of vibration. But, the variation of damping with eccentric moment is found to be trivial, with peak value at lower frequency. Similar pattern has been also observed in all the other pile groups.

Stiffness and damping of various modes of vibration for different pile groupings has been studied along with frequencies for a particular eccentric moment as illustrated in Fig. 5. Figure 5a shows that, as the number of piles in a group increases, there observed an increase in horizontal stiffness, which is subsequently declining with increasing frequency. This decrement in stiffness can be interpreted due to the loosing of soil-pile bond at higher frequency. However, it is found that horizontal damping is maximum at lower frequency and rapidly decreasing, with an increase in frequency. This exemplifies that, for an increased frequency, frequency independent material damping is changing into frequency dependent viscous damping. The figure also expounds that, as the number of piles in group has a direct influence on damping.

From Fig. 5b it is found that, rocking stiffness and damping exhibits same variation as that of horizontal mode of vibration, but in case of rocking stiffness, the response is almost steady for a particular pile group, in spite of frequency increment.

### *3.2 Group Efficiency Ratio (GER)*

Group Efficiency Ratio (GER) is a term used to get a better intuition for understanding the impacts of pile-soil-pile interaction on dynamic stiffness and damping in case of group piles. GER can be expressed as following:

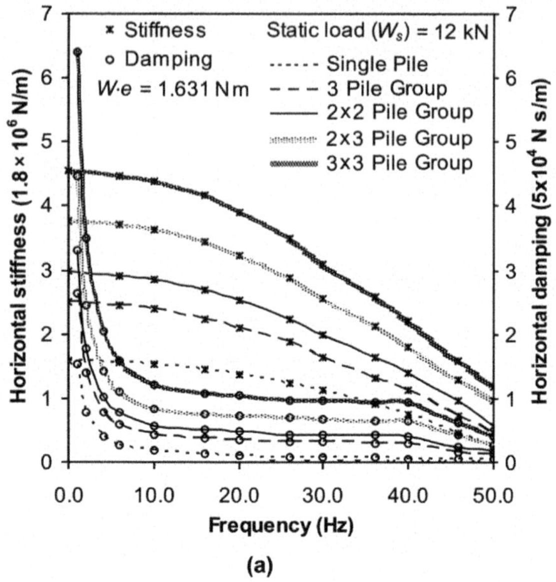


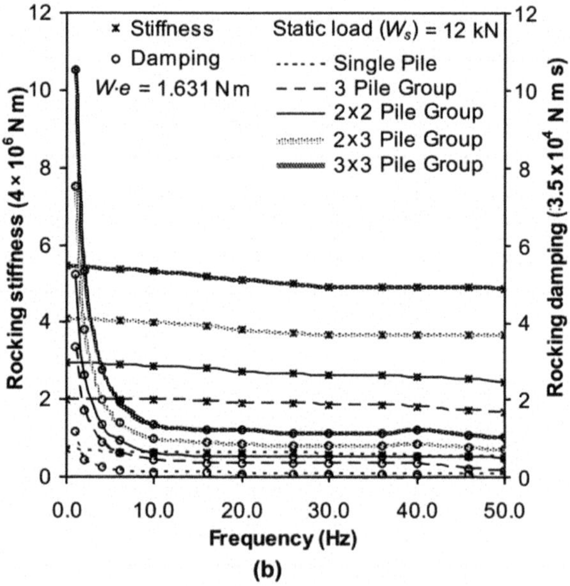


**Fig. 5** Variation of soil-pile system stiffness and damping with frequency for various pile groupings **a** horizontal mode **b** rocking mode

$$GER = \frac{\text{Stiffness of pile group}}{\text{Sum of stiffness of individual piles}}$$

The variations on GER for horizontal stiffness/damping and rocking stiffness/damping with frequency are shown in Fig. 6. For both modes, plot exhibits decrement in GER with an increase in number of piles. From the plot, it is found that, GER for horizontal stiffness is exhibiting a value less than 1, which depicts that soil-pile interaction has significant impact on the stiffness of pile groups for horizontal motion. However, with regard to GER for rocking stiffness, it is found that values are close to 1, for the frequency range considered in the analysis.

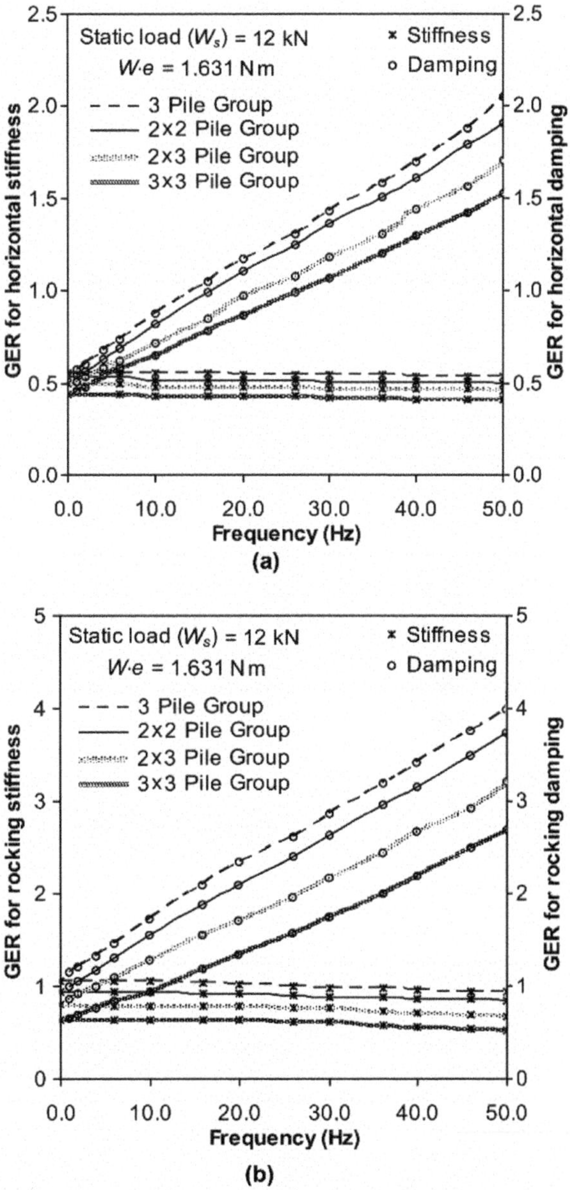


**Fig. 6** Variation of GER with frequency for different pile groups horizontal mode **b** rocking mode

For an increase in frequency, GER of stiffness is reducing and GER of damping is increasing for horizontal as well as rocking vibration, whereas the GER for damping in both cases is increasing to a value higher than 1, which infers that the damping of the group system is increased by pile-soil-pile interaction.

## 4 Conclusions

This paper has studied broadly about the dynamic response of various pile arrangements under coupled vibration along with its impedance functions. Field experiments and analytical studies to mimic the same response has been performed by Choudhary et al. [3] and Choudhary et al. [4]. Additionally, this small scale tests details can be used for analysing responses of full scale piles by using appropriate scaling laws available. Further, this particular study has concentrated on understanding the behaviour of piles arranged in different pattern. Some of the conclusions are being expounded below:

- Stiffness and damping for both modes of vibration is observed to be decreasing with an increase of eccentric moment. Stiffness reduction can be inferred owing to the loosening of soil-pile bond at immense vibration.
- For an increased frequency, stiffness has been exhibiting a gradual reduction and damping showcased a sudden reduction, with peak value at lower frequency. This can be inferred as the frequency independent material damping conversion into frequency dependent viscous damping.
- The stiffness and damping of the soil-pile system improves with the number of piles in a group for both horizontal and rocking modes of vibrations.
- GER concept is introduced to estimate the out-turn of pile-soil-pile interaction in group piles and it is found that GER for both impedance function decreases with an increase in number of piles in group.
- GER for horizontal stiffness is observed to be less than 1, due to intense soil-pile interaction on group piles, whereas GER for rocking stiffness is found close to 1.
- In case of GER for damping in both modes, it is noted that, with an increase in frequency, GER is increasing to a value more than 1.

## References

1. Biswas S, Manna B (2018) Experimental and theoretical studies on the nonlinear characteristics of soil-pile systems under coupled vibrations. J Geotech Geoenviron Eng 144(3):04018007. https://doi.org/10.1061/(asce)gt.1943-5606.0001850. Author F (2016) Article title. J 2(5):9–110
2. Biswas S, Ralli R, Manna B, Choudhary SS, Datta M (2022) Vertical nonlinear response of single and 2 × 2 group pile under strong harmonic excitation. Sadhana—Acad Proc Eng Sci 47(1). https://doi.org/10.1007/s12046-021-01789-9

3. Choudhary SS, Biswas S, Manna B (2020) Experimental and numerical study of pile foundations subjected to rotating machine-induced coupled excitations. Int J Geotech Eng 14(6):614–625. https://doi.org/10.1080/19386362.2019.1620536
4. Choudhary SS, Biswas S, Manna B (2021) Effect of pile arrangements on the dynamic coupled response of pile groups. Geotech Geol Eng 39(3):1963–1978. https://doi.org/10.1007/S10706-020-01599-6
5. El-marsafawi BH, Han YC, Novak M (1992) Dynamic experiments on two pile groups. 118(4):576–592
6. Elkasabgy M, El Naggar MH (2013) Dynamic response of vertically loaded helical and driven steel piles. Can Geotech J 50(5):521–535. https://doi.org/10.1139/cgj-2011-0126
7. Goit CS, Saitoh M, Mylonakis G (2016) Principle of superposition for assessing horizontal dynamic response of pile groups encompassing soil nonlinearity. Soil Dyn Earthq Eng 82:73–83. https://doi.org/10.1016/j.soildyn.2015.11.001
8. Han YC, Sabin GCW (1995) Impedances for radially inhomogeneous viscoelastic soil media. J Eng Mech 121(9):939–947.https://doi.org/10.1061/(asce)0733-9399(1995)121:9(939)
9. Kaynia and Kausel (1982) Dynamic behaviour of pile groups. In: Proceedings of 2nd international conference on numerical methods of offshore pilings, Austin, Texas, pp 509–532
10. Manna B, Baidya DK (2010) Dynamic nonlinear response of pile foundations under vertical vibration-theory versus experiment. Soil Dyn Earthq Eng 30(6):456–469. https://doi.org/10.1016/j.soildyn.2010.01.002
11. Novak' M, Asce M, Sheta M (1980) Approximate approach to contact effects of piles
12. Sharnouby BE, Novak M (1986) Flexibility coefficients and interaction factors for pile group analysis. Canadian Geotech J 23(4)
13. Novak M, Fakhry A-E (1978) Impedance functions of piles in layered media. J Eng Mech Divis 641–652
14. Novak N, Mitwally H (1990) Random response of offshore towers with pile-soil-pile interaction. J Offshore Mech Arct Eng 112(1):35–41. https://doi.org/10.1115/1.2919833
15. Sinha SK, Biswas S, Manna B (2015) Nonlinear characteristics of floating piles under rotating machine induced vertical vibration. Geotech Geol Eng 33(4):1031–1046. https://doi.org/10.1007/s10706-015-9885-5

# Effect of Fin Inclination on Different Shapes of Fin Under Axial Load

**Poonam P. Gawande and A. I. Dhatrak**

## 1 Introduction

In case of deep foundation piles are more commonly used where loose soil and large loads comes in frame. Piles are more significant in taking vertical loads. The numerous structures such as bridge abutment, ocean engineering structures, etc. are subjected to huge amount of vertical load. In case of circular pile, vertical resistance offered are insufficient as required, that's why there is need of improvement in foundation techniques.

Spin fin pile is best alternatives over these traditional circular piles. These fin pile foundation was first time developed by PND Engineers in 1983. The representations of fin pile are such as circular pile welded with steel plates known as fins. These fins are generally fitted at an upper or bottom portion of circular piles making slight angle with pile periphery. "The geometry of these fin pile is such that when it is seen from bottom to top, the top of one fin meets the bottom of the adjacent fin, thus giving 360 degree coverage. These fins are fitted along the circumference of the circular pile foundation" [6]. Both traditional impact and vibratory hammers, along with templates and attachments, can successfully drive spin fin piles. These piles are more effective in taking loads as compare to circular pile because, when it is pushed into earth fill it actually rotates like screw. This rotation effect is caused by fins. These results in formation of strong bond with nearby soil, fins making it firmly fix into the soil. A soil cone is generated when the pile is pushed in compression, thus increasing the pile's axial capacity.

Fin piles provide predominantly higher ultimate lateral capacity and lateral resistance when compared with a regular circular reference pile [5, 10–12] . The different shapes of fin such as triangular [2, 11, 15] , rectangular [3, 11, 14, 17, 18], tapered

P. P. Gawande (✉) · A. I. Dhatrak
Civil Engineering Department, GCoE, Amravati 444602, India
e-mail: poonamgwnd25@gmail.com

 
B. T. Jose et al. (eds.), *Proceedings of the Indian Geotechnical Conference 2022 Volume 2*, Lecture Notes in Civil Engineering 477,
https://doi.org/10.1007/978-981-97-1741-5_9

[7] are effective in carrying. They found that increase in fin area influences capacity of pile.

Position of fins near the pile top provides higher resistance than pile bottom for lateral loading condition [7, 11]. Fins attached at the bottom of spin fin pile provide higher vertical load capacity in comparison with conventional pile [17, 18]. Also, fins placed at the pile end have a significant impact in increasing the uplift load capacity [15].

However, these studies of fin piles are limited up to straight fin for different loading only. Fin inclination effect on vertical load needs attention.

## 2 Numerical Modeling

Numerical analysis was done by using MIDAS GTS—NX 3D software. The various Geotechnical Engineering problems such as tunneling, piling etc. are solved by MIDAS software package. Investigation was done by using a 3D finite element model for spin fin pile having triangular shape and trapezoidal shape. In this analysis, linear elastoplastic sand was assumed as a foundation material. The simple and precise Mohr–Coulomb model was used to replicate the nonlinear behavior of sand. Drained conditions was preferred for modeling of fin and pile. Five essential inputs are needed for numerical modeling, which is cohesion (c), frictional angle ($\varphi$), young's modulus ($E$), Poisson's ratio ($\mu$), and dilatancy ($\Psi$). The direct shear test was used for finding frictional angle and cohesion, and unit weight was determined from modified proctor test for sands.

Sand properties used for analysis are shown in Table 1. Spin fin piles were analyzed to allow comparisons of the fin inclinations with triangular shape fin and trapezoidal shape fin. Fins are placed at $0^0$ to the pile and $90^0$ to the pile; this represents straight orientation and inclined orientation of fin with circumference of pile as presented in Figs. 1 and 2. The model fin pile used in analysis was made up of hollow circular mild steel pile having outer diameter ($D$) 1.2 m with wall thickness 75 mm. At the bottom of pile, four numbers of fins were welded to each other. Because of axial load fins were welded onto the bottom of pile. The geometry of spin fin pile is same for both the shapes of fin. In case of trapezoidal fin, the ratio of top width of fin to bottom width of fin ($B_{tf}/B_{bf}$) was kept 0.5. Hollow portion of piles was filled with sand having same properties as shown in Table 1. The dimensions of the spin fin piles used in the numerical analyses are given in Table 2.

Assumption was made that fin pile to be driven in normally consolidated soil having $Ko = 0.42$. In order to reduce the estimated strength of structural members, when they come into contact with one another, a strength reduction factor is used. The strength reduction factor should be given in between interface surface element, which is sand–spin fin pile interaction. The value of strength reduction factor is considered 0.65. This factor is directly taken from sand-steel interfaces. The relations between interface properties, soil strength parameters, and strength reduction factor of soil layer are as follows:

**Table 1** Sand properties

| Property | Loose sand | Medium dense sand |
|---|---|---|
| Relative density (%) | 40 | 55 |
| Unit weight (kN/m$^3$) | 16.33 | 16.5 |
| Young's modulus (MPa) | 20 | 27 |
| Poission's ratio | 0.3 | 0.3 |
| Friction angle | 34 | 37.88 |
| Cohesion (kN/m$^2$) | 1 | 2 |

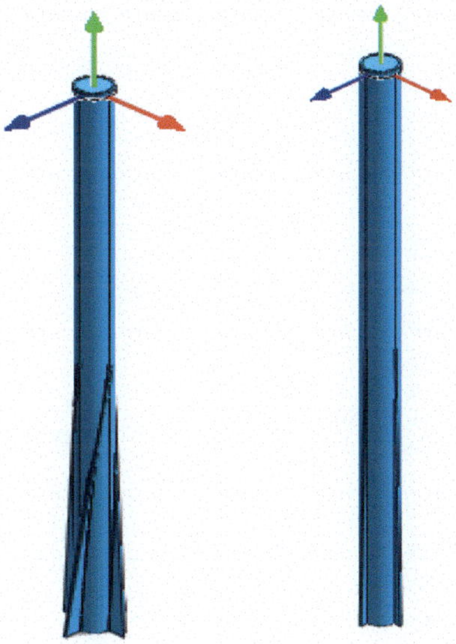

**Fig. 1** Triangular shape fin pile in 3D view with **a** inclined fin, **b** straight fin

$$\tan\phi^i = R_{\text{int}} \tan\phi\prime C_i = R_{int} C\prime$$
$$\Psi_i = 0 \text{ if } R_{\text{int}} < 1 \text{ otherwise } \Psi_i = \Psi,$$

where $\phi^i$, $\Psi$, and $C_i$ are the frictional angle, dilatancy angle of interface and cohesion of soil, respectively.

## 3 3D Geometry of Fin Pile and Meshing

Three-dimensional geometrical model of two different shapes was developed in MIDAS GTS NX software for the analysis. Triangular and trapezoidal shapes of fin pile were modeled with soil block as a foundation material. "A soil bed with sides 22.5 times the fin pile's diameter and a depth 2.5 times its length was chosen"

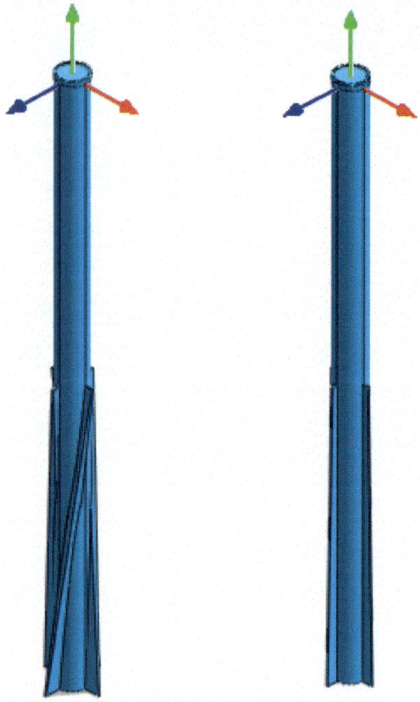

**Fig. 2** Trapezoidal shape fin pile in 3D view with **a** inclined fin. **b** straight fin

**Table 2** Spin fin pile properties

| Property | Value |
|---|---|
| Pile diameter, D (m) | 1.2 |
| Pile thickness (m) | 0.075 |
| Fin thickness (m) | 0.075 |
| Pile length (m) | 24 |
| Slenderness ratio | 20 |
| Fin length/Pile length | 0.5 |
| Fin width (m) | 0.6 |
| No. of piles | 1, 3, 4, 5, 6 |
| Pile spacing, S(m) | 3*D |
| Size of Pile cap (m) | S + 2D + 0.3 |
| Depth of Pile cap (m) | 0.6 |
| Material | Mild steel |
| Poisson's ratio | 0.3 |
| Young's modulus (GPa) | 200 |
| Unit weight (kN/m$^3$) | 78 |

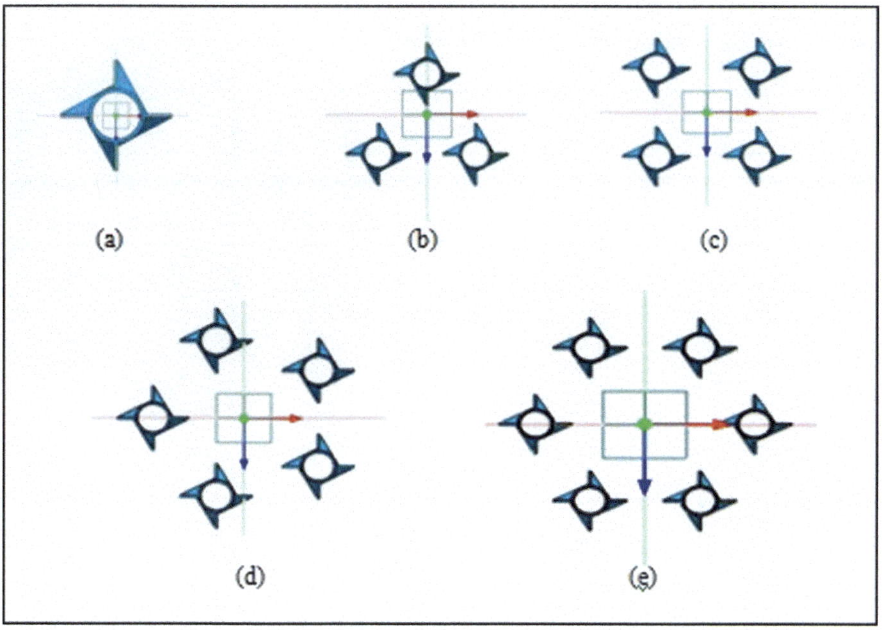


**Fig. 3** Plan view of group of triangular shape inclined fin pile arrangement for single; **b** three; **c** four; **d** five, and **e** six

[12]. Depth pile cap was kept constant 0.6 m for analysis and area was varying according to pile group. The axial capacities for triangular and trapezoidal shape fin piles were analyzed in comparison with fin inclinations. For comparison purpose, the cross-sectional area of inclined fin and straight fin pile was kept same, but the cross-sectional area of trapezoidal fin and triangular fin was different.

The investigations were executed by changing different fin shape for single and group of pile. Pile group arrangement as presented in Fig. 3. Three-dimensional view of four group fin pile embedded in sand is presented in Fig. 4. Boundary conditions and loading can be applied to geometric shapes, nodes, and elements also.

## 4 Results and Discussions

The load settlement curves for triangular shape and trapezoidal shape spin fin pile is plotted. As per criteria set into IS: 2911–2013 (Part-IV) [8] , the ultimate axial load of pile was taken. Axial load for different shapes of fin pile was compared with selected parameters such as density and pile group. Load–settlement (*P–Y*) curves extracted from numerical analyses for spin fin pile with loose sand is shown in Figs. 5, 6, 7, 8, and 9. The same patterns are followed for medium dense spin fin pile. But, some of the curves are shown due to space restriction.

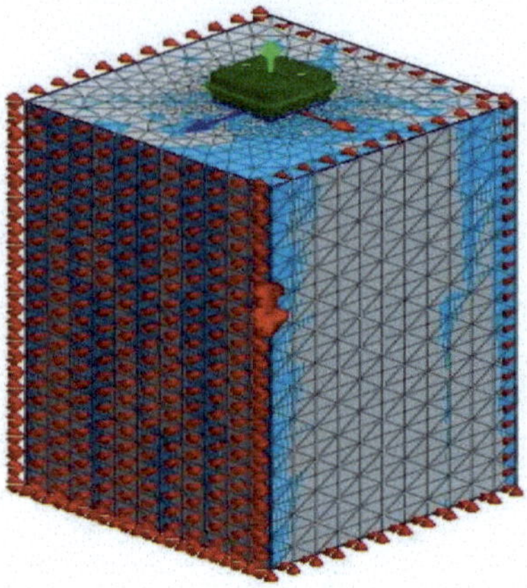

**Fig.4.** Three-dimensional view of four group of fin pile embedded in soil

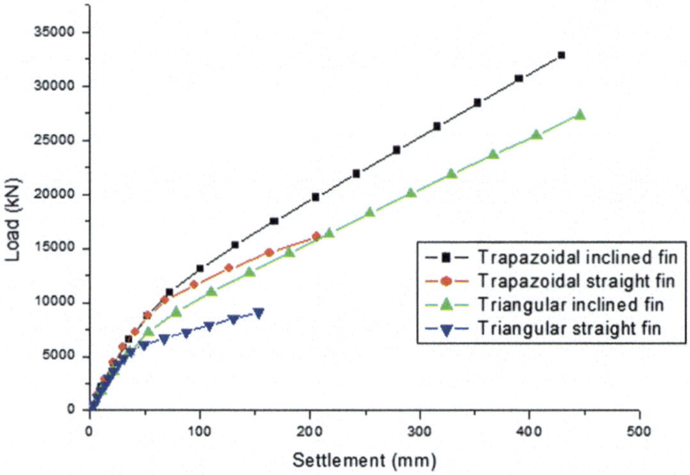


**Fig. 5** P–Y curves for single fin pile

## *4.1 Effect of Fin Shape on Vertical Capacity*

The comparison between rectangular fins and triangular fins with different area, but same width for lateral loading was made by Nasr [10], Sakr et. al. [14] and Bariker et. al. [11]. To ascertain how fin affects performance, analyses were performed on triangular and trapezoidal fins for vertical loading. Figures 5, 6, 7, 8 and 9 depicted the variations of vertical load vs vertical deflection for different fin shapes. It is prominently observed that the vertical load of the pile improves significantly when fins of any shape are used, as the inclusion of fins at the pile's base increases soil resistance [17]. It follows that the axial load of piles with trapezoidal-shaped fins increased by around 26% compared to heaps with triangular-shaped fins. This is

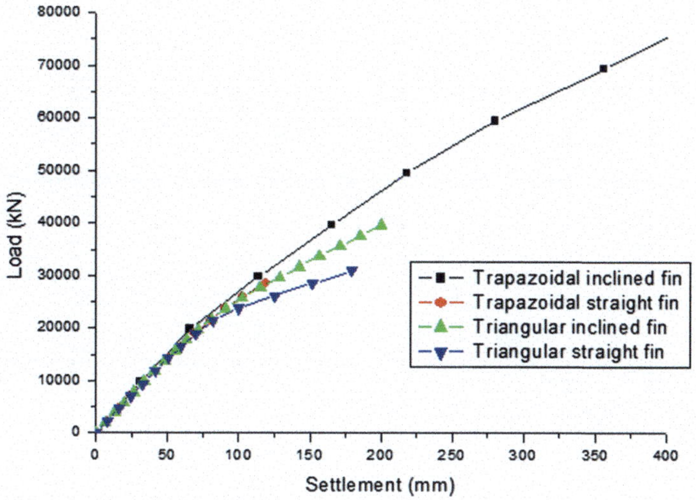


**Fig. 6** P–Y curves for group of three fin pile

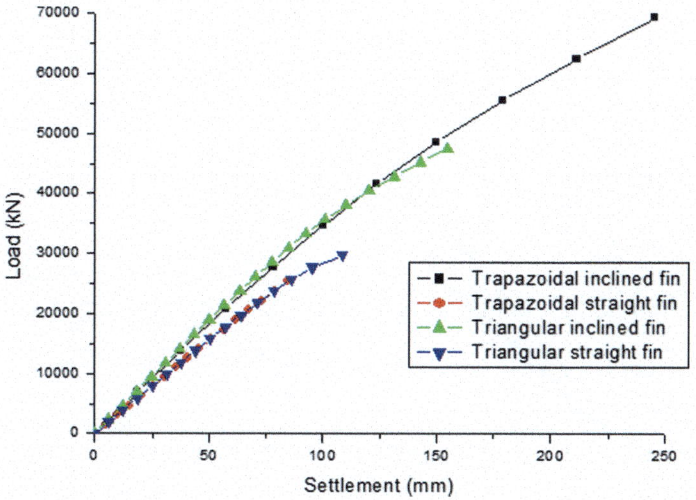


**Fig. 7** P–Y curves for group of four fin pile

explained by the larger area of soil resistance present in a trapezoidal finned pile, which raises the ultimate vertical load even more.

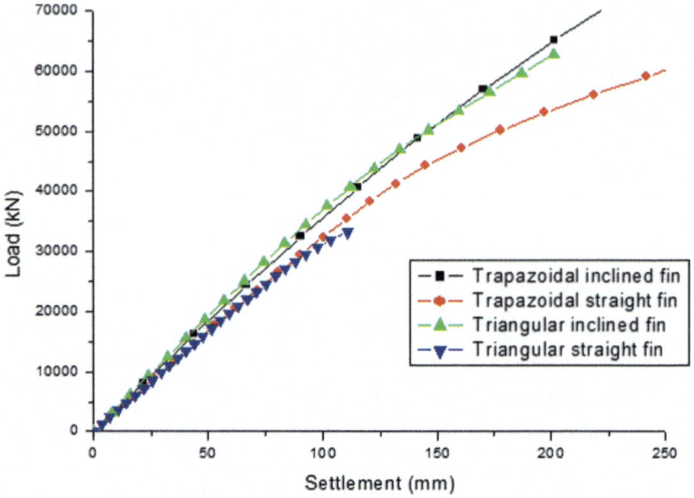


**Fig. 8** P–Y curves for group of five fin pile

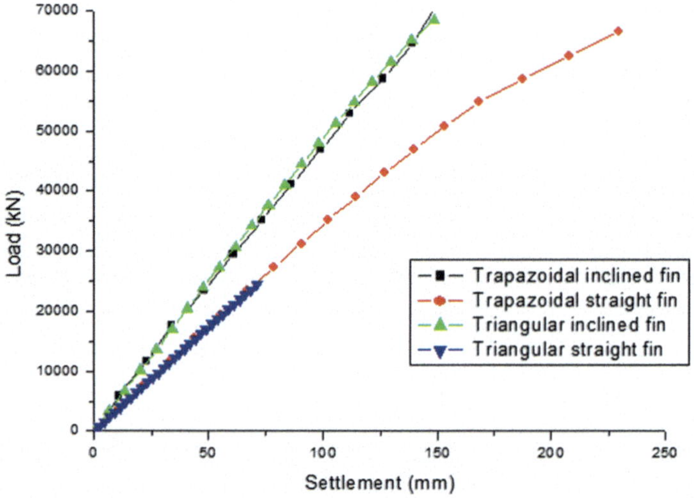


**Fig. 9** P–Y curves for group of six fin pile

## *4.2 Effect of fin Inclination on Vertical Capacity*

From Figs. 10 and 11, it is evident that piles with inclination fins offer a significantly stronger vertical resistance and stiffer behavior than that of straight fin for both the shape of fins. Based on the outcomes for slenderness ratio of $L/D = 20$ the axial load of incline finned pile is increased by about 43 and 45%, over that of a straight finned pile for both the shape of fins, i.e., trapezoidal shape of fin and triangular shape fin

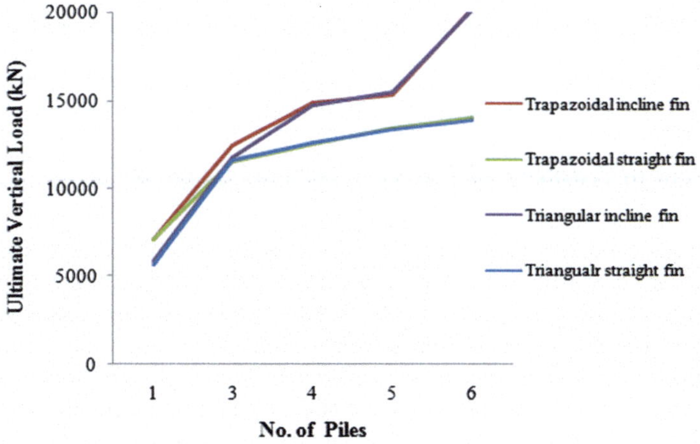


**Fig.10** Variation of no. of piles versus ultimate vertical load for loose sand

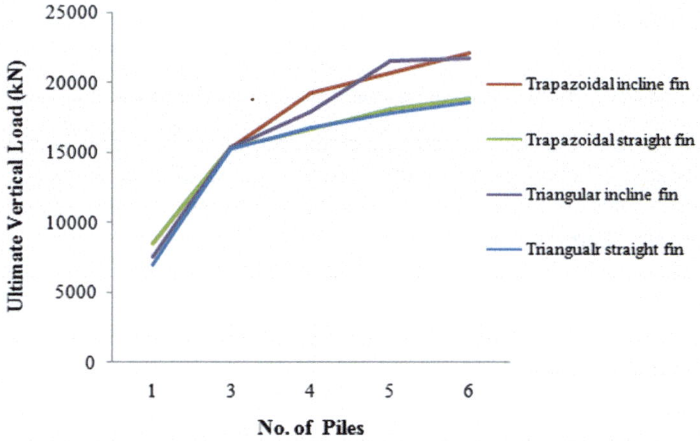


**Fig.11** Variation of no. of piles versus ultimate vertical load for medium dense sand

pile, respectively. The inclined finned pile has the more vertical resistance capacity because it covers whole peripheral surface of the pile.

### *4.3 Effect of Sand Density on Vertical Capacity*

Figures 10 and11 confirm that there is increment in ultimate vertical load with sand density for trapezoidal shape of fin and triangular shape fin pile. As sand density rises, the stiffness and shear strength of the sand also increases. In case of triangular shape fin pile, as sand density went from 40 to 55%, the increment in the axial load

# Undrained Uplift Capacity of Under-reamed Pile in Layered Clays

**Rajat Rathore, Ritesh Ingle, Yogesh D. Patil, and Jitesh T. Chavda**

## 1 Introduction

Conventional cylindrical piles will not be able to provide the sufficient uplift resistance required for the tall structures like transmission line towers, patrolling towers, chimneys, tall and slender structures, etc.; therefore, in such cases, the under-reamed piles are used to withstand the uplift forces. In case of piles in clay, the cohesion strength may vary linearly throughout its depth, or the soil deposit may have layered clay as soft over hard clay and vice-a-versa. Therefore, it is necessary to assess the undrained uplift capacity of the under-reamed pile in layered clayey soil.

The research that has already contributed to the understanding on behavior of under-reamed pile is reasonably addressed [1–7]. According to Martin and Stephen [1], a twin bulb under-reamed pile can be used as an alternate foundation for over-consolidated hard clay. In terms of dimensional stability and homogeneity for the piles, Peter et al. [2] reported that self-compacting concrete performs better. Contrarily, piles formed of vibrated concrete exhibit comparable compression and uplift outcomes. George and Hari [5] used finite element analysis to assess the compressive and uplift capacity of the under-reamed pile in homogeneous clay. The uplift capacities of conventional pile, single under-reamed pile, and double under-reamed pile in clay with linearly increasing cohesion of soil with depth were determined using finite element limit analysis by Khatri et al. [7]. With reference to other areas of research on under-reamed pile, the undrained uplift capacity of the

R. Rathore (✉) · Y. D. Patil · J. T. Chavda
Department of Civil Engineering, Sardar Vallabhbhai National Institute of Technology, Surat, Gujarat, India
e-mail: rajatrathore21081997@gmail.com

R. Ingle
Department of Civil Engineering, National Institute of Technology, UT of Jammu & Kashmir, Srinagar, Jammu and Kashmir, India



B. T. Jose et al. (eds.), *Proceedings of the Indian Geotechnical Conference 2022 Volume 2*, Lecture Notes in Civil Engineering 477,
https://doi.org/10.1007/978-981-97-1741-5_10

under-reamed pile in layered clay is still not completely understood. In this regard, the undrained uplift capacity for the layered clay as soft over hard clay and vice-a-versa with a cohesion ratio of 2 is evaluated in the study for varying clay thickness ratios as $h/H = 0$ to 1 and different $L/D$ ratios of pile.

## 2 Problem Statement

In the study, the under-reamed pile with single bulb in soft clay overlain hard clay and hard clay overlain soft clay is simulated to determine the failure load for various $h/H$ ratios of layered clay and varying $L/D$ ratios of pile. For this, the finite element software Plaxis 2D is used. The codal provision (IS 2911 Part III [8]) is used to determine the geometry (Fig. 1, $D =$ diameter of shaft, $Du = 2.5$ times diameter of bulb) of the under-reamed pile. The pile length is kept constant as $L = 4.5$ m, and the $L/D$ ratio is varied as 5, 15, and 20. The diameter of the shaft is varied to vary the $L/D$ ratio. The pile is modeled using Linear Elastic model, while the soil is modeled using Mohr–Coulomb model, and the properties of the soil and pile are provided in Table 1.

The cohesion ratio of soil is taken as 2 in the study, where soft soil is having undrained cohesion value of 15 kPa and hard clay is having the undrained cohesion value of 30 kPa. The layered clay is modeled by considering soft clay overlain hard clay and hard clay overlain soft clay, and the value of undrained uplift capacity is determined. Based on several trials, a fine mesh (Fig. 2) is selected in the study, and it appears to be the most appropriate for this analysis and the water table level

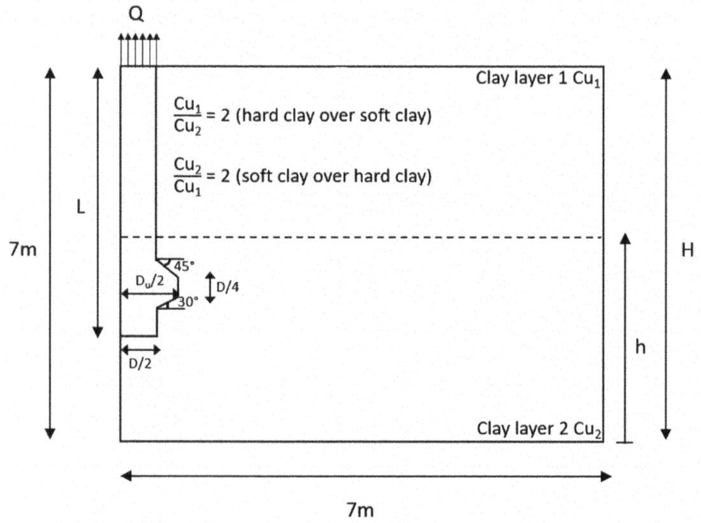


**Fig. 1** Schematic representation of single under-reamed pile in layered clay

**Table 1** Material properties of soil and pile used in study [after George and Hari [3]]

| Parameter | Soil | Pile |
|---|---|---|
| Frictional angle ($\phi$) | 0° | – |
| Unit weight ($\gamma$) | 16 kN/m$^3$ | 27 kN/m$^3$ |
| Young's modulus ($E$) | 15 MPa | 31,000 MPa |
| Poisson's ratio ($\nu$) | 0.35 | 0.15 |

is considered at ground level, i.e., the top boundary of FE model. The soil has the following values: friction angle = 0°, unit weight = 16 kN/m$^3$, Young's modulus = 15 MPa, and Poisson's ratio = 0.35. The pile is modeled using Linear Elastic with the following values: unit weight = 27 kN/m$^3$, Young's modulus = 31,000 MPa, and Poisson's ratio = 0.15.

In this study, the $h/H$ ratio is varied by considering the variation in $h$ from the bottom of FE domain (see Fig. 3), where the overall height of model $H = 7$ m which is fixed and the thickness of top clay layer is varied from 0 to 7 m to have the variation in $h/H$ ratio. For each fixed value of the $L/D$ ratio, i.e., 5, 15, 20, the undrained uplift capacity of the under-reamed pile in both the cases of soft clay overlain hard clay and hard clay overlain soft clay is evaluated. Then, considering both cases, i.e., soft overlain hard and vice-a-versa, a design chart is prepared.

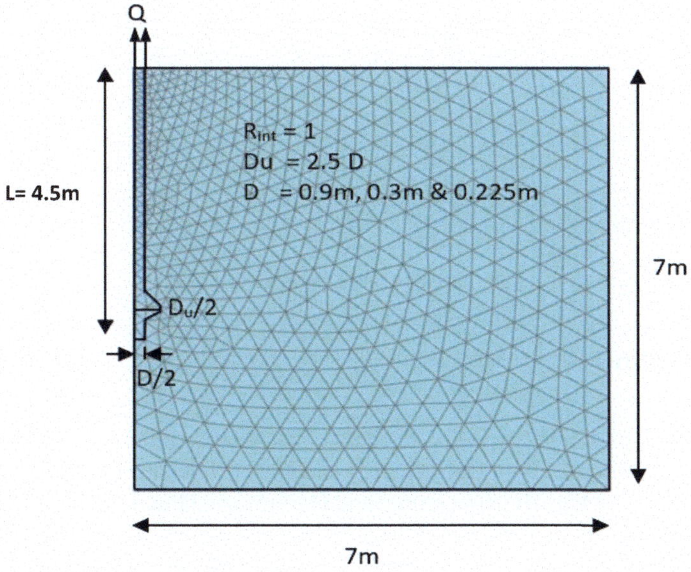


**Fig. 2** FE model of single under-reamed pile in layered clay

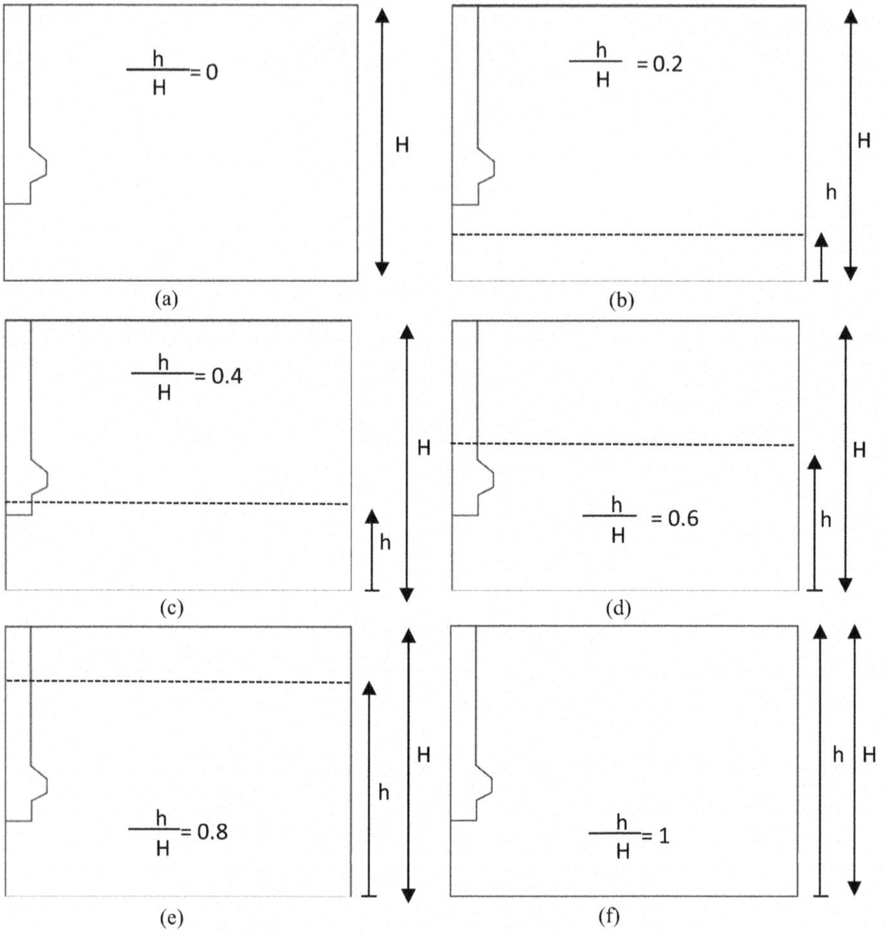


**Fig. 3** Representation showing the variation in h/H ratio from 0 to 1

## 3 Validation

In the study, the undrained capacity of single under-reamed pile is evaluated using FE analysis considering the data from George and Hari [5], Khatri et al. [7], and Codal guidelines (IS 2911 Part III [8]). George and Hari [5] used Plaxis 3D and Khatri et al. [7] used Optum G2 in their studies. The geometry and model characteristics are same as used by George and Hari [5], which are listed in Table 2. Based on present FE analysis, the load-deformation plot is obtained, and then the FE results are compared with the solution of George and Hari [5], Khatri et al. [7], and IS 2911 Part III [8] (see Fig. 4). The present study results match well with the value obtained from IS 2911 Part III and reasonably well with George and Hari [5] and Khatri et al. [7]. Then using the same validated model, the layered clay is introduced with varying

**Table 2** Properties of soil and pile as per George and Hari [5]

| Parameter | Soil | Pile |
|---|---|---|
| Frictional angle ($\phi$) | 1° | – |
| Unit weight ($\gamma$) | 16 kN/m$^3$ | 27 kN/m$^3$ |
| Cohesion ($c$) | 15 kPa | – |
| Young's modulus ($E$) | 15 MPa | 31,000 MPa |
| Poisson's ratio ($\nu$) | 0.35 | 0.15 |
| Length ($L$) | – | 4.5 m |

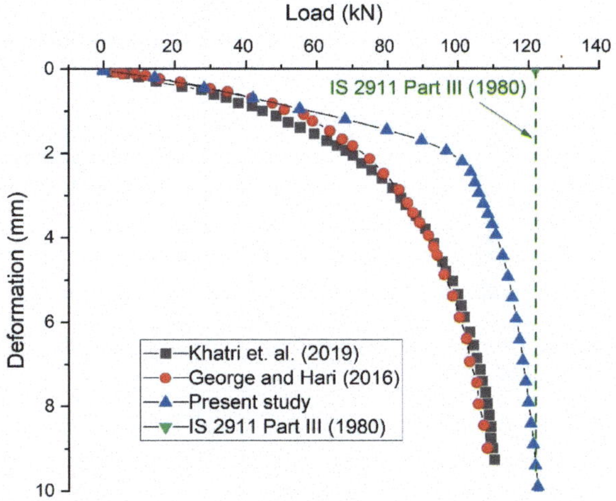


**Fig. 4** Comparison of present study results with literature

h/H ratios for different L/D ratios of under-reamed pile subjected to uplift load in undrained clay.

## 4 Results and Discussion

In the study, the undrained capacity of under-reamed pile in layered clay is evaluated and is normalized by dividing the ultimate failure load of pile in layered clay with the ultimate failure load of pile in single clay layer. This normalized factor is reported as $Fc$. The $Fc$ for under-reamed pile in layered clay with soft clay overlain hard clay and hard clay overlain soft clay is evaluated and depicted in Fig. 5. For each $L/D$ ratio, considering the soft undrained clay ($cu = 15$ kPa) overlain by hard undrained clay ($cu = 30$ kPa), the undrained uplift capacity of under-reamed pile increases with increase in the $h/H$ ratio, indicating that the presence of a thicker hard clay layer below the soft clay layer results in higher undrained capacity. Similarly, the $Fc$ reduces for a thicker soft clay layer below the hard clay layer which results in lower undrained capacity. The results from the study are also provided in Table 3.

7. Kumar A, Khatri VN, Gupta SK (2020) Effect of linearly increasing cohesion on the compression and uplift capacity of the under-reamed pile in clay. SN Appl Sci 2(2):1–17
8. IS: 2911 Part III. 1980. Indian Standard Code of Practice for Design and Construction of Pile Foundations, Under-reamed Piles. Bureau of Indian Standard

# A Review of the Design and Axial Capacity Calculation Practices for Helical Piles

**Arnab Debnath and Vishwajeet. P. Singh**

## 1 Introduction

Helical piles are low displacement deep foundation component which are installed by screwing into the ground. Their utilization has increased over the last few decades because of their several advantages over conventional piles and the development of high torque generating machines [21].

Traditionally, helical piles were used as foundation for transmission towers to resist uplift loads or to resist smaller compressive forces, e.g., from residential buildings. But when scaled-up, they can also be used to resist heavy loads. They are now widely used for residential buildings, machine foundation, telecommunication towers, and boardwalks in environmentally sensitive areas. In addition, they are also used as foundation for resisting tensile and lateral loads, e.g., guy anchors in poles and towers, tiebacks for retaining walls, foundation tie-down, offshore structures, tilt-up braces, and for slope stabilization.

They have various advantages over conventional pile foundation systems. They are easy and quick to install, produce less vibration, can be used for remedial works, and can be loaded immediately after installation. Moreover, they are environmentally friendly as they do not require concrete, do not produce spoils of excavation, and can be removed, reused, and recycled.

This paper discusses the design aspects, failure mechanism, and the various factors influencing the capacity of helical piles. The most common methods used for the prediction of axial capacity of helical piles have been presented. A 2D finite element

A. Debnath (✉) · Vishwajeet. P. Singh
Motilal Nehru National Institute of Technology, Prayagraj 211004, India
e-mail: arnab.2020rce04@mnnit.ac.in

Vishwajeet. P. Singh
e-mail: vps15783@mnnit.ac.in

 
B. T. Jose et al. (eds.), *Proceedings of the Indian Geotechnical Conference 2022 Volume 2*, Lecture Notes in Civil Engineering 477,
https://doi.org/10.1007/978-981-97-1741-5_11

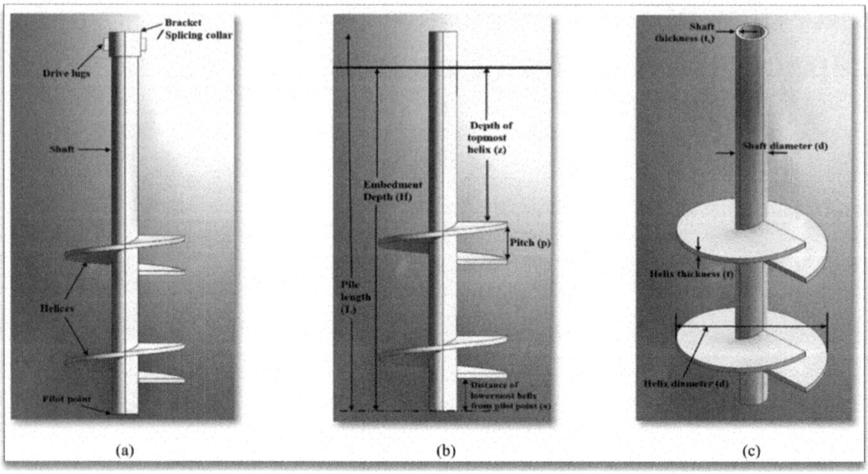


**Fig. 1** Definitions and components of helical pile: **a** Main components of a typical helical pile; **b** and **c** Definitions of a typical helical pile

analysis (FEA) has also been conducted on helical pile, and the results have been compared with field test results obtained from the literature.

## 2 Typical Helical Pile Model

### *2.1 Main Components of Helical Piles*

A standard helical pile is made of three main sections:

- **Shaft**: The shaft provides resistance through skin friction and helps to transfer the superstructure load to the helix.
- **Termination**: The termination is comprised the drive lugs which helps to screw the pile into the soil, a splicing collar or bracket.
- **Bearing elements**: They consist of helices, which are the most important component for providing load resistance in helical piles (Fig. 1).

### *2.2 General Specifications*

The general design considerations and definitions conforming to the International Building Code (IBC 2009), Acceptance Criteria (AC358):2007, Chance Technical Design Manual [5], HAI engineering design manual [10], and some important studies, which are extensively followed for most research works and field practices, are listed below:

- *Steel Grade*: A wide variety of steel is used for manufacturing helical piles depending upon the project requirements. Generally, helical piles are manufactured using Grade 50 steel better.
- *Shaft wall thickness* ($t_s$): $t_s$ is selected to resist loads without buckling or structural failure, depending upon the requirements of a project. $t_s$ is typically taken between 5.5 mm and 9.5 mm.
- *Shaft diameter* ($d$): Generally, round shaft helical pile diameters range from 73 to 305 mm. Square shaft is also used in cases where there is no problem of buckling or bracing.
- *Shaft length* ($L$): The shaft is manufactured in lengths ranging from 1 to 6 m. Additional extension shafts can also be used to connect to the main shaft if the pile needs to be installed at greater depths.
- *Helix diameter* ($D$): Typically, helices of diameters 150–400 mm are manufactured. The helix diameter varies from 2 to 4 times the shaft diameter.
- *Helix thickness* ($t$): The thickness varies from 5.5 to 25.4 mm depending on the project requirements [24].
- *Pitch* ($p$): Helical pile manufacturers use a standard pitch of 3 inches (76.2 mm).
- *Number of helices* ($n$): Helical pile consists of a minimum of one helix. Additional helices may be welded to the shaft to increase the pile capacity. Practically, the maximum number of helix plates used is five in cohesive soils and six in cohesionless soils.
- *Helix spacing* ($s$): The distance between two helices is essential for determining the failure mechanism in helical piles. Helical pile manufacturers generally provide an optimum spacing of three times the diameter of the lower helix to avoid overlapping between the stress zones of the helices and incorporate individual failure mechanisms, maximize capacity, reduce shaft length, and enable the helices to be placed in the same layer, thus increasing the accuracy for bearing capacity calculations.
- *Embedment depth* ($H$): The distance between the topmost helix and the ground surface is another determining factor for the failure mechanism in a helical pile. The minimum recommended embedment depth is five times the largest helix diameter (5D) as per CTDM [5] and 6D as per HAI [10] and Gavin et al. [9] for compressive loading. For tensile loading, the minimum embedment depth is 12D.
- *Spacing between helical piles in the group* ($S$): The minimum recommended center-to-center spacing between adjacent helical piles is five times the diameter of the largest helix with spacing not less than 3 feet (0.91 m).
- *Speed of installation:* Usually, a speed of one pitch per revolution is prescribed by helical pile manufacturers.
- *Installation torque* ($T$): $T$ is directly proportional to the diameter of pile shaft (d). A maximum torque of 10,000–120,000 Nm is recommended for pile having shaft diameters from 76.2 to 323.85 mm [24].

- *Factor of safety* (*FOS*): A FOS of 2 and 1.5 for permanent and temporary structures, respectively, is recommended. A FOS of 2 is considered for all allowable load calculations according to AC358:2007 and IBC:2009. CFEM:2006 recommends the ultimate capacity to be multiplied by a geotechnical resistance factor of 0.4 for compressive loading and 0.3 for tensile loading.

## 3 Failure Mechanism and Load Distribution in Helical Piles

Helical pile may fail due to the collapse of any of its four main elements: Helix Capacity, Bracket Capacity, Shaft Capacity, and Soil Capacity [1]. According to IBC:2009, to avoid any failure, each of the elements must conform to the minimum strength requirements to avoid failure.

From a geotechnical perspective, helical piles experience either deep failure or shallow failure. Deep failure develops locally near the helices in two ways: individual failure or cylindrical failure. In individual failure, the failure plane forms around each independent helix, while in cylindrical failure, the failure plane forms around the soil mass between the uppermost and the lowermost helices [7, 16, 23]. In shallow failure, the failure plane reaches the soil surface. The factors that determine the type of failure occurring in helical piles are discussed in the following section.

Studies show that most of the resistance is provided by the helix [29] and a very small part of the total resistance is offered by shaft friction, especially those having smaller shaft diameters. In their study, Clemence and Lutenegger [16] found that resistance offered by shaft friction was not taken into consideration in over 80% of projects. The lowermost plate provides largest resistance under compression, while the uppermost plate provides maximum resistance under tension in multi-helix piles [13, 26]. It has been observed that the largest soil displacement takes place within a radial distance of 1.5 times the helix diameter from the axis of the shaft. According to different manuals, code books, and research papers, the failure of helical piles is generally considered to be the load at a pile head displacement of 5–10% of largest helix diameter ($D_{max}$) or average helix diameter ($D_{avg}$), provided that no plunging failure develops [13, 28].

## 4 Factors Influencing Helical Pile Capacity

Bearing capacity of helical piles depends on several factors such as the type of connection, dimensions of the pile, soil properties, type of load applied, embedment depth, and soil–pile interaction. These factors which influence the load-carrying mechanism of a helical pile have been briefly described.

## 4.1 Embedment Ratio (H/D)

The embedment ratio H/D is defined as the ratio between embedment depth (i.e., the distance between ground surface and pilot point of shaft) and the helix diameter. Deep failure occurs when the embedment ratio is greater than the critical embedment ratio $(H/D)_{cr}$. Shallow failure occurs when $H/D$ ratio is smaller than $(H/D)_{cr}$ [8, 14, 28]. Mittal and Mukherjee [17] and Sprince and Pakrastins [26] after performing parametric analyses using FEA concluded that the load-carrying capacity in helical piles increases with the increase in H/D ratio under both tensile and compressive loadings.

A helical pile is considered as deep if the depth of the top most helix ($z$) to its diameter ($D$) is greater than 5–7 [5]. According to the uplift tests conducted by Trofimenkov and Maruipolshii [27], $(H/D)_{cr}$ varied between 4 and 5 and between 5 and 6 for clay and sand, respectively. Narasimha Rao and Prasad [19] suggest that shallow failure occurs at $H/D < 2$, a transition phase occurs at H/D between 2 and 4, and deep failure occurs at H/D > 4, in clays.

## 4.2 Spacing Ratio (s/D)

The spacing ratio is defined as the ratio of the distance between two adjoining helices to the diameter of the helix. The transition of failure mechanism from cylindrical to individual depends upon the spacing ratio and is almost independent of the diameter of helix and soil characteristics. Cylindrical failure occurs when the spacing ratio (s/D) is smaller than the critical spacing ratio $(s/D)_{cr}$, while individual failure occurs when the spacing ratio is greater than the critical spacing ratio value, i.e., $(s/D) > (s/D)_{cr}$. From the literature, it was found that the pile-bearing capacity is directly proportional to spacing ratio up to a limiting value (i.e., $s/D_{cr}$) of 3.

## 4.3 Number of Helices

Increase in the number of helices increases the helical pile capacity under both tensile and compressive loadings. Sakr [23] and Polishchuk and Maksimov [22] demonstrated in their finite element analyses that double helix pile had 40% and 30% higher bearing capacity, respectively, compared to a single helix pile. In another study, Papadopoulou et al. [20] found that the pile capacity under compression increased by 40–55% in clayey soil but the tensile capacity remained unchanged. Thus, incorporating an additional helix can provide satisfactory increase in pile capacity at minimal cost.

### 4.4 Helix and Shaft Diameter

The most critical factor in determining the load-bearing capacity of helical pile is the diameter of its helix (D). *D* is directly proportional to the pile capacity. In contrast, diameter of shaft (d) has a relatively minor influence on the pile-bearing capacity [2, 14]. Sprince and Pakrastins [26] claim that the magnitude of the increase in pile capacity with helix diameter varies with the nature of soil.

### 4.5 Pitch

Pitch of the helix and inclination angle is useful during installation processes rather than for the determination of pile capacity. Finite element analyses by George et al. [2] revealed that the compressive capacity of helical piles does not change with the variation of pitch but the tensile capacity reduces slightly.

### 4.6 Installation Disturbance

The installation disturbance generated when helical piles are screwed into the ground affects the pile capacity. This is because the soil properties in the immediate vicinity of the pile are altered. Therefore, the disturbance due to installation should be taken into account in designing of such piles [28] (Nasr 2009). Spagnoli and Gavin [25] in their finite element analyses and laboratory tests revealed that there was an increase in the uplift capacity. This increase was due to the compaction of the soil in the vicinity of the pile as a result of the installation effect. Mendoza et al. [3] and Zapata et al. [29] also found a higher pile capacity in their FEA when the installation disturbance was not considered. It can therefore be concluded that the pile capacity may increase in loose sand due to compaction. However, there might be a decrease in the pile capacity in stiff clay and dense sand due to reduction of soil stiffness after installation. It is also observed that installation disturbance generally affects the tensile capacities of helical piles and not their compressive capacities. This happens since the pile interacts with the disturbed soil above the bottom-most helix under uplift loading [30].

## 5 Determination of Axial Capacity of Helical Piles

Trofimenkov and Maruipolshii [27] proposed one of the first expressions (Eq. 1) for determination of the axial load-carrying capacity of single helix pile.

$$Q_u = R_u A_h + f P_s (L - D), \tag{1}$$

where $R_u$ = uplift resistance pressure, $Q_u$ = ultimate uplift capacity, $f$ = average specific pressure of soil, $A_h$ = area of helix, $P_s$ = perimeter of the shaft, $D$ = diameter of helix, and $L$ = length of shaft.

## 5.1 *Analytical Models Based on Cylindrical Shear Failure Method*

Mooney et al. [18] proposed Eq. 2 for determining the bearing capacity of multi-helix pile in cohesive soil based on cylindrical failure mechanism:

$$Q_u = \pi DLS_f c_u + A_h c_u N_c + \pi d\alpha H_{\text{eff}} c_u, \tag{2}$$

where $c_u$ = undrained shear strength of soil, $S_f$ = spacing factor, $N_c$ = bearing capacity factor, $H_{\text{eff}}$ = effective embedment depth, and $\alpha$ = adhesion factor.

Narasimha Rao and Prasad [19] gave the following expressions (Eqs. 3a, 3b, and 3c) for estimating the value of $S_f$:

$$\text{For}\, s/D < 1.5 \quad S_f = 1.0, \tag{3a}$$

$$\text{For } 1.5 < s/D < 3.5 \quad S_f = 0.863 + 0.069(3.5 - s/D), \tag{3b}$$

$$\text{For } 3.5 < s/D < 4.6 \quad S_f = 0.700 + 0.148(4.6 - s/D). \tag{3c}$$

For the determination of pile capacity under compression in cohesionless soil, the following equation (Eq. 4) was proposed by Mistch and Clemence [16]:

$$\text{Q}_u = \frac{1}{2}\pi D\gamma'\left(H_b^2 - H_t^2\right)K_0 \tan\Phi + \gamma' H A_h N_q + \frac{1}{2}P_s H_{\text{eff}}^2 \gamma' K_0 \tan\Phi, \tag{4}$$

where $H_b$ and $H_t$ = embedment depth of the topmost and bottom-most helices.

For determining the uplift capacity of a helical pile embedded in cohesionless soil, Mistch and Clemence [16] gave the subsequent expressions (Eqs. 5 and 6):

$$\text{For } H/D < (H/D)_{cr,} \quad Q_u = \frac{1}{2}\pi\gamma' D\left(H_b^2 - H_t^2\right)K_0 \tan\Phi + \gamma' F_q H A_h, \tag{5}$$

$$\text{For } H/D > (H/D)_{cr}, \quad Q_u = \frac{1}{2}\pi\gamma' D\left(H_b^2 - H_t^2\right)$$

$$K_0 \tan\Phi + \gamma' F_q H A_h + \frac{1}{2}P_s H_{eff}^2 \gamma' K_0 \tan\Phi, \tag{6}$$

where $P_s$ and $F_q$ are the perimeter of shaft and bearing capacity factor, respectively.

Narasimha Rao and Prasad [19] presented the following equation (Eq. 7) for determining the uplift capacity in cohesive soils:

$$Q_u = W_a + \pi S_f D L c_u + A(c_u N_u + \gamma' H) + \pi d \alpha H_{\text{eff}} c_u, \tag{7}$$

where $W_a$ is the total weight of anchor.

### 5.2 Analytical Models Based on Individual Shear Failure Method

Most of the codes and manuals recommend providing sufficient spacing between the helices so as to attain individual shear failure. This is done for achieving a greater efficiency and accurate prediction of the pile capacity. CFEM:2006 presents the following equation (Eq. 8) for calculation of the load-carrying capacity of helical pile in c-Φ soil when its helices are kept at a minimum spacing of $3D_{\text{max}}$:

$$Q_u = \Sigma Q_h + Q_s, \tag{8}$$

where $Q_h$ = individual bearing capacity for each helix expressed as (Eq. 9).

$$Q_h = A_h(N_c c_u + \gamma D N_q + 0.5\gamma B N_\gamma). \tag{9}$$

For a helical pile having a shaft diameter less than 100 mm, the shaft friction ($Q_s$) may be ignored. For pile having $d > 100$ mm, $Q_s$ can be calculated as (Eq. 10) [4].

$$Q_s = \Sigma(\pi D f_s \Delta L_f), \tag{10}$$

where $L_f$ is the pile length over which $f_s$ and $\pi D$ are constant and $f_s$ is the sum of the adhesion friction between the pile and soil.

### 5.3 Empirical Models for Determining Helical Pile Capacity

To calculate the load-carrying capacity of pile using installation torque (T), Hoyt and Clemence [11] presented the following equation (Eq. 11):

$$Q_t = K_T T, \tag{11}$$

where $T$ and $K_T$ are the final installation torque (kNm) and resistance-to-torque ratio ($\text{m}^{-1}$), respectively. $K_T$ is influenced by the shape of pile shaft, direction of loading (tension or compression), dimensions of the pile, and characteristics of the soil.

Torque-capacity correlation is a quick method for helical pile capacity calculation and is commonly used as a verification tool as its results are not very reliable.

### *5.4 Numerical Modeling*

The analytical pile load capacities are often validated against field load tests which are relatively time-consuming and expensive. Therefore, to study the behavior of helical piles in different design considerations, finite element method (FEM) is employed. FEM can simulate the actual geometry of a spatial structure such as the helical blade at a microlevel. For most of the previous FEA, the piles were considered to be wished-in-place into a soil body consisting of a meshed continuum. Although they give fairly accurate insights into the behavior and capacity of screw piles, they do not usually consider the installation effects of installation on the behavior of the soil. In 2D FEM analysis, helix has mostly been considered as a plane plate for simplification purposes, which does not represent the actual geometry. The actual helix model can be designed and realized only in a three-dimensional FEA. For FEM analysis on helical piles, a strength reduction factor ($R_{\text{int}}$) is used at the contact surfaces to simulate the decrease in the shear strength of the soil at the interface [20]. Some other software studies on helical piles have also been carried out in applications such as LPile which is based on p-y analysis, Particle Flow Code (PFC) 3D which is based on discrete element analysis (DEM), etc. [8].

## 6 Case Study

George et al. [2] conducted a 3D FEA on helical piles in Plaxis 3D and validated it against field load test. A 2D FEM analysis was conducted on one of the problems presented in the paper. The results have been compared and presented in this paper. The details of the design and other considerations are provided in Table 2.

To conduct the FEM, the pile and soil were first modeled in the Plaxis 2D. The model was then discretized into small elements, known as meshing. Material properties were then assigned and load was applied in the last step. The model was then analyzed by executing the calculation tool. Figures 2 and 3 show the helical pile model and the mesh as designed in Plaxis 2D.

**Table 2** Pile and soil properties considered for case study

| Pile dimensions considered | |
|---|---|
| Shaft diameter (d) | 0.273 m |
| Helix diameter (D) | 0.61 m |
| Embedment depth (H) | 5.5 m |
| *Model boundary* | |
| Width | 10 D (7 m approximately) |
| Depth | 5 D (9 m approximately) |
| *Pile properties* | |
| $E$ | 200 GPa |
| $\mu$ | 0.2 |
| *Soil properties* | |
| Density | 20 kN/m$^3$ |
| $c$ | 0 |
| $\Phi$ | 30° |
| $E$ | 80 MPa |
| $\mu$ | 0.3 |
| Rint | 0.4 |
| Constitutive model | Mohr–Coulomb |
| Maximum load | 250 kN |
| Loading type | Compressive |

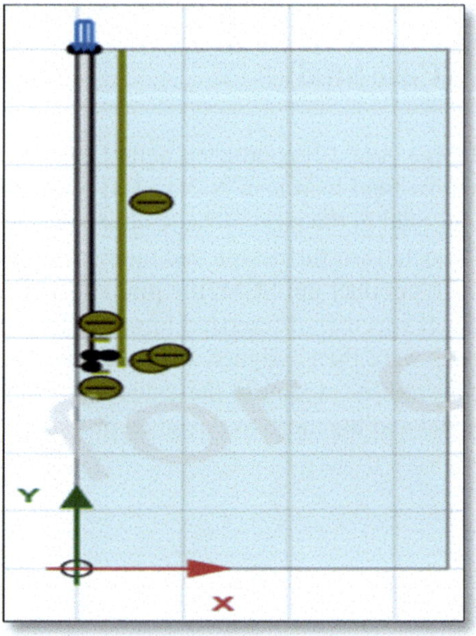


**Fig. 2** Pile model created in Plaxis 2D

**Fig. 3** Meshing of the model

After the analysis was complete, the results were studied and compared. Figures 4a and b show the stress distribution developed around the pile. From the figures, it is clear that the maximum stress developed in the soil body occurs around the helix and up to a distance of about 4D below the pile.

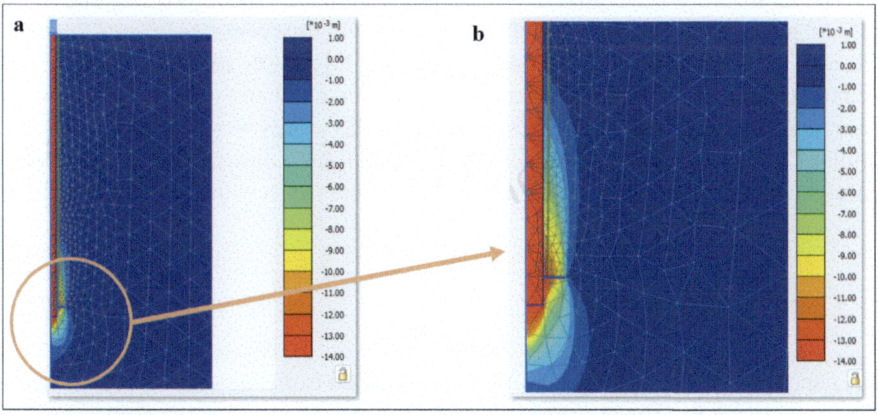


**Fig. 4** **a** Stress zones developed after FEA. **b** Magnified view of the stress zone

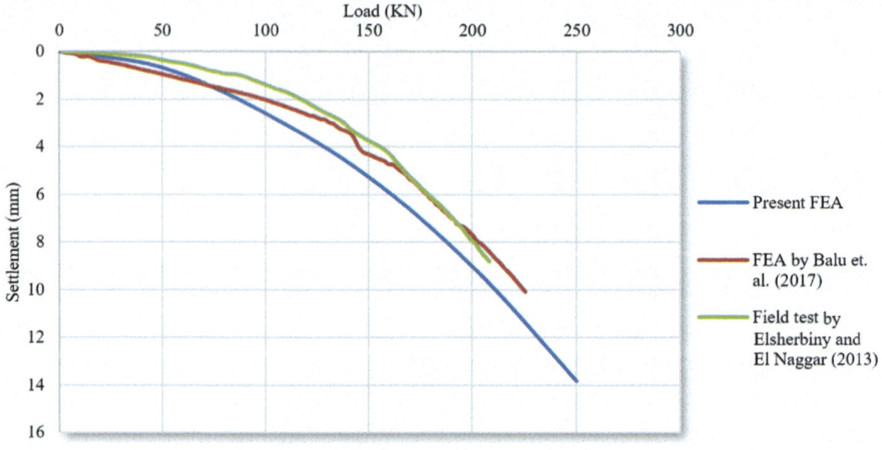


**Fig. 5** Comparison of load versus settlement curves

**Table 3** Comparison of load capacity corresponding to 0.1D settlement

| Calculation method | FEA by George et al. [2] | Present FEA study | Settlement by field load test (10% of D) |
|---|---|---|---|
| Load at carrying capacity (kN) | 179 | 163 | 176 |

A load vs. displacement graph was plotted from the results obtained after the analysis. Figure 5 shows a comparison of the load versus settlement graph obtained from the present study with the previous study.

The graph shows that a similar trend and the ultimate capacity corresponding to 0.1D of settlement obtained from the FEA were within acceptable limits (Table 3). The results obtained using 2D FEA provided slightly lower capacity compared to the field tests and 3D analysis. This also shows that a 3D analysis provides a more accurate result compared to the 2D analysis, as the values were closer to field test results. This may be due to the fact that the actual helix design can be accurately represented in a 3D model with proper pitch and inclination with respect to the shaft.

## 7 Conclusion

This paper presents an aggregated inference obtained from numerous studies, codes, and design manuals. The performance of helical piles under static compressive and tensile loadings has been discussed. From the discussion, the following conclusions can be drawn:

1. Embedment depth, helix spacing, and the soil properties are the main factors influencing the failure mechanism in helical piles.
2. In the absence of maximum allowable displacement, a pile head displacement between 5 and 10% of D may be considered as failure, provided that no plunging failure occurs.
3. An analytical solution has been proposed for calculating the tensile and compressive capacity of helical pile embedded in both cohesionless and cohesive soils.
4. The different factors influencing the bearing capacity of helical piles have been discussed in detail and the following points can be concluded:

    i. Critical embedment depth ($H/D_{cr}$) may be assumed to be 5.
    ii. Critical spacing ratio $s/D_{cr}$ may be considered as 3.
    iii. Use of additional helices increases pile capacity at minimal cost.
    iv. The bearing capacity of the pile increases as the helix and shaft diameters increase. The lowermost and the uppermost helices are the primary contributor to the compression and tensile resistance, respectively.
    v. For helical piles, installation disturbance should be taken into consideration, particularly for stiff soils under tensile loading, as the soil above the bottommost helix will be disturbed.

5. Shaft resistance may be neglected if the diameter of the shaft is less than 100 mm.
6. The 2D FEA conducted on single helix pile showed lower bearing capacity compared to the 3D analysis and field tests. But the results were similar and the difference in bearing capacity was within acceptable range.

## References


1. AC358:2007 (2007) Acceptance criteria for helical foundation systems and devices, ICC evaluation service, Inc.
2. George BE, Banerjee S, Gandhi SR (2017) Numerical analysis of helical piles in cohesionless soil. Int J Geotech Eng. https://doi.org/10.1080/19386362.2017.1419912
3. Mendoza CC, Cunha R, Lizcano A (2015) Mechanical and numerical behavior of groups of screw (type) piles founded in a tropical soil of the Midwestern Brazil. Comput Geotech 67:187–203. ISSN 0266–352X
4. Canadian foundation engineering manual (2006) In: Canadian geotechnical society. 4th edn. BiTech Publishers Ltd., Richmond, B.C.
5. Chance Technical Design Manual (2018) 4 edn, Hubbell Power Systems, Inc.
6. Clemence SP, Lutenegger AJ (2015) Industry survey of state of practice for helical piles and tiebacks. J Deep Found Inst 9(1):21–41. https://doi.org/10.1179/1937525514Y.0000000007
7. Das BM (1990) Earth anchors. Elsevier, Amsterdam, the Netherlands
8. Debnath A, Singh VP (2022) Analysis and design methods of helical piles: a critical review with emphasis on finite element method. Arab J Geosci 15:1496. https://doi.org/10.1007/s12517-022-10774-3

9. Kenneth G, Paul D, Ali T (2014) Field investigation of the axial resistance of helical piles in dense sand. Canadian Geotech J 51:1343–1354. https://doi.org/10.1139/cgj-2012-0463
10. HAI Engineering Manual (2014) (EDM Rev 02), Helical Anchors Inc.
11. Hoyt RM, Clemence SP (1989) Uplift capacity of helical anchors in soil. In: Proceedings of the 12th international conference on soil mechanics and foundation engineering, Rio de Janeiro, Brazil
12. Kurian N, Shah S (2009) Studies on the behaviour of screw piles by the finite element method. Can Geotech J 46:627–638
13. Ben L, Naggar E, Hesahm M (2008) Axial testing and numerical modelling of square shaft helical piles under compressive and tensile loading. Canadian Geotechn J 45:1142–1155
14. Merifield R (2011) Ultimate uplift capacity of multiplate helical type anchors in clay. J Geotech Geoenviron Eng 137:704–716. https://doi.org/10.1061/(ASCE)GT.1943-5606.0000478
15. Meyerhof GG (1976) Bearing capacity and settlement of pile foundations. ASCE J Geotech Eng 102(GT3):197–224
16. Mitsch MP, Clemence SP (1985) The uplift capacity of helix anchors in sand. In: Clemence SP (ed) Proceedings of uplift behavior of anchor foundations in soil, Detroit, Mich., 24 October 1985. American Society of Civil Engineers (ASCE), New York. pp 26–47
17. Satyendra M, Sanjeev M (2015) Behaviour of group of helical screw anchors under compressive loads. Geotech Geol Eng 33. https://doi.org/10.1007/s10706-015-9841-4
18. Mooney JS, Adamczak S, Clemence SP (1985) Uplift capacity of helix anchors in clay and silt. American Society of Civil Engineers, pp 48–72
19. Narasimha Rao S, Prasad YVSN (1993) Estimation of uplift capacity of helical anchors in clays. J Geotech Eng 119(2):352–357
20. Konstantina P, Harry S, Vasileios P (2014) Finite element analyses and experimental investigation of helical micropiles. Geotech Geol Eng 32. https://doi.org/10.1007/s10706-014-9771-6.
21. Perko HA (2009) Helical piles: a practical guide to design and installation. Wiley and Sons, Hoboken
22. Inna P, Maksimov F (2017) Numerical analysis of helical pile–soil interaction under compressive loads. In: IOP conference series: materials science and engineering. vol 262. pp 012099. https://doi.org/10.1088/1757-899X/262/1/012099
23. Mohammed S (2009) Performance of helical piles in oil sand. Canadian Geotech J 46:1046–1061. https://doi.org/10.1139/T09-044
24. Schmidt R, Nasr M (2004) Screw piles: uses and considerations. Struct Magazine 29–31
25. Giovanni S, Kenneth G (2015) Helical piles as a novel foundation system for offshore piled facilities. https://doi.org/10.2118/177604-MS
26. Andina S, Leonids P (2009) Helical pile behavior analysis in different soils. Scient J Riga Tech Univer. Construct Sci 10:121–130
27. Trofimenkov JG, Maruipolshii LG (1965) Screw piles used for mast and tower foundations. In: Proceedings of the 6th international conference on soil mechanics and foundation engineering, vol 2. pp 328–332
28. Wang D, Merifield RS, Gaudin C (2013) Uplift behaviour of helical anchors in clay. Can Geotech J 2013(50):575–584. https://doi.org/10.1139/cgj-2012-0350
29. Zorany Z, Cristina T, José S, Thorel L (2015) Discussion of "Field investigation of the axial resistance of helical piles in dense sand. Canadian Geotech J 52. https://doi.org/10.1139/cgj-2015-0212
30. Zhang DJY, Chalaturnyk R, Robertson PK, Sego DC, Cyre G (1998) Screw anchor test program (Part I & II): instrumentation, site characterization and installation. In: Proceedings of the 51st Canadian geotechnical conference, Edmonton

# Connected Versus Disconnected Piled Raft Systems: A Comparative Experimental Assessment

**Prasun Halder and Bappaditya Manna**

## 1 Introduction and Background

Globally, the concept of connected piled raft (CPR) foundation is a very popular among the geotechnical engineers [1–11]. However, as an alternative solution, Wong et al. [12] proposed to structurally disconnect raft from piles with a cushion which pioneered the concept of disconnected piled raft (DPR) system. Few numbers of research have been conducted in this domain so far [2, 13–20, 22–24]. However, very fewer physical experiments have been conducted in this domain, and as a result, the load transfer process of DPR is not yet fully understood. Therefore, this paper presents some experimental results which would clarify the difference in the load transfer process between CPR and DPR. This study also focusses on the stiffness and load-sharing aspects of the raft and piles in these two different foundation types.

### *1.1 Materials for Experiment*

Yamuna sand (YS) is used in the model tests as bed material. A graded granular material CM is used as cushion. Table 1 shows the properties of these materials.

For model pile, a 1 m long aluminum pipe (25.4 mm diameter and 2 mm thick) is used. To model the raft, a 300 × 300 × 25 mm aluminum plate is used. The elastic modulus and Poisson's ratio of the aluminum are obtained as 69,230 N/mm$^2$ and 0.33, respectively. Using the scaling law of bending rigidity [25], a prototype pile of

P. Halder (✉)
Indian Institute of Technology Jammu, Jammu and Kashmir 181221, India
e-mail: prasun.siliguri@gmail.com

B. Manna
Indian Institute of Technology Delhi, New Delhi 110016, India

 
B. T. Jose et al. (eds.), *Proceedings of the Indian Geotechnical Conference 2022 Volume 2*, Lecture Notes in Civil Engineering 477,
https://doi.org/10.1007/978-981-97-1741-5_12

**Table 1** Granular material properties

| Parameters | YS | CM |
|---|---|---|
| Mean particle size, $d_{50}$ (mm) | 0.3 | 2 |
| Min dry unit weight (kN/m$^3$) | 13.8 | 14 |
| Max dry unit weight (kN/m$^3$) | 16.8 | 17 |
| Friction angle (Degree) | 40.7 | 47 |

15.5 m length and 600 mm diameter can be represented with this model pile when the scaling factor is 15.5.

### 1.2 Test Program

All the strain-controlled (0.02 mm/second) model tests are performed in a 2 × 2 × 1.8 m steel tank (schematic diagram as Fig. 1) where sand bed having the $R_d$ of 70% is prepared with tamping method. The granular platform is placed over the sand bed covering the pile heads. About 350 Ω strain gauges are attached on the pile to measure the axial loads. All the tests are terminated when the raft settles up to 30 mm [3, 11, 26, 29].

As the cost, time, and space requirements for a full-scale load test are high, the reduced-scale 1-$g$ model laboratory tests are preferred in geotechnical engineering. Therefore, keeping the limitations (i.e., stress level) of model 1-$g$ tests in mind, an attempt is made to predict the response of DPR under vertical load and have an idea about the load transfer mechanism involving the key effect of granular cushion.

## 2 Results and Discussion

### 2.1 Load-Settlement Performances of CPR and DPR

Table 2 shows the loads carried by unpiled raft (UR), 2 × 2 CPR, and 2 × 2 DPR foundations at 30 mm settlement. The granular cushion is observed to have a decremental impact on the capacity of 2 × 2 DPR.

The settlement performance of CPR and DPR is quantified using a non-dimensional quantity termed as 'Settlement Efficiency' [13, 27, 28] and it is defined as:

$$\eta = \frac{w_{UR} - w_{DPR}}{w_{UR}} \times 100\%, \tag{1}$$

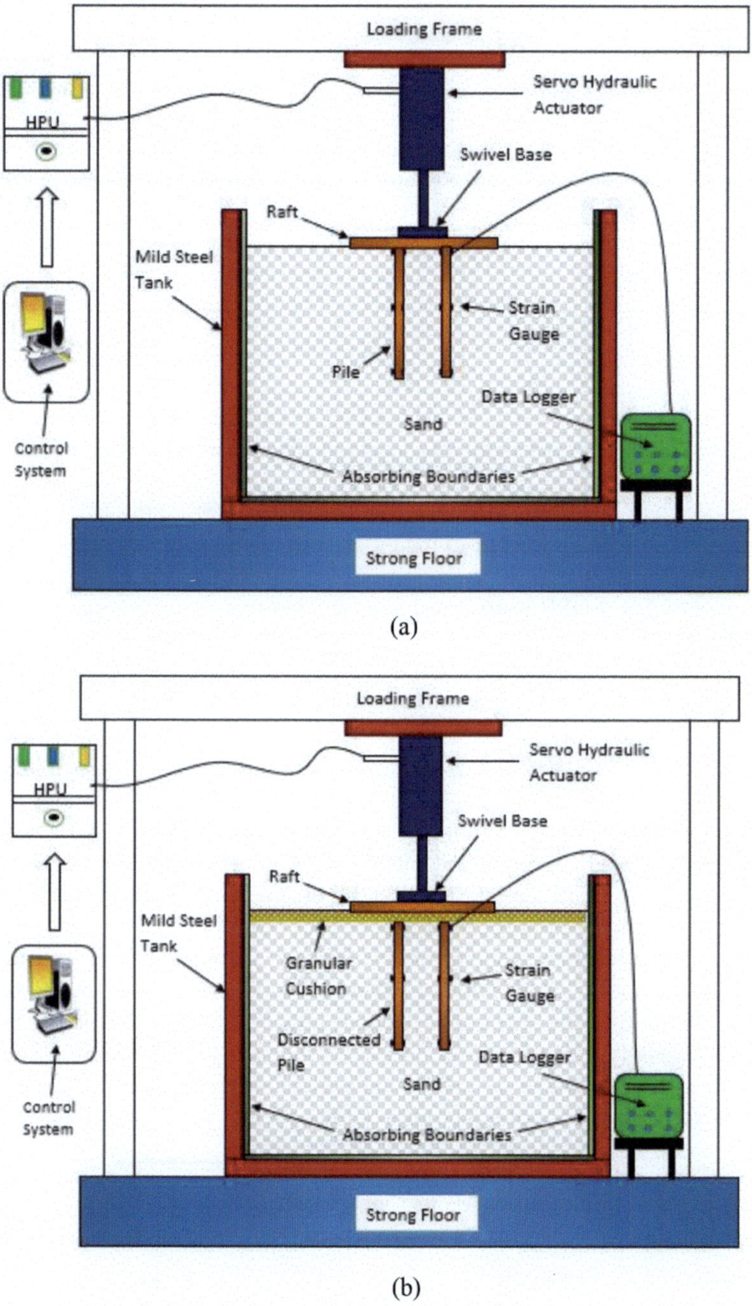


**Fig. 1** Schematic of 2 × 2 foundation system with loading arrangement: **a** CPR; **b** DPR

**Table 2** Load responses of different foundations

| Cases | Load (kN) |
|---|---|
| UR | 15 |
| 2 × 2 CPR | 17.8 |
| 2 × 2 DPR (*h/d* = 3) | 15.9 |

**Table 3** Variation of maximum $\eta$ for CPR and DPR

| Cases | $\eta_{max}$ |
|---|---|
| 2 × 2 CPR | 30% |
| 2 × 2 DPR (*h/d* = 3) | 11% |

where $w_{UR}$ and $w_{DPR}$ are the UR and DPR settlements, respectively. In case of CPR, $w_{CPR}$, i.e., CPR settlement has been considered for calculation of settlement efficiency.

As $w_{UR} > w_{DPR}$ and $w_{UR} > w_{CPR}$ and ($0 \leq \eta \leq 100\%$), the efficiency of the DPR and CPR at a given load level increases with higher value of '$\eta$' [27]. Table 3 highlights the variation of maximum $\eta$ value for CPR and DPR at 30 mm settlement. The $\eta$ value of 2 × 2 DPR is observed to have a lower value as compared to that of 2 × 2 CPR.

## 2.2 *Foundation Stiffness*

Table 4 shows the variation of foundation stiffness for 2 × 2 CPR and 2 × 2 DPR under vertical load. Here, the representation has been made in terms of the ratio of the stiffness of 2 × 2 CPR or 2 × 2 DPR to the stiffness of UR. From Table 4, it can be seen that the ratio of the stiffness of 2 × 2 CPR to UR is 1.42 as the piles attract the higher load at the beginning. Interestingly, this ratio is found to achieve an average value of 1.2 due to the introduction of structurally connected piles beneath the raft. However, the ratio of the stiffness of the raft in 2 × 2 CPR to UR is observed to increase nonlinearly from an initial value of 0.6, indicating the non-uniform raft–soil contact, to an average value of 0.95 as the settlement progresses. So, the piles are found to lessen the stress under the raft in CPR system.

**Table 4** Variation of foundation stiffness for CPR and DPR

| Foundation stiffness ratio | Initial value | Average value |
|---|---|---|
| Stiffness ratio of 2 × 2 CPR to UR | 1.42 | 1.2 |
| Stiffness ratio of raft in 2 × 2 CPR to UR | 0.6 | 0.95 |
| Stiffness ratio of 2 × 2 DPR to UR | 1.25 | 1.03 |
| Stiffness ratio of raft in 2 × 2 DPR to UR | 1.2 | 1.06 |

**Table 5** Pile–raft load sharing in CPR and DPR

| Foundation components | Initial load sharing (%) | Final load sharing (%) |
|---|---|---|
| Piles in 2 × 2 CPR | 90 | 26 |
| Raft in 2 × 2 CPR | 10 | 74 |
| Piles in 2 × 2 DPR | 4 | 10 |
| Raft in 2 × 2 DPR | 96 | 90 |

Table 4 also depicts that the ratio of the stiffness of 2 × 2 DPR to UR attains an initial value of 1.25 and then decreases to an average value of 1.03 with increasing settlement highlighting the higher load bearing ability of 2 × 2 DPR than UR. Due to large over burden stress under the raft provided by the granular cushion platform, the average value of the stiffness ratio of the raft in 2 × 2 DPR to UR is found to be 1.06. Therefore, it can be concluded that, in DPR, the piles enhance the raft stiffness and act as soil reinforcements.

### 2.3 Pile–Raft Load Sharing

Table 5 shows the gradual distribution of load between raft and piles. In 2 × 2 CPR, the piles attract the load initially due to higher stiffness and then the raft share enhances with settlement. However, in DPR, the raft attracts the load initially and then the piles take part in load sharing.

From this table, it can also be noticed that the raft in DPR shares 90% of the total load when the settlement reaches 30 mm. The pile load share is very small. This happens because of the less no. of the piles used in this study. However, with higher number of piles beneath the raft, the pile load share would increase up to 35–40%. However, this pile load-sharing percentage in DPR would be lesser than the pile load share in case of a conventional CPR system. It is noteworthy that the relative compaction of the cushion layer must be around 90–95% for better result of the whole DPR system.

## 3 Limitations and Future Scope of the Present Study

The followings are some of the limitations and future scopes of the present study:

- This study entirely focusses to understand how a granular cushion layer beneath the raft alters the load transfer process of a vertically loaded disconnected piled raft system from that of the conventional connected piled raft system. To establish the relationship between raft thickness and cushion thickness, further detailed parametric studies need to be done by varying raft thickness and cushion thickness. However, this aspect does not fall in the scope of the present study.

- Under seismic condition, the piles would be subjected to base shear, bending as well as axial force. However, this present study is done under static vertical loading condition only. The seismic assessment of DPR system and the related design aspects are in the future scope of this study.
- With the introduction of the granular cushion between raft and piles for a disconnected piled raft, the interaction between piles, cushion, and raft becomes complex and dependent on settlement. This particular interaction study needs a thorough research and extended experimental works which are in the future scope of this study.

# 4 Conclusion

From the experimental results, it can be concluded that the compacted granular cushion platform beneath the raft holds the key to alter the load transfer process of a DPR from that of a conventional CPR. The $2 \times 2$ CPR shows higher load bearing capacity as well as higher settlement efficiency as compared to $2 \times 2$ DPR. However, in a DPR, the raft stiffness gets enhanced by the soil reinforcing piles which is different from the case of a CPR. Additionally, the raft is found to take the initial load in a DPR which is exactly opposite from the established pattern of a CPR where the piles take the load initially.

# References

1. Clancy P, Randolph MF (1993) An approximate analysis procedure for piled raft foundations. Int J Numer Anal Methods in Geomech 17:849–869
2. Clancy P, Randolph MF (1996) Simple design tools for piled raft foundations. Geotechnique 46(2):313–328
3. Horikoshi K, Randolph MF (1998) A contribution to optimum design of piled raft. Geotechnique 48(3):301–317
4. Katzenbach R, Choudhury D (2013) ISSMGE combined pile–raft foundation guideline. In: TC212 design guideline, Darmstadt, Germany, pp 1–23
5. Katzenbach R, Arslan U, Moorman C (2000) Piled raft foundation projects in Germany. Design applications of raft foundation, Thomas Telford, London, pp 323–391
6. Katzenbach R, Leppla S, Choudhury D (2016) Foundation systems for high-rise structures. CRC Press, Boca Raton, FL
7. Kumar A, Choudhury D (2017) Load sharing mechanism of combined pile-raft foundation (CPRF) under seismic loads. Geotech Eng J Southeast Asian Geotech Soc (SEAGS) and Assoc Geotech Soc Southeast Asia (AGSSEA) 48(3):95–101
8. Kumar A, Patil M, Choudhury D (2017) Soil-structure interaction in a combined pile-raft foundation—a case study. Proc Institut Civil Eng—Geotech Eng 170(2):117–128
9. Poulos HG, Davis EH (1980) Pile foundation analysis and design. Wiley, New York
10. Randolph MF (1994) Design methods for pile groups and piled rafts. In: Proceedings of 13th ICSMFE, New Delhi, vol 5, pp 61–82

11. Sanctis LD, Mandolini A (2006) Bearing capacity of piled raft on soft clay soil. J Geotech Geoenviron Eng 132(12):1600–1610
12. Wong IH, Chang MF, Cao XD (2000) Raft foundations with disconnected settlement reducing piles. Design application of raft foundations and ground slabs, Chap. 17, Thomas Telford, London, pp 469–486
13. Cao XD, Wong IH, Chang MF (2004) Behavior of model rafts resting on pile-reinforced sand. J Geotech Geoenviron Eng 130(2):129–138
14. Dezfouli M, Dehghani M, Asakereh A, Kalantari B (2018) Behavior of geogrid reinforced and unreinforced non-connected pile raft foundation. Int J Civil Eng. https://doi.org/10.1007/s40999-018-0362-4
15. Fioravante V, Giretti D (2010) Contact versus noncontact piled raft foundations. Can Geotech J 47:1271–1287
16. El Sawwaf M (2010) Experimental study on eccentrically loaded raft with connected and disconnected shorts piles. J Geotech Geoenviron Eng 136(10):1394–1402
17. Ata A, Badrawi E, Nabil M (2015) Numerical analysis of unconnected piled raft with cushion. Ain Shams Eng J 6:421–428
18. Azizkandi A, Rasouli H, Baziar MH (2018) Load sharing and carrying mechanism of piles in non-connected pile rafts using a numerical approach. Int J Civil Eng. https://doi.org/10.1007/s40999-018-0356-2
19. Eslami A, Veiskarami M, Eslami MM (2012) Study on optimized piled-raft foundations (PRF) performance with connected and non-connected piles-three case histories. Int J Civil Eng 10(2):100–111
20. Hor B, Song MJ, Jung MH, Song YH, Park YH (2015) A 3D FEM analysis on the performance of disconnected piled raft foundation. In: Proceedings of 15th Asian regional conference on soil mechanics and geotechnical engineering, South Korea, vol 2(34). pp 1238–1243
21. Liang FY, Chen IZ, Shi XG (2003) Numerical analysis of composite piled raft with cushion subjected to vertical load. Comput Geotech 30(6):443–453
22. Pecker A (2004) Design and construction of the Rion Antirion Bridge. In: Proceedings of geo-trans, geotechnical engineering for transportation projects, Los Angeles, California, ASCE, GSP 126, pp 216–240
23. Jamiolkowski MB, Ricceri G, Simonini P (2009) Safeguarding Venice from high tides: site characterization and geotechnical problems. In: Proceedings of international conference on soil mechanics and geotechnical engineering, vol 4. Alexandria, pp 3209–3230
24. Mattsson N, Simon C, Menoret A, Ray M (2013) Case study of a full-scale load test of a piled raft with an interposed layer for a nuclear storage facility. Geotechnique 63(11):965–976
25. Wood DM (2004) Geotechnical modeling. Spon Press, Taylor and Francis Group
26. Halder P, Manna B (2021) A new model for the prediction of load sharing in piled raft system—an experimental investigation. Arab J Sci Eng. https://doi.org/10.1007/s13369-021-05481-2
27. Tradigo F, Pisano F, di Prisco C, Mussi A (2015) Non-linear soil-structure interaction in disconnected piled raft foundations. Comput Geotech 63:121–134
28. Halder P, Manna B (2021) Large scale model testing to investigate the influence of granular cushion layer on the performance of disconnected piled raft system. Acta Geotech 16:1597–1614
29. Halder P, Manna B (2021) Load transfer mechanism for connected and disconnected piled raft: a comprehensive study. Acta Geotech. https://doi.org/10.1007/s11440-021-01409-0

# Study on Mechanism of Load Transfer in Pile Raft Foundation

**Vijaykumar, S. K. Prasad, and L. Chethan**

## 1 Introduction

Pile foundations are the most popular forms of deep foundations used for both onshore and offshore structures. They are often used to support high-rise buildings resting on weak soil layers. In the recent days, the concept of pile raft foundation has become very popular for tall towers. Petronas Tower in Kuala Lumpur, Burj Khalifa in Dubai, and Messeturm building in Frankfurt are a few of the landmark tall towers built on pile raft foundations. The research study indicates that pile raft foundation is a more economical design approach compared with pile foundation. In pile foundations, total load from the superstructure is transferred through pile cap to soil through bearing piles. In the piled raft foundation, the pile cap itself acts as raft and rests on the ground. Hence, some amount of load is transferred through raft to the soil, and the remaining load is transferred through piles. Therefore, the load carried by piles is reduced. The main purpose of providing pile raft foundation system is to improve the overall stability of the system, increase the load carrying capacity, and decrease the amount of settlement of the foundation system. This study concentrates the load transfer mechanism from superstructure to soil through the pile raft foundation system.

Many researchers have tried to present their thoughts on pile raft foundation. Irfan and Irshad [1] discussed combined piled raft foundation (CPRF) considering both raft and piles to take their share of the total load applied. Kumar and Choudhury [2] showed the effect of pile head connection condition on the behavior of combined pile raft foundation (CPRF) by using finite element-based geotechnical program PLAXIS-3D. Azad Hameed and Ola Adel [3] conducted three-dimensional

Vijaykumar (✉)
Department of Civil Engineering, Government Engineering College, Raichur 584135, India
e-mail: vijay.kp21@gmail.com

S. K. Prasad · L. Chethan
Department of Civil Engineering, Vidyavardhaka College of Engineering, Mysuru 570002, India

 
B. T. Jose et al. (eds.), *Proceedings of the Indian Geotechnical Conference 2022 Volume 2*, Lecture Notes in Civil Engineering 477,
https://doi.org/10.1007/978-981-97-1741-5_13

linear finite element investigation on pile raft foundation system and studied the load-settlement behavior. They argued that raft size and pile length influence the performance of pile raft foundation system. Raj and Srijit [4] reviewed the performance of pile raft foundation system in terms of serviceability and load carrying capacity and investigated the settlement behavior of a square-piled raft in a layered soil using numerical analysis. Venkataraman and Huang [5] studied the performance of combined piled raft foundation. According to them neither the pile group alone nor the raft by itself ensure the safety and the serviceability of the structure their support. It is the combined system that ensures the safety and serviceability of the structure. Hence a complete knowledge of the contribution of each element namely the pile, raft, and the soil on the overall behavior is essential to generate an effective design. In spite of many attempts on the development in the area of pile raft foundation, there is scope for increased studies, especially to understand how the forces are transferred from superstructure to soil through the complex medium of pile raft foundation which includes raft, individual piles, and their interaction among themselves and with the soil they are in contact with [5].

## 2 Problem Statement

The focus of this paper has been to identify the mechanism of load transfer from superstructure to soil through medium of raft and piles. For this purpose, a pile raft foundation system comprising of square raft of 9m side and thickness 0.5m along with 9 number of circular piles at equal spacing of 3m center to center both ways and length 20m is considered in a homogeneous clayey sand of thickness 75m with unit cohesion of 10kN/m$^2$, friction angle of 30°, and unit weight of 17 kN/m$^3$. The pile raft foundation is subjected to gradually increasing load from zero to failure. The displacements are restrained at the boundaries. The width of the pile raft foundation system is modeled to be 75m based on the consideration of boundary effect. It was found that any increase in width beyond 75m did not appreciably change the parameters such as stresses and deformations in pile raft foundation or adjoining soil. Further, convergence check showed that the 10-noded element size is sufficient. The effect of water table is not considered. Further, piles and raft are considered to be rigid and elastic, and soil is modeled according to Mohr–Coulomb criteria. The loading is applied gradually on top of the raft till failure. Stresses in soil adjoining raft and pile and settlement in soil are studied for bringing out the mechanism of load transfer.

To validate the numerical analysis in the present study, plate load test was simulated using circular plate. The ultimate load carried by the plate was assessed using Terzaghi's theory and compared with the present numerical tool. The results are given in Table 1. It can be observed that this is a good assessment between the two results (Fig. 1).

The contributions of pile and raft in carrying the total load are based on finite element approach using the software PLAXIS-3D. The three-dimensional model

**Table 1** Plate load test comparison findings for circular plates

| Sl. no. | Problem description | Ultimate load from Terzaghi theory in kN | Ultimate load from PLAXIS in kN |
|---|---|---|---|
| 1 | C = 35 kN/m$^2$ | 53 | 58 |
| | $\Phi = 0^0$ | | |
| | ϒ = 17 kN/m$^3$ | | |
| | B = 0.5 m | | |
| | D = 0.5 m | | |
| 2 | C = 10 kN/m$^2$ | 152 | 161 |
| | $\Phi = 40^0$ | | |
| | ϒ = 17 kN/m$^3$ | | |
| | B = 0.305 m | | |
| | D = 0.5 m | | |

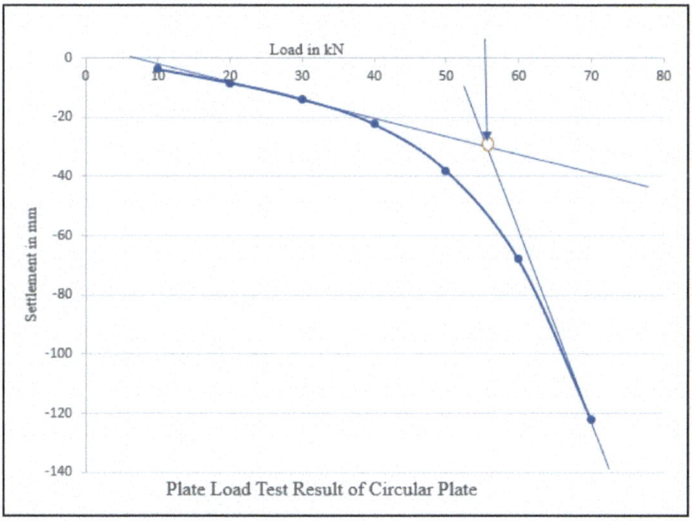


**Fig. 1** Plate load test result of circular plate

has been developed to consider the infinite soil system comprising of piles and raft. The objectives are to identify the stress in soil in different regions and settlement characteristics of pile raft foundation system.

Figures 2 and 3 present the geometric model of three-dimensional pile raft foundation system. Figure 4 represents a three-dimensional soil element (10-noded tetrahedrons) which is used for the present work, and Fig. 5 represents the plan of the pile raft foundation indicating the locations of the pile and raft. It can be observed that there are three different types of pile, namely pile type A, pile type B, and pile type C. Pile type A is in the four corners, each having interaction with 3 piles. Pile type B is the four piles in the middle on the outer side, each having interaction with neighboring 5 piles, and only one pile type C at the center which has a maximum interaction with 8 neighboring piles. It can be seen that the 9 piles have center-to-center spacing of 3 m both ways. Figure 6 presents the results of stress contour in different directions and locations. Table 2 gives the data values used in the present study to model soil, pile and raft. In the present study, no interface element is considered between pile and soil. Hence it is idealized that there is no slip between soil and pile surface during loading. For this purpose, interface strength reduction factor is considered as 1.

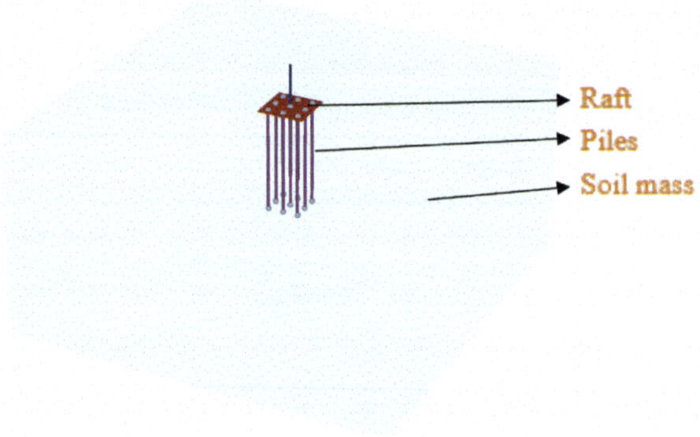


**Fig. 2** PLAXIS-3D geometrical model

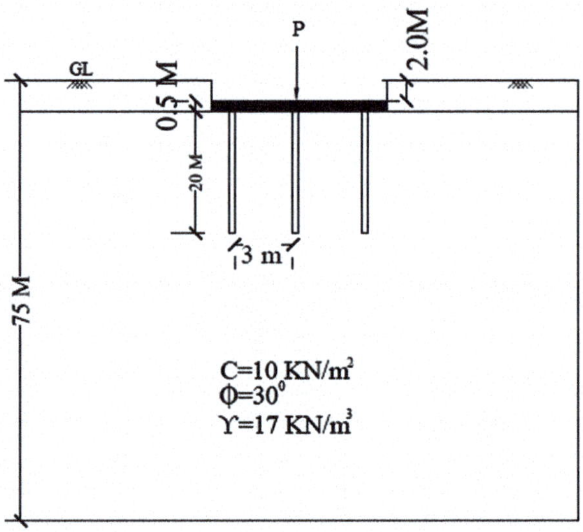


**Fig. 3** Geometrical model of pile raft foundation system in homogeneous soil

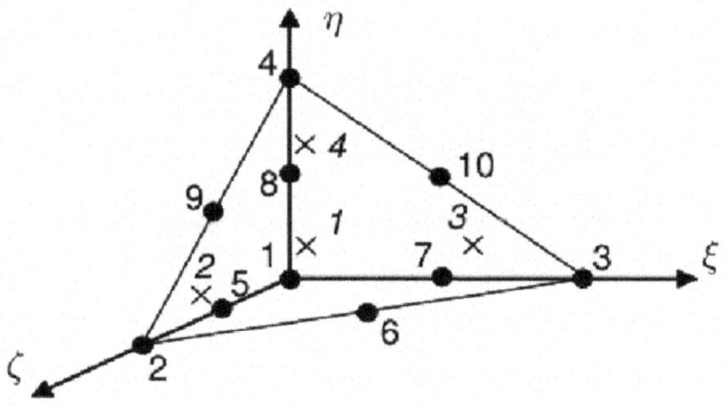


**Fig. 4** Three-dimensional soil elements (10-noded tetrahedrons)

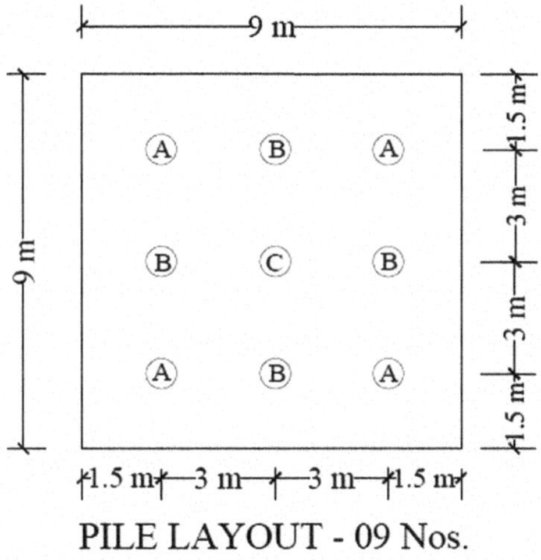


**Fig. 5** Plan of pile raft foundation system with pile types A, B, and C

## 3 Mechanism of Load Transfer

Pile raft foundation is a complex foundation system in which the interactions between raft and soil, pile and soil, pile and pile, and raft and pile have to be seriously considered. Further, when the load is applied from the superstructure, the transfer of load through the foundation to the soil happens in a complex manner. To understand the mechanism of load transfer from superstructure to soil below, the stresses developed in soil and settlement experienced by foundation at different location under different loading conditions are carefully analyzed in the present paper. For this purpose, the soil around the pile raft foundation is classified in to three regions as detailed in Fig. 7. Region 1 is the zone exactly below the raft. The soil in this region is subjected to direct compression from top, and hence the normal stress in vertical direction will be predominant. Region 2 is the zone of soil surrounding the piles along their length. The soil in this region is subjected to shear because of the interaction with pile. The challenge is whether the shear is fully mobilized and in which region the shear is

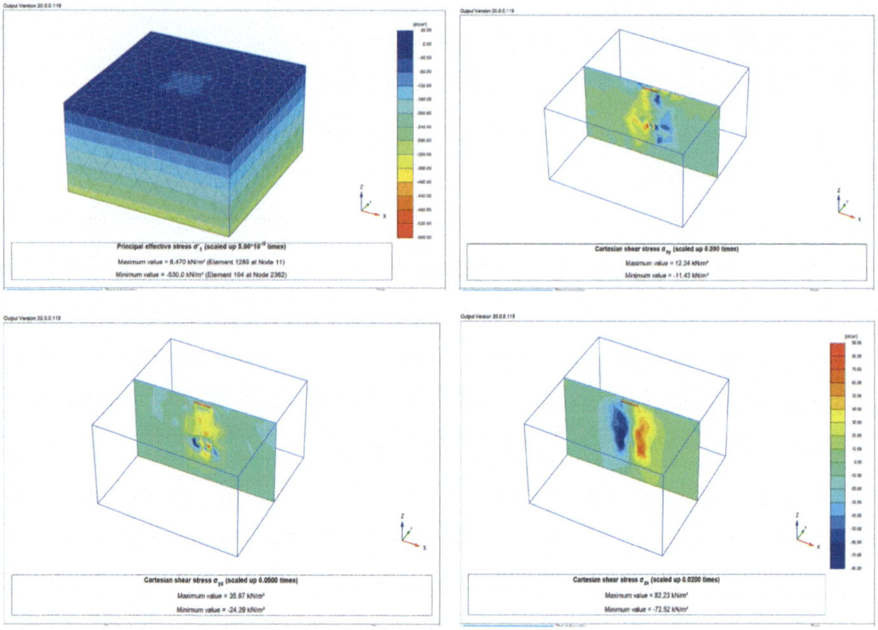


**Fig. 6** PLAXIS-3D effect of stress in XY, YZ, and XZ-directions

fully mobilized. Further, Region 3 is the soil below the bottom of piles which is also subjected to compression under the pile. The same is presented in Table 3.

### Stresses in Soil in Region 1

In order to make sure about the stress development in soil below the base of raft, Fig. 8 indicates the sections along which the vertical normal stress and corresponding vertical settlement are plotted. Sections A-A to E-E are in Y-direction, and sections 1-1 to 5-5 are in X-direction. Figure 9 indicates the distribution of normal stress from sections A-A to E-E. It can be seen that the stresses start to increase from the ends and are transferred toward the center. With increase in load the stresses increase near the end, and the amount of increase near the center is nearly insignificant. At the final load, stresses are close to zero near the center and maximum along the edges. If at all failure happens, soil yields from near the edge toward the center. Figures 10 and 11 indicate the distribution of settlements along different sections in both the directions. It can be seen that the settlement is maximum near the center and reduces toward the edges. The trend is similar in both the directions. Further, the central sections C-C and 3-3 experience the maximum settlement, and the settlement reduces toward the edges. It may be seen that the stresses and displacement are not symmetric even though the loading and geometry are symmetrical. This variation is because the points chosen for stress and displacement determination are not exactly at the same elevations and locations on the two sides.

**Table 2** Properties of soil, piles, and raft used in the present analysis [6]

| Soil properties | | |
|---|---|---|
| Soil type = | Clayey sand | |
| Material model = | Mohr–Coulomb | |
| Density $\gamma$ = | 17 | $kN/m^3$ |
| Young's modulus E = | 3000 | $kN/m^2$ |
| Poisson's ratio $\mu$ = | 0.3 | |
| Cohesion C = | 10 | $kN/m^2$ |
| Angle of friction $\Phi$ = | 30 | o |
| Interface strength reduction factor (rigid) | 1.0 | |
| *Pile properties* | | |
| Pile diameter = | 1.0 | m |
| Young's modulus E = | $20 \times 10^6$ | $kN/m^2$ |
| Length of pile = | 20 | m |
| Number of piles = | 9 | Nos |
| *Raft properties* | | |
| Raft size = | $9.0 \times 9.0$ | m |
| Young's modulus E = | $20 \times 10^6$ | $kN/m^2$ |
| Density $\gamma$ = | 25 | $kN/m^3$ |
| Thickness = | 0.5 | m |
| Poisson's ratio $\mu$ = | 0.3 | |

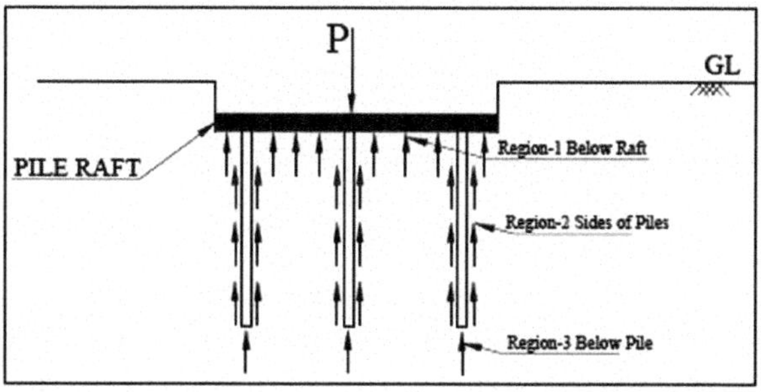


**Fig. 7** Pile raft foundation system

**Table 3** Different zones in soil where the stress variations under increased loading are studied

| | |
|---|---|
| Region 1 | Portion of the soil just below the pile raft foundation |
| Region 2 | Portion of the soil adhering to the surface of the pile along the length of the pile experiencing shear stress |
| Region 3 | Portion of the soil below the bottom of the pile |

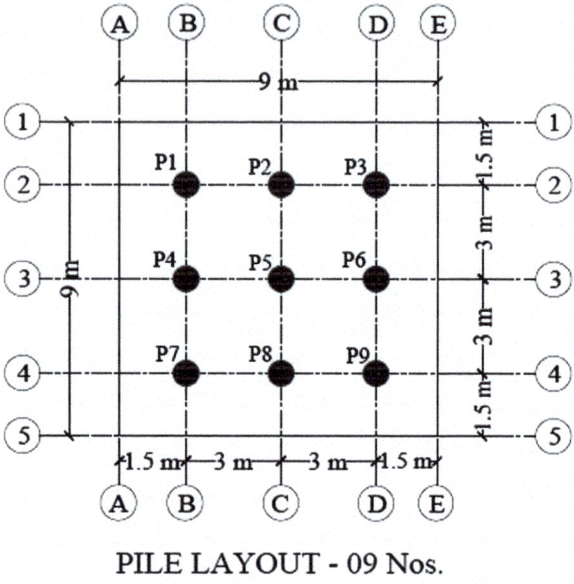


**Fig. 8** Plan of pile layout with sections 1-1 to 5-5 and A-A to E-E

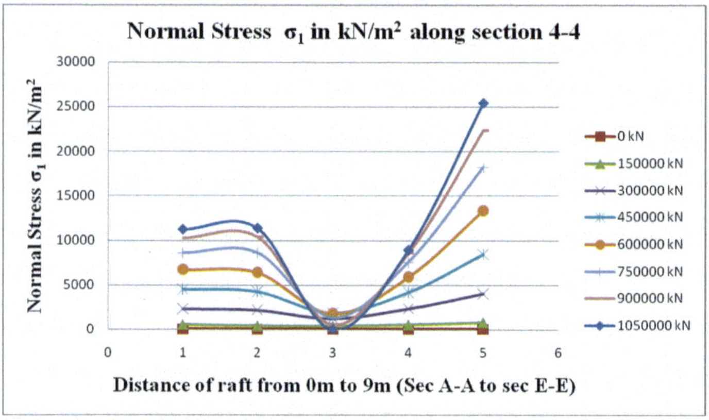


**Fig. 9** Effect of normal stress below the raft at section 4-4 for different loading conditions

## Stresses in Soil in Region 2

Figures 12, 13, and 14 are plotted for pile types A, B, and C, respectively. It is interesting that the maximum shear stress is experienced in the soil surrounding edge pile of type A which has minimum interaction with neighboring piles (only 3 neighboring piles interact). As seen in Fig. 12, the stresses start mobilizing from top, reach a maximum value near the top, and transfer the peak stress toward the bottom with increase in load. The maximum shear stress develops near the bottom toward failure stage. The shear stress pattern is changed near the center of pile along the

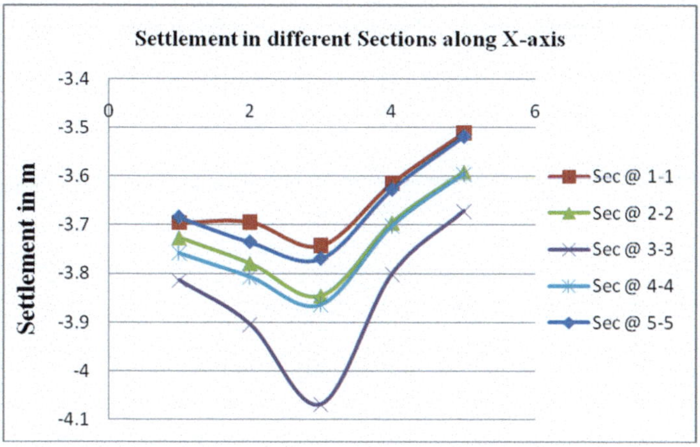


**Fig. 10** Maximum settlement below the raft for different sections along X-direction

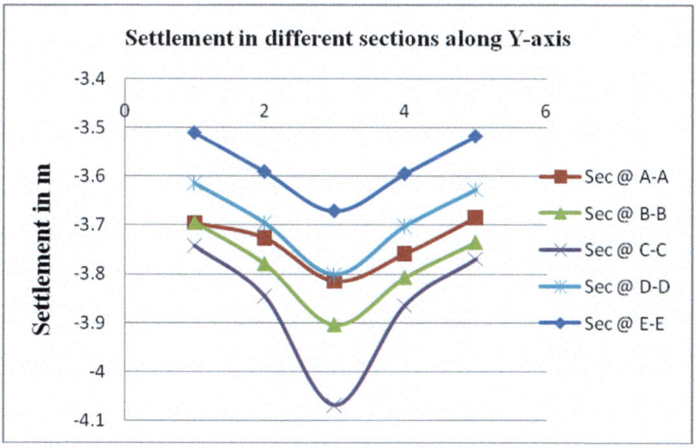


**Fig. 11** Maximum settlement in different sections along Y-direction

length. Figure 13 shows the variation in shear stress along the length of pile with increase in load for pile type B. It is interesting that there is a reduction in peak value (magnitude) of shear stress as compared to those of pile type A. Shear stresses are more near the bottom compared to the case of pile type A. But the peak is near the top.

Figure 14 shows the variation in shear stress in soil along the length of pile of type C at the center with increase in load. This pile has maximum interaction with all piles. However, because it is at the center, the interaction is well balanced, and hence the peak shear stress is much smaller than that observed for cases of pile types A and B. The two ends are subjected to least values, and the maximum is near the top. The signs of shear stress are different at top and bottom.

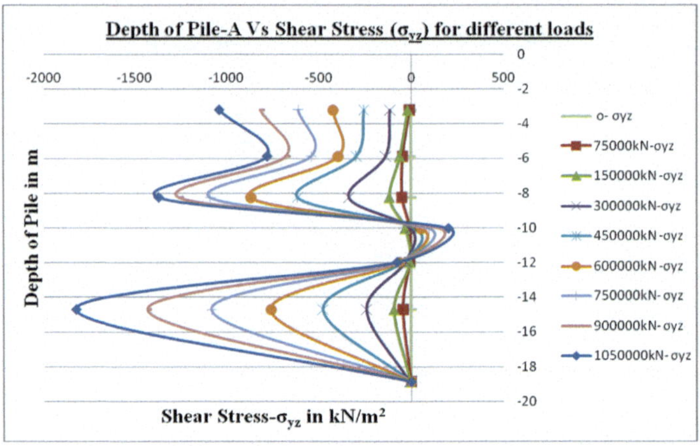


**Fig. 12** Effect of shear stress in YZ-direction for pile type A

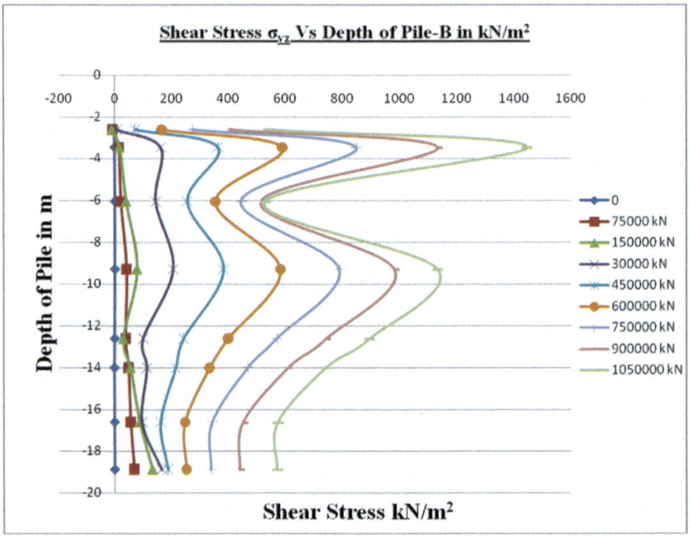


**Fig. 13** Effect of stress in YZ-direction for pile type B

## Stresses in Soil in Region 3

Figure 15 presents the normal stress distribution at the pile tips with increasing loads for all the three types of piles A, B, and C. With increase in load the magnitude of normal stress increases. The rate of increase is the maximum in the beginning and gradually decreases toward the end. Middle pile is subjected to maximum normal stress, and the edge piles are subjected to lower values.

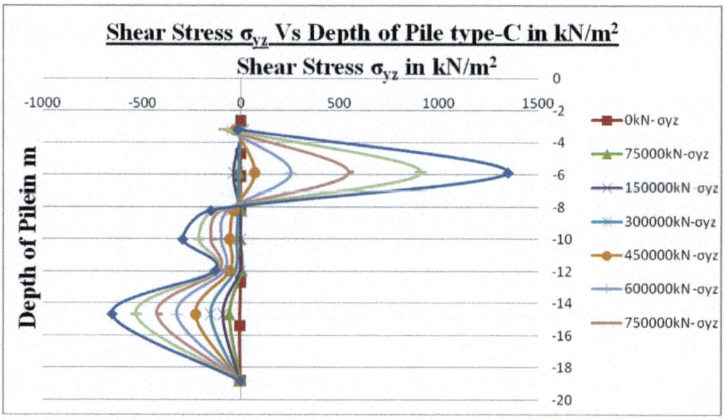


**Fig. 14** Effect of stress in YZ-direction for pile type C

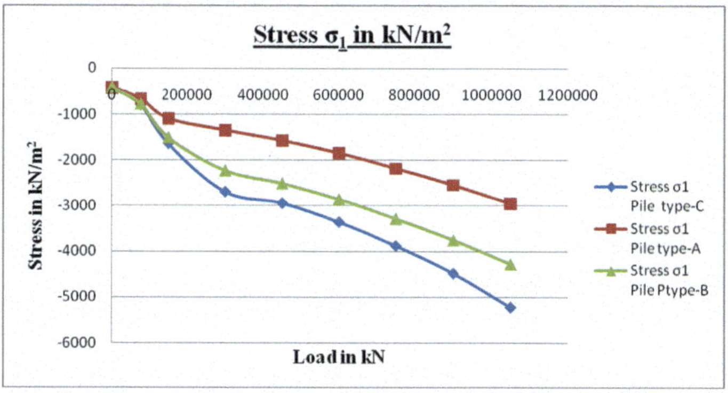


**Fig. 15** Effect of stress $\sigma_1$ for pile types A, B, and C

From the present study the following understanding can be made with regard to transfer of forces from foundation to the soil.

The pile raft foundation is sufficiently stiff compared to the surrounding soil. The load transfer first takes place from the raft to the soil around. Simultaneously the bottom of the pile also transfers the stresses to the soil. The rate at which the force is transferred at the base of pile is more in the beginning, and gradually it decreases with increasing load. Then, the load transfer happens along the surface of pile. This transfer is mainly due to shear, and it appears that the transfer happens near the top at the start, and gradually it is transferred toward the bottom. The biggest challenge is whether the shear is completely mobilized. The effectiveness of pile depends on the mobilization of shear.

## 4 Concluding Remarks

The load transfer mechanism from foundation to soil in pile raft foundation is a complex phenomenon because of the involvement of many foundation components in different regions. It should be noted that the interaction between raft and soil, pile and soil, pile and pile, and pile and raft should be carefully understood. Further, the performance with increase in load can be very different because of nonlinear behavior of soil, changes in interaction in different components, and many other factors. An attempt is made to explain the load transfer mechanism from pile raft to soil in a homogeneous soil. The foundation is subjected to increasing concentric vertical load from zero till failure. The foundation is considered to be rigid and elastic, soil is deformable, and behavior of soil is nonlinear. From the present work, the following inferences are made.

- The load transfer takes place due to compression at the top near the raft and at bottom near the bottom tip of piles.
- The rate of load transfer is more in the beginning near the bottom tip of pile, and gradually it decreases.
- The stresses developed are more near the edge than the center in the soil below the raft. Hence load transfer gradually increases from edges toward center.
- The load transfer is mainly through shear along the surface of piles. The transfer starts from the top and gradually increases toward the bottom.
- The soil in the zones near the central pile is subjected to less shear stress compared to that near the piles at the ends.
- However, the compressive stress in soil at the base is more near the central pile than that at the edge piles.

The performance of pile raft foundation system completely depends on how the shear is mobilized along the length of pile. The overall load carrying capacity and settlement control completely depend on this aspect. Hence, the number of piles used, pile size, and length of pile also significantly influence the overall performance.

## References

1. Irfan J, Irshad A (2022) Analysis and design of piled raft foundation taking into account interaction factors. Adv Civil Eng 2022(1334136):11. https://doi.org/10.1155/2022/1334136
2. Kumar A and Choudhury D (2015) Behaviour of combined pile-raft foundation (CPRF) under static and pseudo-static conditions using PLAXIS3D. In: 6th international conference on earthquake geotechnical engineering, 1–4 November 2015, Christchurch, New Zealand
3. Azad Hameed R, Ola Adel Q (2020) Behavior of pile rafts with raft size, number of piles and different pile length. In: AIP conference proceedings 2213, 020196. https://doi.org/10.1063/5.0000211, Published Online: 25 March 2020
4. Raj B, Srijit B (2020) Settlement behaviour of a pile raft subjected to vertical loadings in multilayered soil. ISSN: 1748–6025 (Print) 1748–6033 (Online) homepage: https://www.tandfonline.com/loi/tgeo20

5. Venkataraman B, Huang M et al (2019) Behavior and performance of piled raft—a comprehensive study. In: Proceeding of 16th Asian regional conference on soil mechanics and geotechnical engineering, Oct 14–18 2019, Taipei, Taiwan. http://seags.ait.asia/16arc-proceedings/16arc-proceedings/
6. Vijaykumar B, Prasad SK (2022) Critical comparison of load carrying capacity of pile, raft and pile-raft foundation systems. J Adv Civil Eng Technol Res (TJPRC: JACETR) 12(2):1–10 © TJPRC Pvt. Ltd.

# Behavior of Soil in OWT Monopile Foundation Subjected to Torsion

**J. Santhoshkumar and K. Muthukkumaran**

## 1 Introduction

India's need for energy has increased significantly as a result of its consistently strong and rapid economic growth. The offshore wind industry has been growing as a result of the large coastline and adherence to international climate agreements. As a renewable energy source, wind energy is growing in popularity and has a wide range of possible applications across the globe. The estimated lifespan of an offshore wind turbine is 20–30 years, whereas foundations are typically built for a longer design life. One of the most popular offshore foundations for wind turbines is the monopile. It is difficult to design foundations for OWTs because these are dynamically delicate structures. The basic goal of a foundation is to safely and within permissible deformations transfer all of the loads from the wind turbine structure to the ground. The main tower, three blades, and nacelle are the components of a wind turbine. The many electrical and electronic components required for the effective and secure conversion of wind power to electrical energy were housed inside the nacelle and the tower. If the yaw system fails, the OWT monopile foundation experiences a torsional effect. The resulting torsional effects then have an impact on the nearby soil, to investigate the response of the surrounding soil (sandy) in the offshore wind turbine monopile foundation when it is impacted torsional owing to the nacelle's yaw mechanism failing. This study will help to understand how the OWT's sandy soils degrade when subjected to torsion.

J. Santhoshkumar (✉) · K. Muthukkumaran
Department of Civil Engineering, National Institute of Technology, Tiruchirappalli 620015, India
e-mail: j.santhoshcivil70@gmail.com



B. T. Jose et al. (eds.), *Proceedings of the Indian Geotechnical Conference 2022 Volume 2*, Lecture Notes in Civil Engineering 477,
https://doi.org/10.1007/978-981-97-1741-5_14

### 1.1 Numerical Study

This study used a finite element application to understand the response of rigid piles embedded in dry homogeneous sand under torsional loading. The finite element software PLAXIS 3D has been to analyze the behavior of monopile. Tamil Nadu was the place selected for this investigation (8.500 °N, 78.300 °E). The Ministry of New and Renewable Energy provided the geotechnical, wind, wave, pile, turbine, and tower data. Based on the reference [1] and [2], various loads acting on the OWT were calculated. From the reference [3], numerical input parameters were collected. With the exception of near the top of the pile, the earth deforms little as it interacts with the pile under lateral loading. Therefore, in this case, the HS small strain model (HS-Small) is the chosen model. Based on Mindlin's theory, the monopile and tower in sandy soil were simulated in PLAXIS 3D using a linear elastic plate element as a hollow steel pipe. A very fine stage of mesh generation was used to divide the geometry into plate elements and volume elements.

## 2 Results

When sandy soil was subjected to environmental loading and static torsion, the variation of deformation with respect to dynamic time showed that the cyclic loading increased the structure's deformation. Cyclic secant stiffness was estimated using the load variation versus deformation graph.

Figure 1 shows the curve with more cycles, and the total displacement decreases. For the first few cycles, the displacement amplification grew more quickly before slowing down. This is a result of the sand around the pile becoming denser. The grains close to the pile are rearranged by the accumulated displacement. With each loading cycle, the hysteresis loop's shape flattened. It exhibits a reduction in loop area, indicating that the behavior is hardening.

### 2.1 Variation of Cyclic Secant Stiffness with Increasing Number of Cycles

The fluctuation in cyclic secant stiffness with respect to the number of cycles for sandy soil subjected to external loading and static torsion is shown in Fig. 2. The active and passive zones in the sand were created by the lateral load on the monopile. Under each cycle of loading, the sand mass develops resistance to applied lateral load with some permanent displacement on the soil monopile system. With more load cycles, the cyclic secant stiffness gradually increases. The decrease in intergranular voids is too responsible for this. Sand surrounding the pile densifies as a result of repetitive loading, raising the sand's shear modulus.

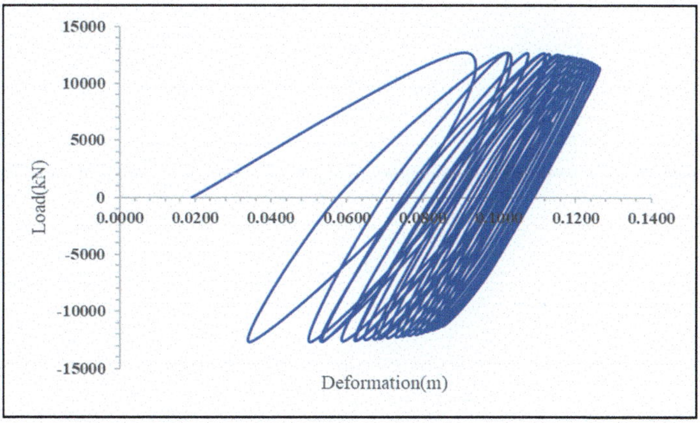


**Fig. 1** Load versus deformation curve

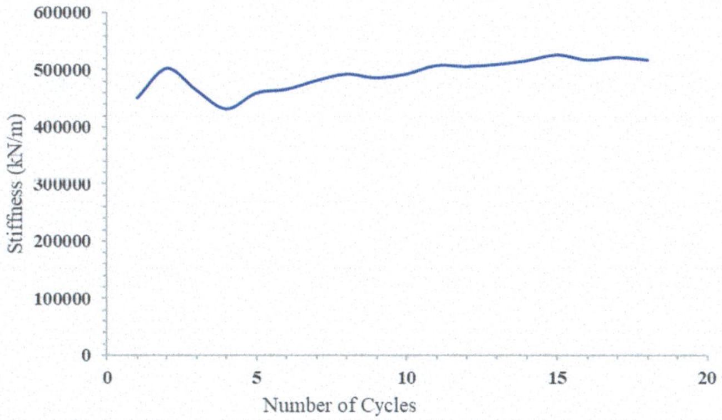


**Fig. 2** Stiffness versus number of cycles curve

## 2.2 *Variation of Shear Stress due to the Application of Lateral Loads in Sand*

We can see from Fig. 3, the shear stress rises as the depth increases. In the initial cycle of cyclic lateral loading in the sand, the monopile creates irrecoverable displacement because it minimizes friction between the pile and soil. The void will decrease as the number of cycles increases, causing friction that raises the shear strength and goes along with the reduction of the rate of increasing displacement.

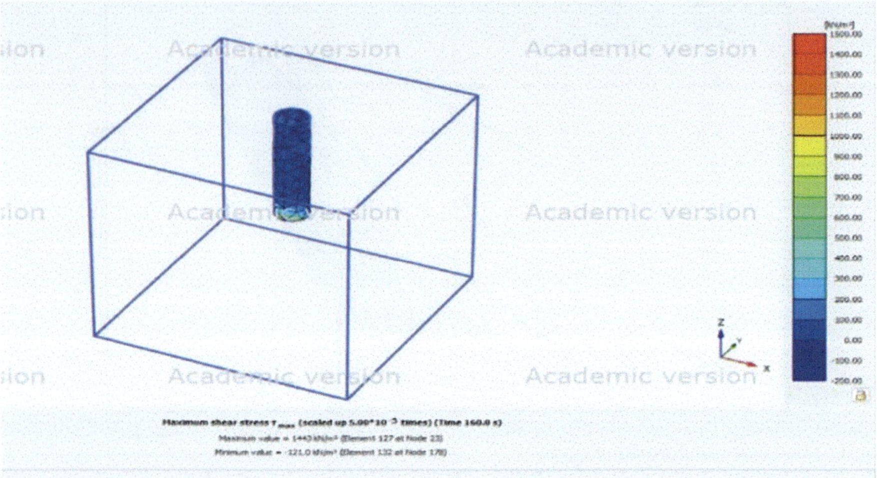


**Fig. 3** Shear stress variation due to the torsion and environmental loading

## *2.3 Variation of Shear Stress due to the Application of Pure Static Torsion in Sand*

Figure 4 shows the fluctuation in shear stress brought on by using static torsion solely. The shear stress contours surrounding the pile under pure static torsion are also used to study the rise in shear stresses over the frictional face of the pile. We can see from the graphs that the torsional load distribution diminishes along the monopile's depth and the shear stress increases with regard to the monopile's embedded length.

# 3 Parametric Study

## *3.1 Effect of Embedded Length of the Monopile in Sandy Soil*

For this investigation, the embedding length was 20, 25, 30, 35, and 40 m (Fig. 5). The sandy ground beneath the pile. The pile was rotated strictly in static motion. The above graph (Fig. 6a–e) illustrates the relationship between torque and torsional deformation for various embedded lengths in sandy soil layer. As the monopile's depth increased, the torsional load's magnitude reduced, because of the frictional resistance provided by the surrounding soil. Due to the impact of overburden soil, the frictional resistance provided by the soil increased with respect to depth. So the torsional load magnitude was decreased and the deformation caused by the torsional load also decreased along the depth of the monopile..

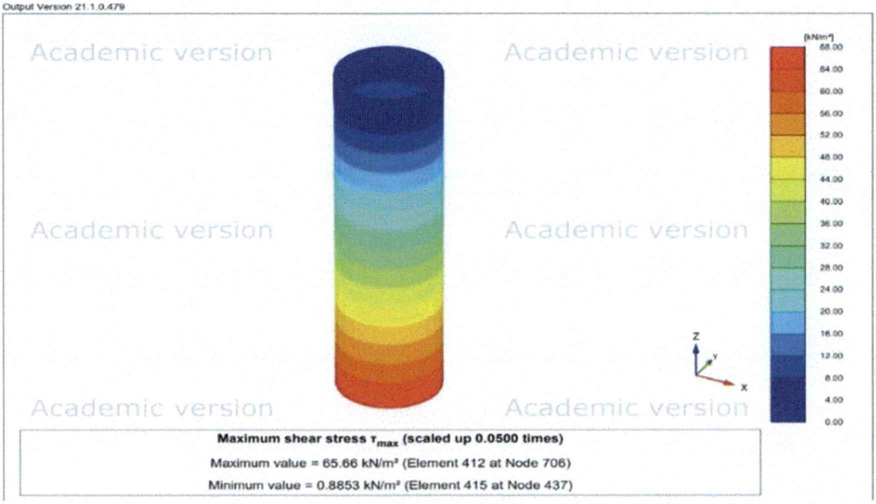


**Fig. 4** Maximum shear stress variation contour for sandy soil stratum subjected to the static torsion

**Fig. 5** Schematic diagram of the OWT in the field

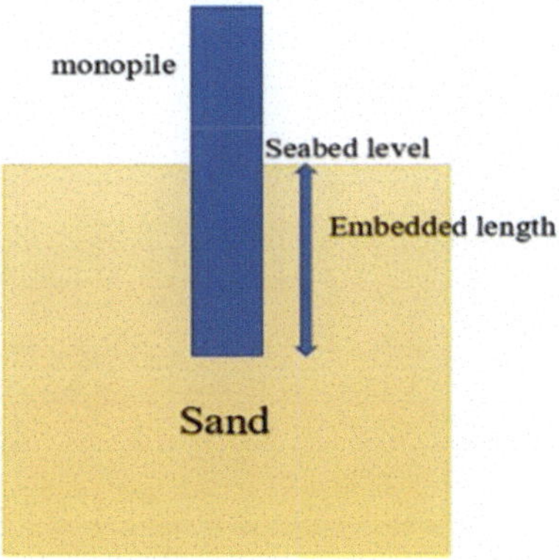


## 3.2 *Effect of Various Soil Strata*

Figure 7a–d show the variance in shear stress as well as the monopile's depth of embedment in various soil strata. We can see from the curve that the shear stress variation follows a linear pattern. With respect to the monopile's depth, the shear stress was rising linearly. The depth of the soil layer that affects shear resistance is the key factor in the variance. When fine-grained soil (clay) and a layer of coarse-grained soil (sand) are combined, the angle of friction and cohesiveness both rise. Shear–stress profiles for piles in clay and sand typically show that the stress rises linearly to a peak value and then sharply declines curvilinearly. For layered soil, the layering was the source of segmental variation in shear stress.

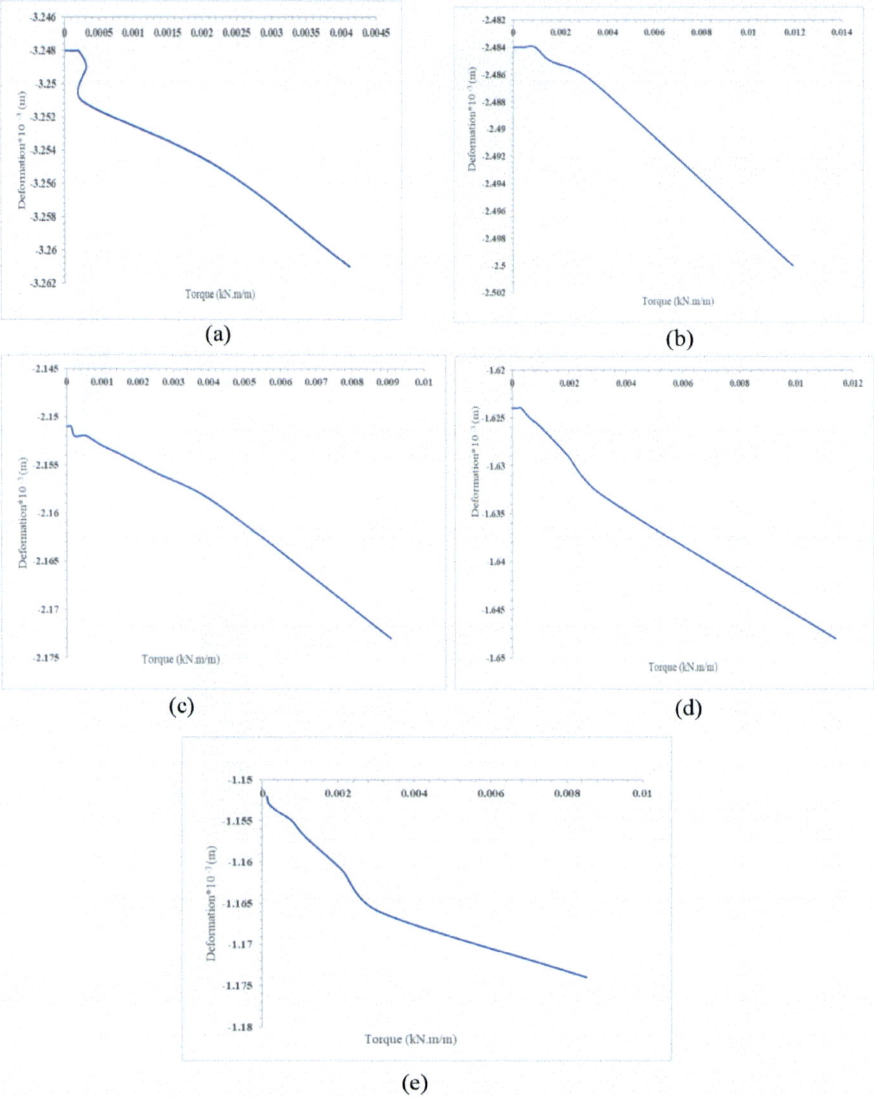


**Fig. 6** **a–e** Torque versus torsional deformation for embedded length of **a** 20, **b** 25, **c** 30, **d** 35, **e** 40 m

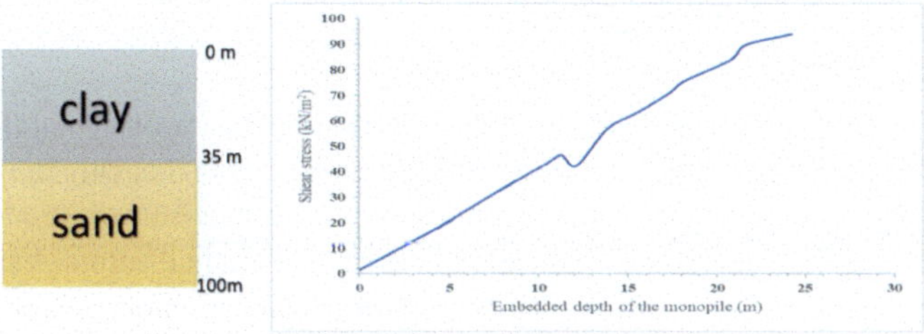


(a)

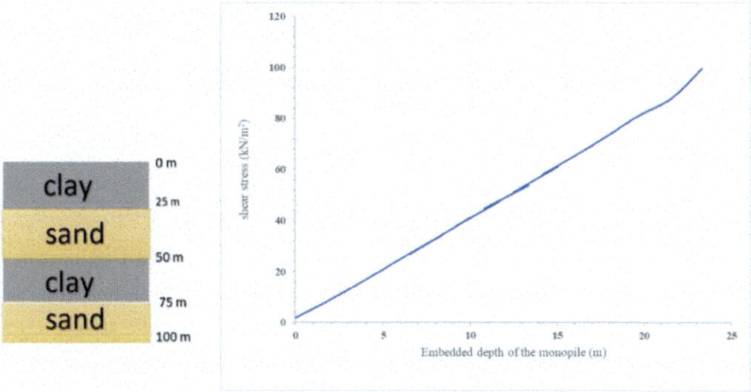


(b)

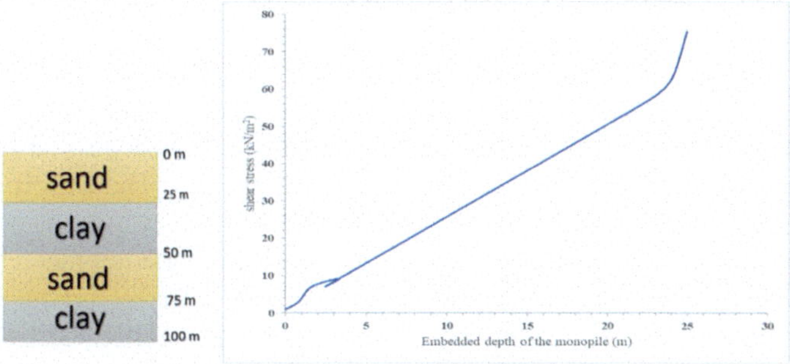


(c)

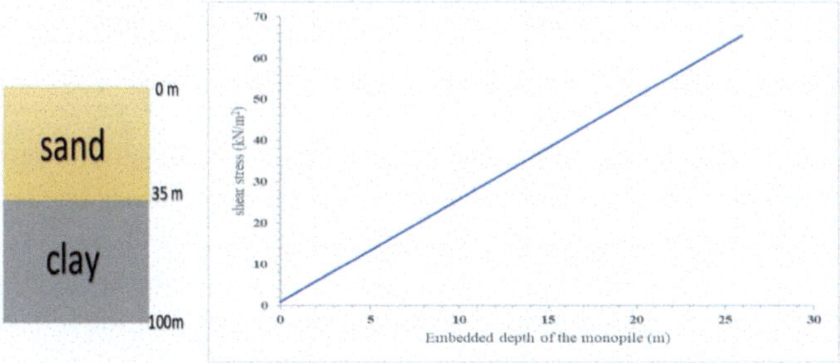


(d)

**Fig. 7 a–d** Shear stress variation along the embedded depth of the monopile

## 4 Experimental Investigation

The precise result can be controlled by doing a field test on a prototype pile, but due to the restricted field test setup facilities and high costs, it is difficult to implement precise loading on a monopile with a large diameter. Because of these above difficulties, which are also quite easy to arrange in the experimental setup. A model pile placed in a soil profile that has been artificially prepared can undergo a scaled test.

*Soil*

The sand used in the model tests was extracted from the Cauvery Riverbed, Tiruchirappalli. The sieve analysis was done as per IS 2720-part 4 (1985). From the curve, the sand is found to be poorly graded sand (SP).

*Test tank*

A small-scale model monopile was used as the subject of laboratory experiments. In a rectangular steel tank, various tests on the small-scale monopile were conducted. The test tank is 1.5 m in length, 1 m in width, and 1 m in height. The test tank's wall has a significant influence on the boundary effect in this model test. In order to prevent the bottom boundary effect in all tests, a minimum soil cushion at the bottom of the pile toe was maintained at 3 times the pile diameter. In order to reduce the possibility of friction between the sidewalls and sand, the tank was polished and smoothed.

*Sand raining device*

To execute the test in dry sand circumstances, the sand must be deposited at a specific density. The pluviation technique can be used to achieve this. The uniform sand bed was created using the sand draining apparatus. To achieve consistent density, the sand was poured into the test tank through the sand raining equipment. To ensure proper sand dispersion, the sand was pluviated in layers, with the vertical pipe being straight. In numerous tests, the device's configuration was calibrated by adjusting the height of the fall in order to obtain a good, consistent soil density. With the aid of a reference steel rod, the height of the fall was kept constant by making sure that, during the pluviation process, the bottom tip of the reference rod only contacts the surface of the sand bed. The photographic image of sand filling in the necessary density is shown in Fig. 8.

Begum and Muthukkumaran [4] and Gandhi and Selvam (2010) varied the free fall height of the sand particles to generate various sand densities (1997). A relationship between the height of fall and relative density for various heights of fall was discovered. From that work, to keep the height of fall constant during the pluviation process, the device was risen in respect to the reference rod. For the relative densities of loose conditions (Id = 30%), medium dense conditions (Id = 50%), and dense conditions (Id = 70%), the height of fall was selected as 4.7, 31, and 72 cm.

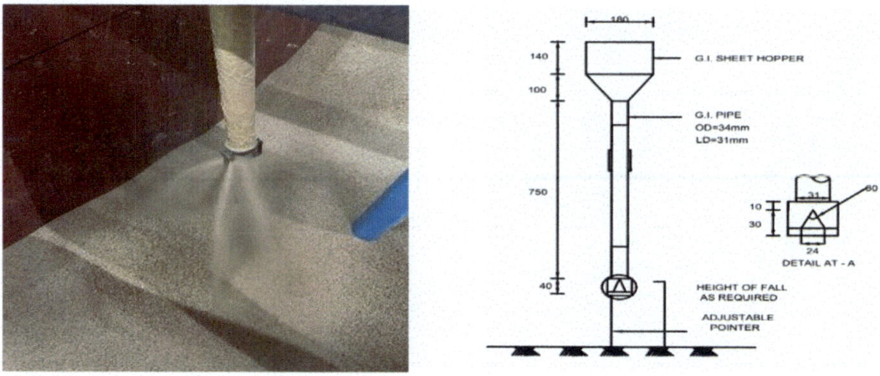


**Fig. 8** Sand raining device

*Model pile*

In this work, the behavior of a monopile was determined using a smooth aluminum model pile. An almost rigid pile was modeled using an open-ended aluminum pipe with a wall thickness of 2.5 mm and a diameter of 63.5 mm. The pile's embedded length was 533 mm and its overall length was 1000 mm. The pile was placed vertically on the sand bed in the center of the testing tank, and the sand cushion was created at the bottom for a depth of 3B. During the sand filling, the preinstalled model pile temporarily gets support from the field vane shear instrument. The layers of sand were filled in until the necessary embedment length was reached. In this experiment, the impact of driving a model monopile was disregarded.

## *4.1 Experimental Program*

A total of nine small-scale torsional load tests are performed to investigate and formulate the effect on piles. Nine tests are carried out on a model pile with an embedded length of 533 mm in the sand with varying relative densities (30, 50, 70%). Then field vane shear instrument was used for the torsional load application. The model pile of length 1m was placed at the required position in the tank. The vane shear apparatus and model pile were connected by the holder arrangement. The sand was poured into the tank to attain a required relative density of 30, 50, and 70% with the help of the sand raining device developed on basis of the air pluviation technique. The torsional load was applied at the pile head using the field vane shear instrument. The experimental setup is shown in Fig. 9. The torsional deformation of soil was measured using a dial gauge attached to the field vane shear instrument. After the failure of the pile, it will rotate freely. The dial gauge readings are reduced because there is no resistance offered by the surrounding soil. The applied torque was calculated by using spring stiffness and angular movement.

ii. As the number of cycles increases, the monopile foundation experiences a larger reversal of accumulated displacement at relatively low load amplitude. The monopile was subjected to recoverable displacement with increasing cycles under serviceability and fatigue loading conditions after initial cycles of irrecoverable displacement.
iii. As loading cycles increased, the hysteresis loop's shape became flatter. It exhibits a reduction in loop area, which indicates that hardening behavior is involved.
iv. As the number of load cycles increases, the stiffness of the cyclic secant increases gradually. This results from the decrease in intergranular voids. Also, The lower load amplitude causes lesser displacement and higher stiffness. This indicates that the lateral cyclic secant stiffness is highly dependent on the load amplitude.. This demonstrates how closely the load amplitude affects the lateral cyclic secant stiffness.
v. The sand surrounding the pile densifies as a result of repetitive loading, increasing the sand's shear modulus.
vi. The performance of the structure may be negatively impacted by an increase in the cyclic secant stiffness of the soil–pile system with an increasing number of load cycles. However, this behavior increases the soil–pile system's natural frequency, which could result in the resonance phenomena.
vii. As the number of cycles increases, there will be a reduction in the void, which causes friction and increases the shear strength while reducing the rate of increasing displacement.
viii. The variation in shear stress is only caused by the application of static torsion, and it increases with regard to the monopile's embedded length, whereas the distribution of the torsional load reduces as the depth of the monopile increases.
ix. The torsional load's magnitude reduced in relation to the depth of the OWT, because of the frictional resistance provided by the surrounding soil. Due to the impact of overburdening the soil, the frictional resistance provided by the soil increases with respect to depth. As a result, both the torsional load magnitude and the deformation it induced decreased, along with the monopile's depth.
x. The shear stress variation follows a linear pattern as the soil layer is changed. With respect to the monopile's depth, the shear stress was increasing linearly. The depth of the soil layer that contributes to shear resistance is the key factor influencing the variation.
xi. The variation of the shear stress is mainly based on the depth of the soil layer contributing to the shear resistance. The addition of the layer of coarse-grained soil (sand) with the fine-grained soil (clay) leads to an increase in the angle of friction and a decrease in cohesion.
xii. The shear–stress profiles for monopile in clay and sand indicate that the stress increases linearly to a peak value followed by a sharp curvilinear decrement. For layered soil, segmental variation of shear stress is originated from the layering.

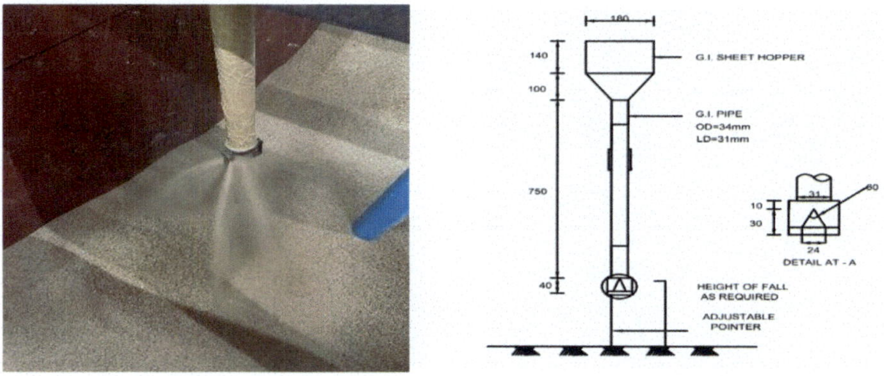


**Fig. 8** Sand raining device

*Model pile*

In this work, the behavior of a monopile was determined using a smooth aluminum model pile. An almost rigid pile was modeled using an open-ended aluminum pipe with a wall thickness of 2.5 mm and a diameter of 63.5 mm. The pile's embedded length was 533 mm and its overall length was 1000 mm. The pile was placed vertically on the sand bed in the center of the testing tank, and the sand cushion was created at the bottom for a depth of 3B. During the sand filling, the preinstalled model pile temporarily gets support from the field vane shear instrument. The layers of sand were filled in until the necessary embedment length was reached. In this experiment, the impact of driving a model monopile was disregarded.

## 4.1 Experimental Program

A total of nine small-scale torsional load tests are performed to investigate and formulate the effect on piles. Nine tests are carried out on a model pile with an embedded length of 533 mm in the sand with varying relative densities (30, 50, 70%). Then field vane shear instrument was used for the torsional load application. The model pile of length 1m was placed at the required position in the tank. The vane shear apparatus and model pile were connected by the holder arrangement. The sand was poured into the tank to attain a required relative density of 30, 50, and 70% with the help of the sand raining device developed on basis of the air pluviation technique. The torsional load was applied at the pile head using the field vane shear instrument. The experimental setup is shown in Fig. 9. The torsional deformation of soil was measured using a dial gauge attached to the field vane shear instrument. After the failure of the pile, it will rotate freely. The dial gauge readings are reduced because there is no resistance offered by the surrounding soil. The applied torque was calculated by using spring stiffness and angular movement.

ii. As the number of cycles increases, the monopile foundation experiences a larger reversal of accumulated displacement at relatively low load amplitude. The monopile was subjected to recoverable displacement with increasing cycles under serviceability and fatigue loading conditions after initial cycles of irrecoverable displacement.
iii. As loading cycles increased, the hysteresis loop's shape became flatter. It exhibits a reduction in loop area, which indicates that hardening behavior is involved.
iv. As the number of load cycles increases, the stiffness of the cyclic secant increases gradually. This results from the decrease in intergranular voids. Also, The lower load amplitude causes lesser displacement and higher stiffness. This indicates that the lateral cyclic secant stiffness is highly dependent on the load amplitude.. This demonstrates how closely the load amplitude affects the lateral cyclic secant stiffness.
v. The sand surrounding the pile densifies as a result of repetitive loading, increasing the sand's shear modulus.
vi. The performance of the structure may be negatively impacted by an increase in the cyclic secant stiffness of the soil–pile system with an increasing number of load cycles. However, this behavior increases the soil–pile system's natural frequency, which could result in the resonance phenomena.
vii. As the number of cycles increases, there will be a reduction in the void, which causes friction and increases the shear strength while reducing the rate of increasing displacement.
viii. The variation in shear stress is only caused by the application of static torsion, and it increases with regard to the monopile's embedded length, whereas the distribution of the torsional load reduces as the depth of the monopile increases.
ix. The torsional load's magnitude reduced in relation to the depth of the OWT, because of the frictional resistance provided by the surrounding soil. Due to the impact of overburdening the soil, the frictional resistance provided by the soil increases with respect to depth. As a result, both the torsional load magnitude and the deformation it induced decreased, along with the monopile's depth.
x. The shear stress variation follows a linear pattern as the soil layer is changed. With respect to the monopile's depth, the shear stress was increasing linearly. The depth of the soil layer that contributes to shear resistance is the key factor influencing the variation.
xi. The variation of the shear stress is mainly based on the depth of the soil layer contributing to the shear resistance. The addition of the layer of coarse-grained soil (sand) with the fine-grained soil (clay) leads to an increase in the angle of friction and a decrease in cohesion.
xii. The shear–stress profiles for monopile in clay and sand indicate that the stress increases linearly to a peak value followed by a sharp curvilinear decrement. For layered soil, segmental variation of shear stress is originated from the layering.

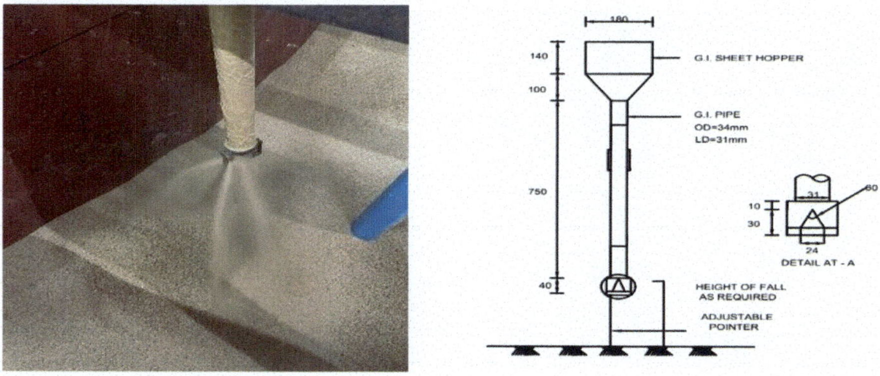


**Fig. 8** Sand raining device

*Model pile*

In this work, the behavior of a monopile was determined using a smooth aluminum model pile. An almost rigid pile was modeled using an open-ended aluminum pipe with a wall thickness of 2.5 mm and a diameter of 63.5 mm. The pile's embedded length was 533 mm and its overall length was 1000 mm. The pile was placed vertically on the sand bed in the center of the testing tank, and the sand cushion was created at the bottom for a depth of 3B. During the sand filling, the preinstalled model pile temporarily gets support from the field vane shear instrument. The layers of sand were filled in until the necessary embedment length was reached. In this experiment, the impact of driving a model monopile was disregarded.

## 4.1 Experimental Program

A total of nine small-scale torsional load tests are performed to investigate and formulate the effect on piles. Nine tests are carried out on a model pile with an embedded length of 533 mm in the sand with varying relative densities (30, 50, 70%). Then field vane shear instrument was used for the torsional load application. The model pile of length 1m was placed at the required position in the tank. The vane shear apparatus and model pile were connected by the holder arrangement. The sand was poured into the tank to attain a required relative density of 30, 50, and 70% with the help of the sand raining device developed on basis of the air pluviation technique. The torsional load was applied at the pile head using the field vane shear instrument. The experimental setup is shown in Fig. 9. The torsional deformation of soil was measured using a dial gauge attached to the field vane shear instrument. After the failure of the pile, it will rotate freely. The dial gauge readings are reduced because there is no resistance offered by the surrounding soil. The applied torque was calculated by using spring stiffness and angular movement.

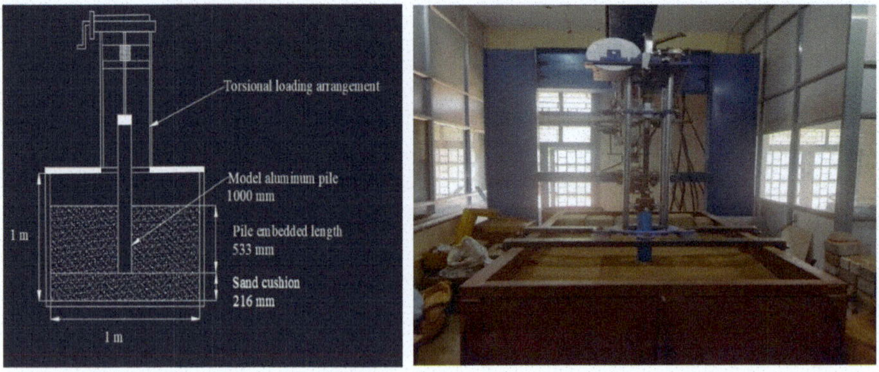


**Fig. 9** Detailed test setup and photographic view of the arrangement for torsional loading

## 4.2 Experimental Results and Validation Shear Strength Calculation

Field vane shear equations were modified based on the pile and soil conditions. In the field vane shear, total torque (T) is the summation of resisting torque at the sides (T1) and resisting torque at the top and bottom (T2). But here the pile is an open-ended hollow pile. So, there is no contribution in the bottom for resisting torque. Here the pile is not fully embedded in the soil, so the top of the pile also does not contribute to resisting torque.

The total torque (T) = resisting torque at the top and bottom (T2).

$$\mathrm{T} = \mathrm{T2} = \int_0^{D/2} S * 2\pi \mathrm{r}\,\mathrm{dr} * \mathrm{r}, \tag{1}$$

$$\text{From the integration}, T = \frac{\pi D^3 S}{12}. \tag{2}$$

The applied torque (T) and diameter of the pile (D) are known parameter. From that data, the shear strength was back-calculated.

$$\text{Shear strength}, S = \frac{12T}{\pi D^3}. \tag{3}$$

The experimental values were checked with the theoretical calculation.

$$\text{Theoretical torque} = \text{skin friction} * \text{perimeter} * \text{diameter}/2. \tag{4}$$

Skin friction = K $\sigma$ tan $\delta$.
Perimeter = $\pi$dl.

**Table 1** Shear strength calculation based on the experiment

| Relative density (%) | Torque (kNm) | Shear strength (kN/m$^2$) |
|---|---|---|
| 30 | 0.002739 | 40.8534 |
| 50 | 0.03423 | 51.067 |
| 70 | 0.00377 | 56.1734 |

**Table 2** Shear strength calculation based on the theoretical calculation and FEM

| Relative density (%) | Shear strength (kN/m$^2$) as per theoretical calculation | Shear strength (kN/m$^2$) as per FEM |
|---|---|---|
| 30 | 48.55 | 47.26 |
| 50 | 66.11 | 59.95 |
| 70 | 86.89 | 70.88 |

K = earth pressure coefficient, σ = vertical stress, D = diameter of the pile, L = length of the pile.

**Experimental calculation**

The torsional deformation was calculated based on the dial gauge reading. The torque was calculated based on the stiffness of the spring attached to the field vane shear test apparatus and the angular movement. The shear strength was back-calculated by using Equation (3). The calculated value is tabulated in the below table.

**Validation**

To verify the accuracy of the model developed, comparisons of the FEM solutions are made with available theoretical (analytical) and experimental (laboratory model tests) results. The analysis is carried out and the interpretations made are sequentially described herein.

The theoretical torque was calculated based on equation (4). Theoretical shear strength was back-calculated by using the theoretical toque. From Tables 1 and 2, shear strength calculated from the experimental and numerical calculation falls within the theoretically calculated shear strength.

## 5 Conclusions

Based on the present work, the following conclusion is drawn.

i. According to the load versus deformation curve, the total displacement decreases as the number of cycles increases. This is a result of the sand around the pile becoming denser.

ii. As the number of cycles increases, the monopile foundation experiences a larger reversal of accumulated displacement at relatively low load amplitude. The monopile was subjected to recoverable displacement with increasing cycles under serviceability and fatigue loading conditions after initial cycles of irrecoverable displacement.
iii. As loading cycles increased, the hysteresis loop's shape became flatter. It exhibits a reduction in loop area, which indicates that hardening behavior is involved.
iv. As the number of load cycles increases, the stiffness of the cyclic secant increases gradually. This results from the decrease in intergranular voids. Also, The lower load amplitude causes lesser displacement and higher stiffness. This indicates that the lateral cyclic secant stiffness is highly dependent on the load amplitude.. This demonstrates how closely the load amplitude affects the lateral cyclic secant stiffness.
v. The sand surrounding the pile densifies as a result of repetitive loading, increasing the sand's shear modulus.
vi. The performance of the structure may be negatively impacted by an increase in the cyclic secant stiffness of the soil–pile system with an increasing number of load cycles. However, this behavior increases the soil–pile system's natural frequency, which could result in the resonance phenomena.
vii. As the number of cycles increases, there will be a reduction in the void, which causes friction and increases the shear strength while reducing the rate of increasing displacement.
viii. The variation in shear stress is only caused by the application of static torsion, and it increases with regard to the monopile's embedded length, whereas the distribution of the torsional load reduces as the depth of the monopile increases.
ix. The torsional load's magnitude reduced in relation to the depth of the OWT, because of the frictional resistance provided by the surrounding soil. Due to the impact of overburdening the soil, the frictional resistance provided by the soil increases with respect to depth. As a result, both the torsional load magnitude and the deformation it induced decreased, along with the monopile's depth.
x. The shear stress variation follows a linear pattern as the soil layer is changed. With respect to the monopile's depth, the shear stress was increasing linearly. The depth of the soil layer that contributes to shear resistance is the key factor influencing the variation.
xi. The variation of the shear stress is mainly based on the depth of the soil layer contributing to the shear resistance. The addition of the layer of coarse-grained soil (sand) with the fine-grained soil (clay) leads to an increase in the angle of friction and a decrease in cohesion.
xii. The shear–stress profiles for monopile in clay and sand indicate that the stress increases linearly to a peak value followed by a sharp curvilinear decrement. For layered soil, segmental variation of shear stress is originated from the layering.

xiii. Torsional load on pile foundations is induced by the action of eccentric horizontal forces on the supporting structures. Such loading not only initiates a twist at the pile head but also reduces its axial capacity significantly.

## References

1. Brian B, Sheil B, McCabe A (2017) Biaxial loading of offshore monopiles: numerical modeling. Int J Geomech 17(2)
2. Basack S, Nimbalkar S (2017) Numerical solution of single pile subjected to torsional cyclic load. Int J Geomech © ASCE 1532–3641
3. Laszlo Arany S, Bhattacharya JM, Hogan SJ (2017) Design of monopiles for offshore wind turbines in 10 steps. Soil Dyn Earthq Eng 92(2017):126–152
4. Begum, N.A and Muthukkumaran, K (2009) Experimental investigation on single model pile in sloping ground under lateral load. Int J Geotechnical Eng 3:133–146

# Evaluation of Nonlinear Load Sharing Ratio of Pile and Raft in Piled Raft Foundation in Cohesionless Soil

**Dinesh Kumar Malviya and Manojit Samanta**

## 1 Introduction

Piled raft foundations are widely preferred as a suitable foundation system for tall buildings and several important structures such as silo, commercial buildings, nuclear power plants, and hospitals [1]. Piled raft foundations consist of a raft, piles, and subsoil. Piled raft effectively improves the bearing capacity and reduces the settlement of the structure to permissible limits [2]. The piles in a small piled raft ($R_w/L_p$ < 1, $R_w$—width of raft and $L_p$—length of pile) improve the load carrying capacity of the foundation, whereas reduce the excessive total and differential settlement in a large piled raft foundation ($R_w/L_p$ > 1). The load on the superstructure is distributed between the raft and pile in a piled raft. The complex soil-structure interaction based on factors such as the geometry of foundation elements and soil conditions governs the load sharing mechanism of piled raft foundations. The optimized piled raft design can be achieved by considering various important governing parameters such as $L_p/D_p$ ratio ($L_p$—pile length, $D_p$—pile diameter), pile spacing, soil–raft stiffness ratio ($K_{rs}$), soil stiffness, which affects the performance of the foundation system [3, 4]. Several laboratory model experiments and centrifuge test has been conducted to understand the behavior of foundations [5]. However, these studies are limited due to high expenses and difficulty in execution. The theoretical investigations such as simplified analysis methods and analytical methods have been widely used to understand the behavior of the piled raft foundations [2, 6]. The rigorous finite element-based numerical analysis of piled raft foundation has been widely used and found

D. K. Malviya · M. Samanta (✉)
CSIR—Central Building Research Institute, Roorkee 247667, India
e-mail: manojit@cbri.res.in

Academy of Scientific and Innovative Research (AcSIR), Ghaziabad 201002, India

D. K. Malviya
e-mail: dinesh.cbri20a@acsir.res.in

 
B. T. Jose et al. (eds.), *Proceedings of the Indian Geotechnical Conference 2022 Volume 2*, Lecture Notes in Civil Engineering 477,
https://doi.org/10.1007/978-981-97-1741-5_15

to be accurate among all methods of piled raft analysis [7]. Several researchers have proposed different models to predict the load sharing in a vertically loaded piled raft foundations. Randolph [8] proposed an empirical relation to estimate the load sharing between the pile and raft based on the results obtained from rigorous numerical analysis of circular raft supported with small pile groups considering the pile–pile and raft–pile interaction. Clancy and Randolph [9] developed an approximate method for estimating the load sharing distribution in piled raft foundation system for large pile groups. Lee et al. [10] developed a model using the normalized nonlinear load–settlement relationship considering the effect of pile–raft interaction to predict the load sharing between pile and raft. Kumar and Choudhury [11] proposed simplified expression for the evaluation of load sharing ratio of piled raft foundation considering the serviceability requirement of the structure.

In this paper, load sharing in a square small piled raft embedded in cohesionless medium and subjected to vertical loading has been presented. The variation of complex pile–pile and pile–raft interaction with foundation settlement has been discussed from experimental investigations. The mechanism of load sharing in a piled raft foundation has been presented using the experimental results of single pile, unpiled raft, pile group, and piled raft foundation. The effect of soil stiffness, soil–raft stiffness, pile group area, and length to pile diameter ratio on load sharing of pile and raft has been investigated using three-dimensional finite element analyses. The parameter of numerical analysis has been considered similar to the experimental studies for comparison purposes of load sharing of pile and raft.

## 2 Experimental Setup

The steel tank of 1600 × 1250 × 1250 mm has been used in the present study. The possibility of volume changes during tank filling and subsequent loading has been eliminated by stiffening it with steel channels at different levels. The test tank has rigid boundaries on all sides except the top surface. The lateral boundaries of the tank are greater than 5 times the raft width, and the vertical boundaries are greater than 10 times the pile diameter from the pile tip. These boundaries are sufficient to minimize the boundary effect [12]. The model piles are close-ended hollow circular pipes with an outer diameter of 20 mm and wall thickness of 1 mm. Two different pile lengths ($L_p$) of 500 mm and 1000 mm have been used in the present study. The $A_g/A_r$ of the tested piled raft is 0.032, where $A_g$ is the area of piles and $A_r$ is the area of raft. The piles are made of the same grade of aluminum alloy as the model raft. They are connected to the raft with nut bolt connection at 6D center to center spacing. The piles are instrumented with strain gauges of 120 Ω gauge resistance and gauge factor of 2.13. The strain gauges are pasted at 0.05, 0.25, 0.50, 0.75, and 0.97 $L_p$ to measure the strain distribution along the pile length. The model raft is 200 × 200 mm in plan and 10 mm in thickness. The rigidity of the raft has been evaluated by the following relationship [4]:

$$K_{rs} = 5.57 \frac{E_r}{E_s} \frac{1 - \upsilon_s^2}{1 - \upsilon_r^2} \left(\frac{B}{L}\right)^{0.5} \left(\frac{t_r}{L}\right)^3,$$

where $E_r$ and $E_s$ are Young's moduli of the raft and the soil, respectively, $\nu_r$ and $\nu_s$ is the Poisson' ratio of the raft and soil, respectively, B is the breadth, L is the length of the foundation, and $t_r$ is the thickness of the raft. Hain and Lee [13] defined the rigidity of the raft based on the $K_{rs}$ ratios ranging from fairly flexible ($K_{rs} = 0.01$) to effectively rigid ($K_{rs} = 10$). The $K_{rs}$ of the raft in present study is estimated to be 2.87 which show the raft to be rigid. Figure 1 shows the complete experimental model test setup and its different components.

Solani river sand, mined locally in Roorkee, has been used in the tests. Figure 2 shows the gradation of the sand used in the present study. The soil has been classified as poorly graded sand. Sand particles are classified as sub-rounded based on morphological characterization. Table 1 shows the properties of sand. Sand mediums are prepared to a relative density of 80%. The angle of internal friction is 38° as obtained from direct shear tests. The amount of soil required has been weighed to achieve the desired density for every 50 mm layer. The soil is poured and compacted manually up to the desired depth. The in situ density has been measured at four corners of the test tank during filling and after test completion.

The dial gauges having least count of 0.01 mm have been mounted over foundation system to measure the settlement. The earth pressure cells have been placed beneath the raft to measure the load carried by raft. The tests have been conducted in a

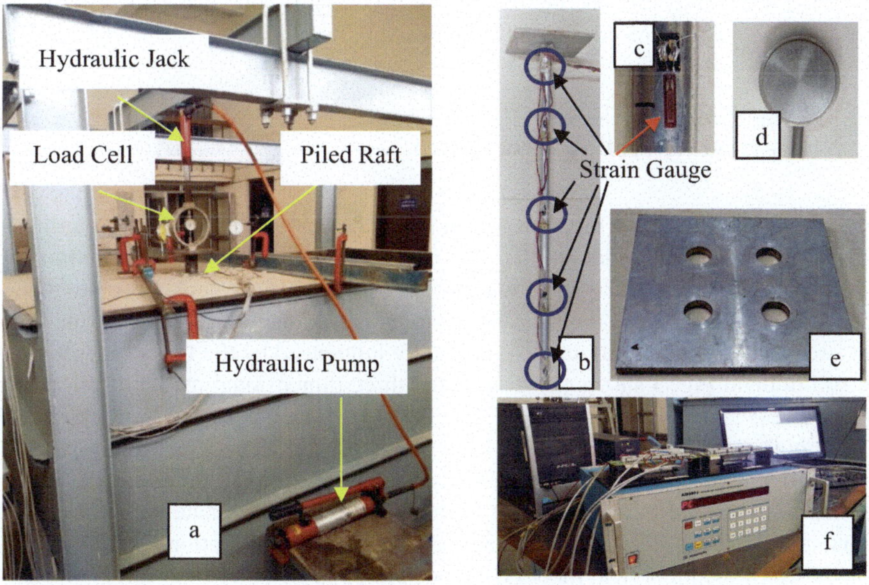


**Fig. 1** Experimental test setup: **a** piled raft in test tank, **b** instrumented pile, **c** strain gauge pasted on pile, **d** earth pressure cell, **e** raft, and **f** data acquisition system

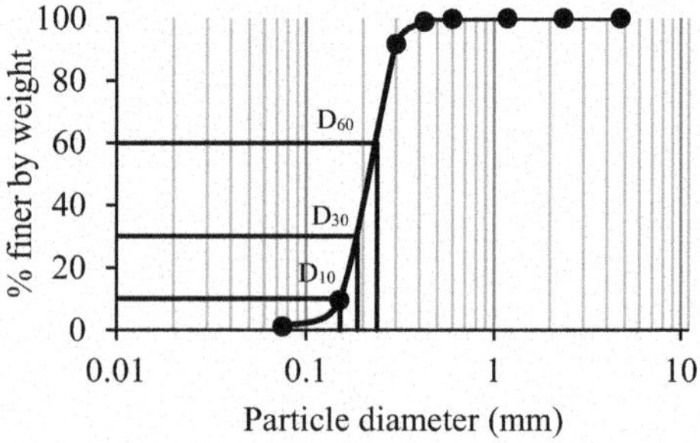


**Fig. 2** Grain size distribution curve

**Table 1** Properties of sand

| | |
|---|---|
| $G_s$ | 2.67 |
| $C_u$ | 1.92 |
| $C_c$ | 1.15 |
| $\rho_{max}$ (g/cc) | 1.66 |
| $\rho_{min}$ (g/cc) | 1.39 |
| IS Classification | SP |
| $\varphi$ (°) | 38 |

load-controlled manner. The vertical load on the piled raft has been applied through a hydraulic jack and measured through a calibrated load cell. The load has been applied in steps, and each load increment of 60 N is applied only when the settlement corresponding to the preceding load becomes very negligible.

## 3 Numerical Modeling Procedure

The behavior of the piled raft foundation has been studied by developing a three-dimensional numerical model in PLAXIS 3D. It consists of the soil continuum, the pile and raft geometry, and the boundary conditions. The trial calculations have been performed by varying the boundaries of numerical domain until the deformations and stresses of the piled raft are insignificantly influenced by the further increase in the soil domain Fig. 3 shows the discretized three-dimensional model of piled raft foundation.

The homogeneous soil layer has been considered in the present analysis. The horizontal and vertical boundaries have been extended up to 10 times the raft width ($R_w$) and 2 times the pile length ($L_p$) from the pile tip to eliminate the undesirable boundary effect [14]. The bottom boundaries are fixed in all directions, while vertical boundaries of the numerical model have been fixed in perpendicular directions and allowed to move in the plane only and the top surface is free to move in all directions. Cohesionless soil is modeled using elastic-perfectly plastic Mohr–Coulomb

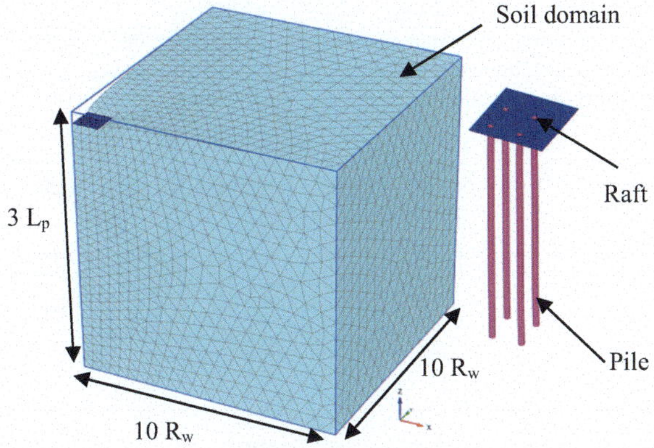


**Fig. 3** Discretized three-dimensional model of piled raft foundation

constitutive model. Drained behavior of the soil medium is considered in the analysis. The strength reduction factor ($R_{int}$) of 0.7 has been considered between pile and soil. The soil continuum has been discretized using 10-noded tetrahedral elements, having three translational degrees of freedom in three perpendicular directions [15]. Several researchers have used this constitutive model to simulate the behavior of sand [16, 17]. Pile and raft are modeled using embedded beam and plate elements, respectively. Table 2 shows the range of soil properties and geometries of the pile and the raft used in the present analysis. The load–settlement response obtained through finite element methodology has been compared with [12]. The piled raft is subjected to vertical loading consists of 3 × 3 pile group rigidly connected with square raft of 10 m width and 1 m thickness. The pile length of 20 and 0.6 m diameter embedded in loose cohesionless soil medium has been considered in the analysis. Figure 4 shows the load–settlement response of piled raft. The results are in close agreement confirming the validation of numerical model used in the present study.

**Table 2** Geotechnical properties of soil and mechanical properties pile and raft

| Parameters | Soil | Raft | Pile |
|---|---|---|---|
| E (kPa) | 8.7–46.4 × $10^3$ | 25 × $10^6$ | 25 × $10^6$ |
| ρ (kN/$m^3$) | 15.5–17 | 25 | 25 |
| ν | 0.28–0.30 | 0.2 | |
| C (kPa) | 0 | – | – |
| φ (°) | 28–33 | – | – |
| Diameter (m) | – | – | 0.5–1.0 |
| Length (m) | – | 7.5 | 20–40 |
| Width (m) | – | 7.5 | – |
| Thickness (m) | – | 0.75–2.5 | – |

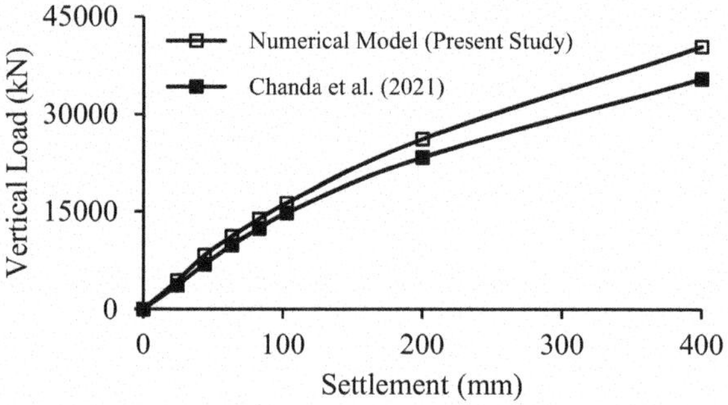


**Fig. 4** Comparison of the load–settlement response obtained by Chanda et al. [12] and present numerical model

# 4 Results and Discussions

## 4.1 *Experimental Investigations*

### 4.1.1 Load–Settlement Performance

The load–settlement behavior of the single pile (SP), unpiled raft (UR), pile group (PG), and piled raft (PR) has been determined from the experimental investigations.

Figure 5 shows the results of a repeat test performed on a single pile to check the reliability and uniformity of the experimental investigation. Figure 6 shows the load–settlement responses of SP, UR, PG, and PR. It can be seen that the pile load on SP increases linearly up to the settlement of 10% of pile diameter and increases nonlinearly up to the settlement of 55% pile diameter. At this settlement level, the pile reaches its ultimate capacity and the load becomes constant with further foundation settlement. In 2 × 2 PG, the load on pile increases up to 10% of pile diameter and increases non-linearly up to 25% of the pile diameter. At this settlement level, the pile reaches its ultimate capacity, and the load becomes constant with further settlement. It can be noted that the load carrying capacity of the 2 × 2 PG is lower than the summation of 4 SP due to the pile–pile interaction which has been discussed in the subsequent section. The load on the UR increases nonlinearly with the foundation settlement. The load on PR increases linearly up to the settlement of 0.002 $R_w$, where $R_w$ is raft width. The load on piled raft increases nonlinearly beyond this settlement level. It can be seen from the load–settlement responses of the UR, PG, and PR that once the pile achieves its ultimate capacity corresponding to settlement of 25% of pile diameter (i.e., 5 mm), further load on the piled raft is carried by the raft. Figure 7 compares strain distribution along piles in SP, PG, and PR at vertical loading of 900N. The strain gradually decreases with the depth of pile in SP, PG, sand PR, respectively. The strains at pile head in PG and SP are 1.54 and 8.94 times the strains in PR, respectively. SP experiences higher strains compared to piles in PG and PR depicting the lowest load carrying capacity of the piles. The piles in PR show lower strains at the pile head depicting the highest load carrying capacity.

**Fig. 5** Load–settlement response of single pile

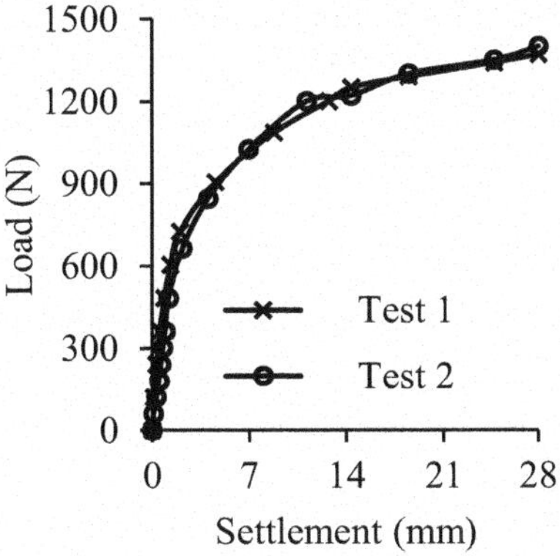


**Fig. 6** Load–settlement response of SP, UR, PG, and PR

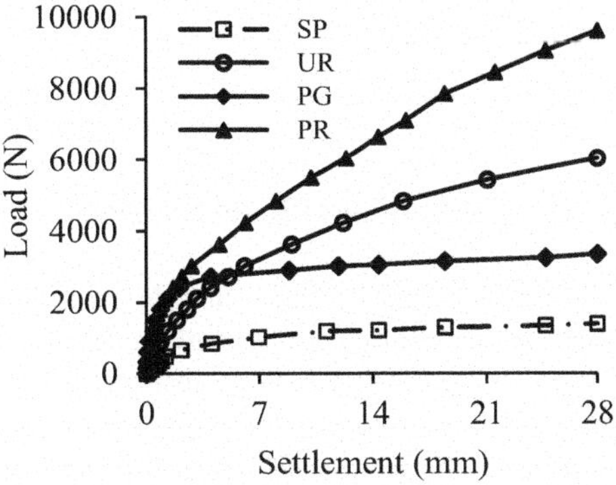


**Fig. 7** Strain distribution along pile in a SP, PG, and PR

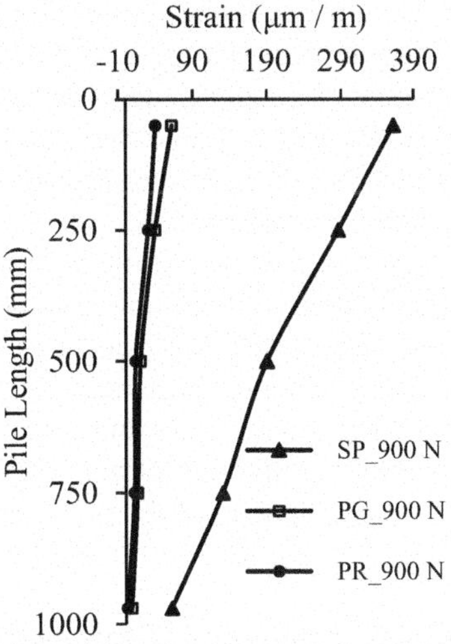

**Fig. 10** Effect of $E_p/E_s$ on load sharing in piled raft

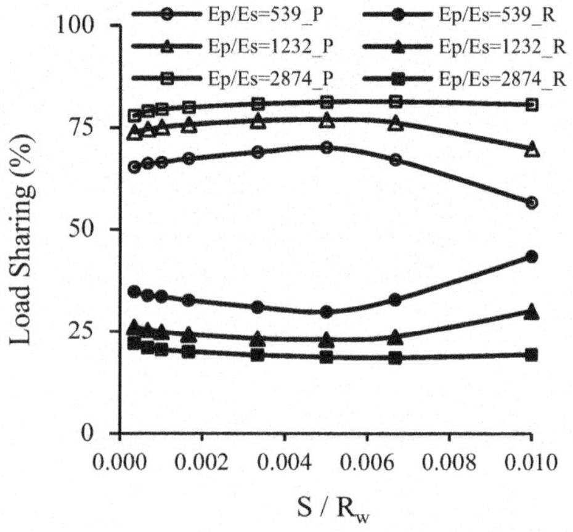


### 4.2.1 Effect of Pile–Soil Stiffness Ratio

The effect of soil stiffness on load sharing of pile and raft has been investigated by varying the soil stiffness from loose to dense cohesionless soil medium. Figure 10 shows the variation of load sharing with foundation settlement in pile and raft with the variation in soil stiffness. The L/D ratio and $A_g/A_r$ of 80 and 0.014 has been considered in the analysis, respectively. The $K_{rs}$ has been varied from 0.36 to 1.90. The load shared by the raft increases with an increase in soil stiffness, whereas the load shared by the piles increases with a decrease in soil stiffness. The pile load sharing increases with foundation settlement up to 0.05 $R_w$ and further decreases gradually with increasing foundation settlement. The load shared by pile and raft at 0.05 $R_w$ foundation settlement is 81.2, 76.9, 70.1, and 18.8, 23.1, 29.9% for $E_p/E_s$ in the range of 2874–539, respectively. The pile load sharing increases to 11 and 6% with increasing $E_p/E_s$ from 539 to 1232 and 1232 to 2874, respectively. The increase in soil stiffness improves the normal pressure between the raft and underlying soil, which is effective in increasing the frictional bearing capacity of the raft [3]. The development of contact pressure with gradually increasing loading consequently increasing foundation settlements and results in increase of raft load sharing in the piled raft foundation. Similar foundation response has been obtained from the experimental investigations in the present study.

### 4.2.2 Effect of Soil–Raft Stiffness ($K_{rs}$)

The effect of soil–raft stiffness has been evaluated by varying the soil stiffness and the raft thickness. Soil–raft stiffness increases with increase in raft thickness and soil stiffness. The L/D ratio and $A_g/A_r$ of 80 and 0.014 has been considered in the analysis, respectively.

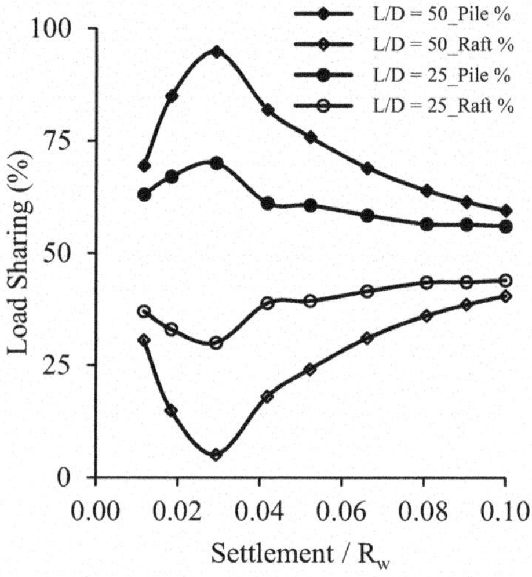


**Fig. 8** Load sharing of pile and raft in a piled raft foundation

### 4.1.2 Evaluation of Load Sharing

The load carried by the four piles in a 2 × 2 PR has been evaluated by summing the load carried by four SP and multiplied with pile–pile interaction factor. The load carried by the raft has been evaluated by multiplying the load carried by raft and pile–raft interaction factor. The estimation of interaction factors has been discussed in subsequent section. Figure 8 shows the nonlinear load sharing of pile and raft in a piled raft foundation with foundation settlement. The pile load sharing increases up to foundation settlement of 0.03 $R_w$ and decreases with further foundation settlement. The pile load sharing at 0.03 $R_w$ is 94 and 70% with slenderness ratios (L/D) 50 and 25, respectively. The raft load increases with foundation settlement greater than 0.03 $R_w$. The raft load share becomes stable approximately at foundation settlement of 0.1 $R_w$. The raft load sharing at 0.1 $R_w$ is 44 and 40% with pile slenderness ratios 25 and 50, respectively.

The raft load sharing in a piled raft increases with the foundation settlement. Horikoshi and Randolph [5] and Lee et al. [10] observed similar results that the raft load sharing increases and pile load sharing decreases with the increasing foundation settlement. The load sharing of the foundation element gradually varies with foundation settlement and becomes constant at settlement of 0.1 $R_w$. Several studies considered the load corresponding to a settlement of 0.1 $R_w$ as the ultimate load carrying capacity of the piled raft foundation [18, 19].

### 4.1.3 Evaluation of Interaction Factor

The piled raft behavior is governed by the complex interaction between the soil–pile–raft interactions and varies with different settlement levels. The pile–pile ($\alpha_{p\text{-}p}$)

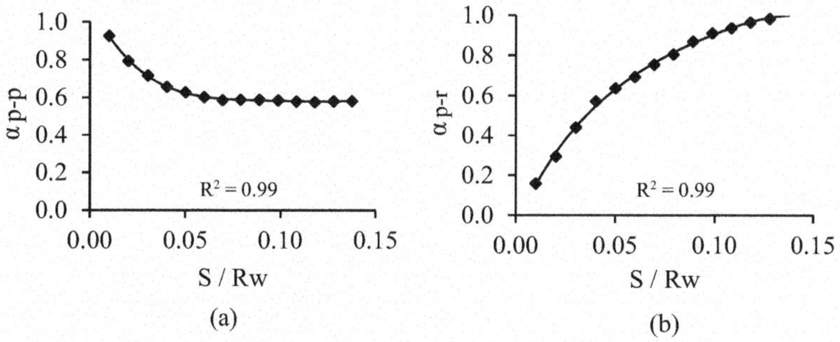


**Fig. 9** Interaction factor: **a** pile–pile ($\alpha_{p\text{-}p}$) and **b** pile–raft ($\alpha_{p\text{-}r}$)

and pile–raft ($\alpha_{p\text{-}r}$) interaction has been estimated using the following relationships:

$$\alpha_{p-p} = L_{PG} / \Sigma L_{SP},$$

$$\alpha_{p-r} = L_R / \Sigma\, L_{UR},$$

where $L_{PG}$—pile group load, $L_{SP}$—single pile load, $L_R$—load of raft in a piled raft, and $L_{UR}$—unpiled raft load. It has been observed that the interaction between the foundation elements of piled raft varies with its settlement. Figure 9 shows the variation of $\alpha_{p\text{-}p}$ and $\alpha_{p\text{-}r}$ with $S/R_w$, where $S$ is the foundation settlement and $R_w$ is raft width.

The pile–pile interaction decreases nonlinearly up to 0.59 corresponding to 0.06 $R_w$ foundation settlements. The pile–pile interaction is constant with further settlement. Lee et al. [10] observed similar variation in pile–pile interaction with foundation settlement. The pile–raft interaction increases nonlinearly with the foundation settlement and becomes constant after 0.10 $R_w$ foundation settlement. The pile load sharing decreases as the pile–pile interaction decreases, whereas the raft load sharing increases as the pile–raft interaction increases.

## 4.2 Numerical Analysis

The influence of different parameters on load sharing performances of the piled raft foundations under vertical loadings has been investigated by three-dimensional finite element analysis. The effect of soil stiffness, length to diameter ratio, area of pile and soil–raft stiffness has been discussed in this section.

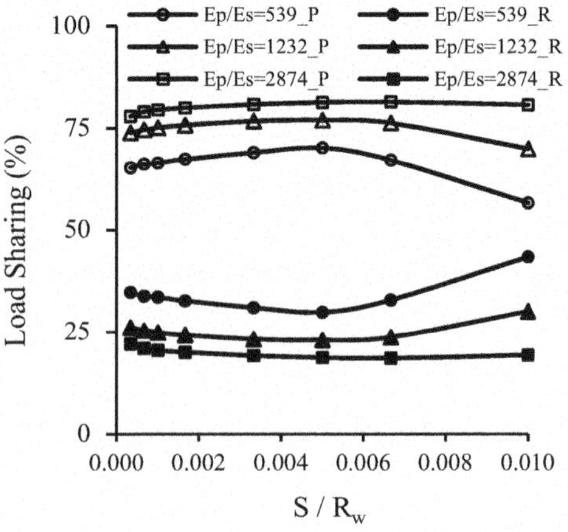


**Fig. 10** Effect of $E_p/E_s$ on load sharing in piled raft

### 4.2.1 Effect of Pile–Soil Stiffness Ratio

The effect of soil stiffness on load sharing of pile and raft has been investigated by varying the soil stiffness from loose to dense cohesionless soil medium. Figure 10 shows the variation of load sharing with foundation settlement in pile and raft with the variation in soil stiffness. The L/D ratio and $A_g/A_r$ of 80 and 0.014 has been considered in the analysis, respectively. The $K_{rs}$ has been varied from 0.36 to 1.90. The load shared by the raft increases with an increase in soil stiffness, whereas the load shared by the piles increases with a decrease in soil stiffness. The pile load sharing increases with foundation settlement up to 0.05 $R_w$ and further decreases gradually with increasing foundation settlement. The load shared by pile and raft at 0.05 $R_w$ foundation settlement is 81.2, 76.9, 70.1, and 18.8, 23.1, 29.9% for $E_p/E_s$ in the range of 2874–539, respectively. The pile load sharing increases to 11 and 6% with increasing $E_p/E_s$ from 539 to 1232 and 1232 to 2874, respectively. The increase in soil stiffness improves the normal pressure between the raft and underlying soil, which is effective in increasing the frictional bearing capacity of the raft [3]. The development of contact pressure with gradually increasing loading consequently increasing foundation settlements and results in increase of raft load sharing in the piled raft foundation. Similar foundation response has been obtained from the experimental investigations in the present study.

### 4.2.2 Effect of Soil–Raft Stiffness ($K_{rs}$)

The effect of soil–raft stiffness has been evaluated by varying the soil stiffness and the raft thickness. Soil–raft stiffness increases with increase in raft thickness and soil stiffness. The L/D ratio and $A_g/A_r$ of 80 and 0.014 has been considered in the analysis, respectively.

**Fig. 11** Effect of $K_{rs}$ on load sharing in piled raft

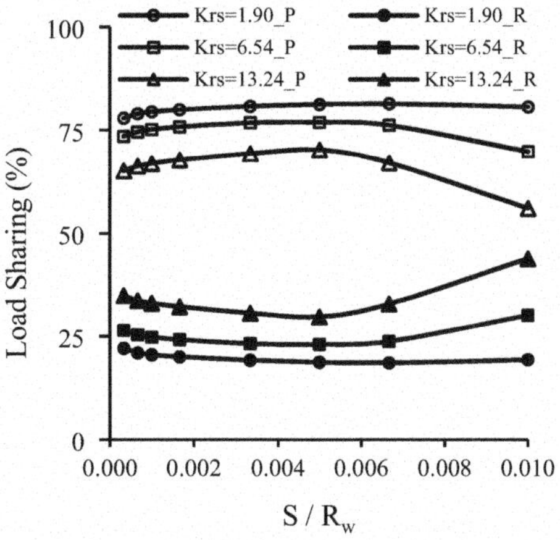


Figure 11 shows the effect of soil–raft stiffness on the load sharing of pile and raft in a piled raft. The raft load share increases and pile load share decreases with the foundation settlement. Similar results have been observed by El-Garhy et al. [20] for piled raft foundations in sand subjected to vertical loading. The raft load share increases by 17 and 24% with varying $K_{rs}$ from 1.90 to 6.54 and 6.54 to 13.24, respectively.

### 4.2.3 Effect of $A_g/A_r$

The $A_g/A_r$ ratio increases with increasing the pile diameter, consequently increasing the group pile area ($A_g$). The relatively stiff soil ($E_s = 46{,}400$ kPa) and flexible raft ($K_{rs} = 0.36$) raft has been considered in the analysis. The slenderness ratio of pile varies in the range of 40–80. Figure 12 shows the influence of $A_g/A_r$ on the load sharing behavior of piled raft. The pile load sharing varies in between 65 and 78% with the variation of $A_g/A_r$ in the range of 0.01–0.05. The pile load sharing increases with an increase in pile diameter. This could be attributed to the improved pile capacity with a larger pile group area. The increase in pile diameter increases the cross-sectional area of the pile tip and surface area of the pile shaft consequently improving the end bearing resistance and skin friction resistance of the pile, respectively.

### 4.2.4 Effect of L/D Ratio

The relatively stiff soil ($E_s = 46{,}400$ kPa) and flexible raft ($K_{rs} = 0.36$) raft have been considered in the analysis. The $A_g/A_r$ has been taken as 0.014. Figure 13 shows the influence of pile length on load sharing of the piled raft. The load sharing of piles increases, and load sharing of raft decreases with increasing the slenderness ratio. The load sharing of pile varies in between 60 and 65% with the variation of L/D in

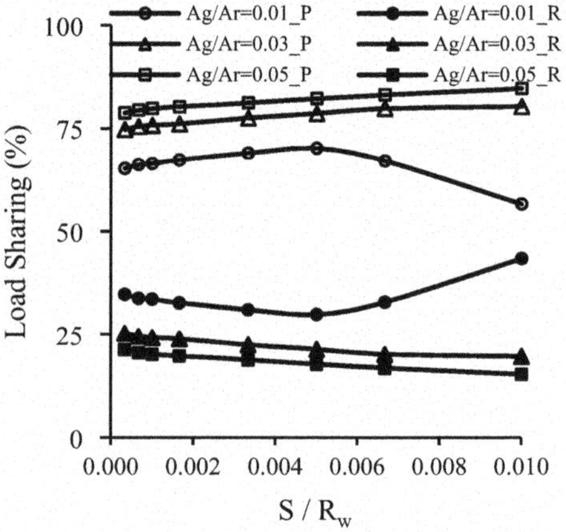


**Fig. 12** Effect of $A_g/A_r$ on load sharing in piled raft

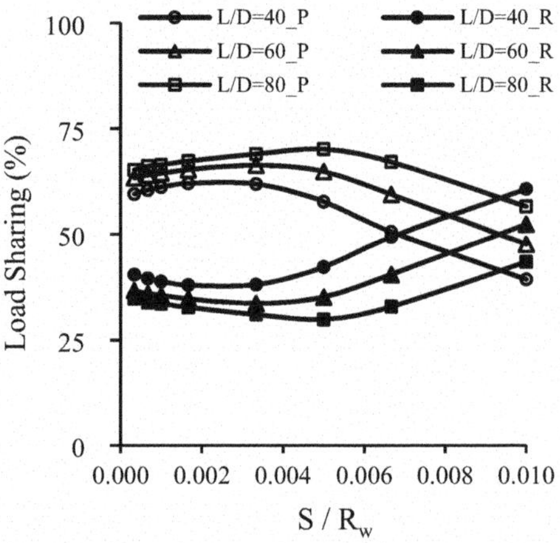


**Fig. 13** Effect of L/D on load sharing in piled raft

the range of 40–80. This increase in load proportion is observed 6 and 3%, when the L/D ratio changes from 40 to 60 and 60 to 80, respectively. The increase in pile length results in higher skin resistance due to larger surface area of the pile shaft attributing larger pile load sharing.

## 5 Conclusion

The experimental and numerical investigation on a vertically loaded small square piled raft in cohesionless soils shows the nonlinear load sharing of pile and raft. The complex pile–soil–raft interaction governs the load sharing between the foundation elements. The experimental and numerical investigations confirm that the load sharing between the pile and raft varies nonlinearly with the foundation settlement.

The increase in foundation settlement results in a decrease in pile–pile interaction, leading to decrease in pile load sharing, whereas pile–raft interaction increases with foundation settlement which results in increase in raft load sharing in a piled raft. The parametric studies show that the load sharing of pile and raft in a piled raft foundation is greatly influenced by the stiffness ratio of the pile and soil, soil–raft stiffness, ratio of pile group area and raft, and the length to diameter ratios of the pile. The pile shares 60–78%, and the raft shares 22–40% of the load applied on the piled raft for the parameters considered in the present study. The understanding of load sharing mechanism obtained from more experimental and numerical studies will be helpful to develop a model for predicting the load sharing in piled raft.

## References


1. Malviya DK, Ansari A, Samanta M (2023) Settlement and load sharing behavior of piled raft foundation: a review. Innov Infrastruc Solut 8(11):1–8. https://doi.org/10.1007/s41062-023-01272-w
2. Poulos HG, Davis EH (1980) Pile Foundation Analysis and Design. Rainbow-Bridge Book Co., New York: Wiley
3. Akbari A, Eslami A, Nikookar M (2021) Influence of soil stiffness on the response of piled raft foundations under earthquake loading. Transportation Infrastructure Geotechnology 8:590–606. https://doi.org/10.1007/s40515-021-00157-8
4. Horikoshi K, Randolph MF (1997) On the definition of raft—soil stiffness ratio for rectangular rafts. Geotechnique 47(5):1055–1061. https://doi.org/10.1680/geot.1997.47.5.1055
5. Horikoshi K, Randolph MF (1996) Centrifuge modelling on piled raft foundations on clay. Geotechnique 46(4):741–752. https://doi.org/10.1680/geot.1996.46.4.741
6. Kitiyodom P, Matsumoto T (2003) A Simplified Analysis Method for Piled Raft Foundation in Non-homogeneous Soils. International Journal of Numerical and Analytical Methods in Geomechanics 27(2):85–109. https://doi.org/10.1002/nag.264
7. Bhartiya P, Chakraborty T, Basu D (2020) Nonlinear subgrade modulus of sandy soils for analysis of piled raft foundations. Comput Geotech 118:103350. https://doi.org/10.1016/j.compgeo.2019.103350
8. Randolph MF (1983) Design of piled raft foundations. International Symposium on Recent Developments in Laboratory and Field Tests and Analysis of Geotechnical Problems, Bangkok, 525–537.
9. Clancy P, Randolph MF (1996) Simple design tools for piled raft foundation. Geotechnique 46(2):313–328. https://doi.org/10.1680/geot.1996.46.2.313
10. Lee J, Park D, Choi K (2014) Analysis of load sharing behavior for piled rafts using normalized load response model. Computer and Geotechnics 57:65–74. https://doi.org/10.1016/j.compgeo.2014.01.003
11. Kumar A, Choudhury D (2018) Development of new prediction model for capacity of combined pile-raft foundations. Computer and Geotechnics 97:62–68. https://doi.org/10.1016/j.compgeo.2017.12.008
12. Chanda D,Nath U, Saha R, Haldar S (2021) Development of Lateral Capacity-Based Envelopes of Piled Raft Foundation under Combined V-M-H Loading. International Journal of Geomechanics, 21 (6). https://doi.org/10.1061/(ASCE)GM.1943-5622.0002023
13. Hain SJ, Lee IK (1978) The analysis of flexible raft pile systems. Geotechnique 28(1):65–83. https://doi.org/10.1680/geot.1978.28.1.65
14. Samanta M, Bhowmik R (2019) 3D numerical analysis of piled raft foundation in stone column improved soft soil. Int J Geotech Eng 13(5):474–483. https://doi.org/10.1080/19386362.2017.1368139

15. Brinkgreve RBJ, Swolfs WM (2012) Plaxis 3D Material Models Manual. PLAXIS BV, Netherland.
16. Kumar A, Choudhury D (2016) Effect of Earthquake on Combined Pile-Raft Foundation. Int J Geomech 16(5):1–6. https://doi.org/10.1061/(ASCE)GM.1943-5622.0000637
17. Manoharan N, Dasgupta SP (1995) Bearing capacity of surface footings by finite elements. Comput Struct 54(4):563–586. https://doi.org/10.1016/0045-7949(94)00381-C
18. Horikoshi K, Randolph MF (1998) A contribution to the optimum design of piled rafts. Geotechnique 48(3):301–317. https://doi.org/10.1680/geot.1998.48.3.301
19. Poulos HG (2001) Piled-raft foundation: design and applications. Geotechnique 51(2):95–113. https://doi.org/10.1680/geot.2001.51.2.95
20. El-Garhy B, Galil AA, Youssef AF, Raia MA (2013) Behavior of raft on settlement reducing piles: Experimental model study. Journal of Rock Mechanics and Geotechnical Engineering 5:389–399

# Settlement-Based Load-Sharing Ratio Variation for Large Piled-Rafts

**Banchiva K. Marak and Baleshwar Singh**

## 1 Introduction

Piled-raft foundation is a hybrid foundation system that is composed of two structural components, that is, the raft and the piles. This foundation system is designed in such a way to utilize the capacity of both the structural components and hence, allow for load sharing between them. This load-sharing mechanism makes this foundation system as the alternative economical option when used in suitable geotechnical conditions, in comparison with the conventional pile foundation in which the capacity of the raft is not taken into consideration. Due to the combined action of both the structural components, the stress as well as displacement fields of the raft and piles overlap, giving rise to various interactions between the raft, piles, and soil, making the overall response of piled-raft foundation very complex [1]. As a result of these interactions, the load-carrying capacity of raft and piles, when combined into piled-raft, becomes different from that of alone unpiled raft (UR) and group piles (GP) [2]. One of the important design components for achieving the optimum piled-rafts is evaluating the load shared between the raft and piles since the contribution of both raft and piles plays an important role in the load-carrying capacity of the piled-raft foundation. This raft and pile load share is generally expressed in terms of load-sharing ratio, defined as the ratio of the proportion of load carried by the piles to the total applied load on piled-raft. The piled-raft foundations are categorized into small piled-rafts and large piled-rafts, according to Viggiani et al. [3]. Small piled-rafts are those in which the width of the raft (*B*r) is small as compared to the length of the piles (*L*) ($Br/L < 1$), with the width of the raft ranging from 5 to 15 m. Large piled-rafts are

B. K. Marak (✉) · B. Singh
Department of Civil Engineering, IIT Guwahati, Guwahati 781039, India
e-mail: k.banchiva@iitg.ac.in

B. Singh
e-mail: baleshwar@iitg.ac.in

 
B. T. Jose et al. (eds.), *Proceedings of the Indian Geotechnical Conference 2022 Volume 2*, Lecture Notes in Civil Engineering 477,
https://doi.org/10.1007/978-981-97-1741-5_16

those in which the width of the raft is comparatively large with respect to the length of piles ($Br/L > 1$), with the width of the raft greater than15 m.

The objective of the present study is to investigate the load-sharing ratio of a large piled-raft foundation in dense sand with respect to average settlement by varying parameters like pile spacing and length and raft thickness. The individual load–settlement responses of the raft and piles in PR are determined initially for all PR configurations, which is then compared with the responses of the unpiled raft and group piles alone.

## 2 Numerical Modeling

This section discusses the numerical modeling procedure and also the validation study of large PR foundations in dense sand. Three-dimensional numerical modeling is carried out with a finite element-based commercial software Plaxis 3D [4]. A suitable mesh size is considered for the analysis on the basis of a mesh convergence study. The numerical model is verified with the centrifuge test results of Park and Lee [1] by modeling the foundation on a prototype scale.

### *2.1 Domain Size and Mesh Convergence*

The size of the soil model is fixed at a lateral distance of 4.5 times the raft width from the edge of the raft on all four sides restraining the horizontal movement and allowing for vertical movement. The vertical depth of the soil domain is taken as seven times the pile length from the base of the raft and the bottom boundary of the soil is fixed; that is, both vertical and horizontal movements are not allowed. The boundaries of the soil model have been selected, keeping in mind that the piled-raft influence zone is well within the soil domain to avoid any undesirable boundary effects. Figure 1 shows the numerical model with the selected boundaries used for the study, along with piled-raft geometry.

Mesh convergence analysis is performed to identify the optimal mesh required to perform the numerical analysis accurately by considering the available five different meshing options in the Plaxis library (such as very coarse, coarse, medium, fine, and very fine). Coarser meshes are unable to take into account of the important soil and structure behavioral characteristics, while very fine mesh takes excessive computational time for the analysis. The element size according to different meshes is identified by non-dimensional element length, which is defined as the ratio of the length of the element to the maximum dimension of the model geometry. The result of the mesh convergence study is shown in Fig. 2. It is found that the load-carrying capacity of piled-raft varies a lot when the mesh is changed from very coarse to fine. However, the results almost converge beyond the fine mesh, and hence, fine meshing is adopted for the present numerical analysis. Meshes are refined locally nearby the structural elements.

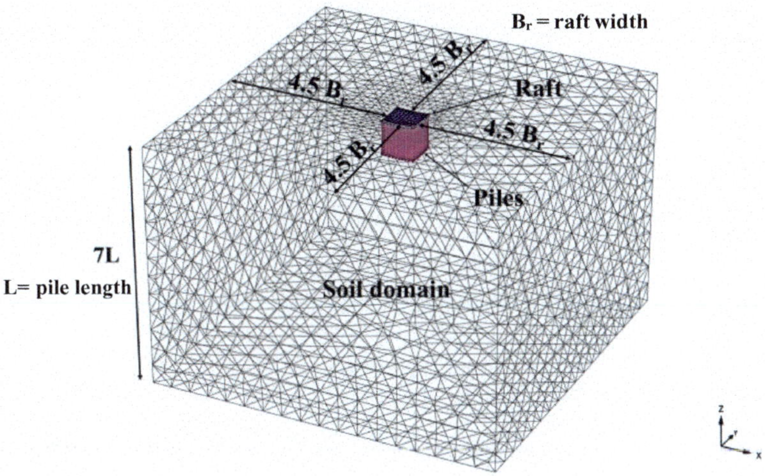


**Fig. 1** Numerical model geometry and meshing

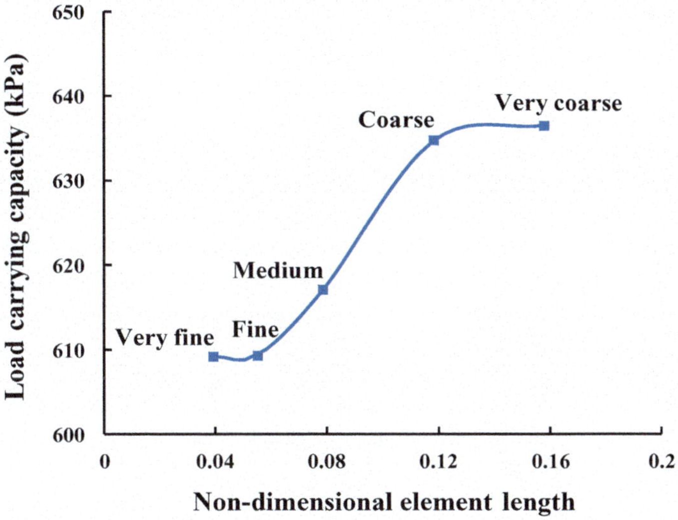


**Fig. 2** Mesh convergence analysis

**Table 1** Input parameters for Plaxis 3D

| Soil properties | Dense sand |
|---|---|
| Relative density (%) | 85 |
| Dry unit weight, $\gamma d$ (kN/m$^3$) | 15.6 |
| Secant Young's modulus, $E_{50}^{ref}$ (MPa) | 37.67 |
| Oedometer stiffness, $E_{oed}^{ref}$ (MPa) | 37.67 |
| Unloading/reloading stiffness, $E_{ur}^{ref}$ (MPa) | 115.2 |
| Friction angle, $\phi$ (°) | 43 |
| Cohesion, $c$ (kPa) | 0 |
| Poisson's ratio, $\upsilon$ | 0.25 |
| Dilatancy angle, $\psi$ (°) | 11 |
| *Raft and pile properties* | |
| Modulus of elasticity, $E$ (MPa) | 30,000 |
| Poisson's ratio, $\upsilon$ | 0.15 |
| Unit weight, $\gamma$ (kN/m$^3$) | 25 |

## 2.2 Material Models

The simulated soil volume consists of 10 node tetrahedral elements, and the behavior of sandy soil considered in the study is simulated using the hardening soil material model. The raft component is modeled considering a plate element which is a six-node triangular element, and the modeling of the pile component is considered with the help of an embedded beam element with a special interface element. Both raft and pile behavior is considered as linear-elastic. The interaction between the soil and the structural components is considered with the help of 12-node interface elements. The reduced shear strength at the soil-structure interface is taken into account by the Interface Reduction Factor ($R_{int}$).

For the present study, the soil properties of the homogeneous dense sand are considered from Nguyen et al. [5] and are given in Table 1. Both the raft and piles are considered to be concrete material having a modulus of elasticity equal to 30,000 MPa with Poisson's ratio of 0.15.

## 2.3 Model Validation

The numerical model used for the present study is validated with that of centrifuge test results performed by Park and Lee [1] by modeling the foundation on a prototype scale with a centrifuge acceleration of 60 g. The piled-raft foundation of a 9 × 9 m raft having a thickness 1.2 m and 4 × 4 pile configuration with spacing of pile 2.4 m and a pile of 0.6 m diameter and 15 m length are considered. Silica sand with 84% relative density having an elastic modulus of 50 MPa and friction angle of 41° has

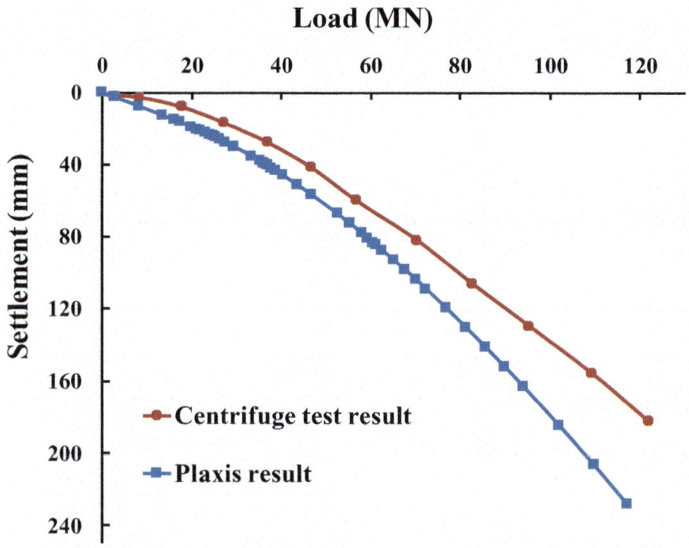


**Fig. 3** Validation of the numerical model with the result of centrifuge test

been used for the analyses. The results of the load–settlement curve for piled-raft foundation obtained from both numerical study and centrifuge test are compared and shown in Fig. 3. The centrifuge test result appears to predict a bit higher stiffness value; however, the overall response shows a reasonably close match with a similar trend of the load–settlement curve. This validated numerical model is then further used to carry out the numerical analysis for the present study.

## 3 Results and Discussions

The evaluation of the load-sharing ratio of a large PR foundation in dense sand with respect to average settlement by varying parameters like pile spacing and length and raft thickness has been carried out. Initially, the individual load–settlement responses of the raft and piles in PR are determined for all the considered PR configurations which are compared with the individual responses of the unpiled raft and group piles, respectively. A large piled-raft foundation of 5 × 5 pile configuration with a raft of size (25 × 25) m, with raft thickness varied as 0.7, 2.0, and 4.7 m corresponding to relative raft–soil stiffness ratio ($K$rs) of 0.4, 10, and 117, respectively [5], and piles of diameter 1 m are considered. The pile spacing and length are normalized with respect to pile diameter. The pile spacing is varied as $s/d$ ratio of 2.5, 3, 4, and 5, while the pile length is varied as $L/d$ ratio of 10, 15, 20, and 25. These various parameters of the structural elements considered for the study are also given in Table 2. The uniformly distributed load of 600 kPa is applied on the foundation. The average settlement

**Table 2** Raft and pile parameters considered for the present study

| Raft parameters | Values |
|---|---|
| Raft width, *B*r (m) | 25 × 25 |
| Raft thickness, *t*r (m) | 0.7*, 2.0, 4.7 |
| Relative raft–soil stiffness ratio, *K*rs | 0.4*, 10, 117 |
| *Pile parameters* | |
| Pile number | 5 × 5 |
| Pile diameter, *d* (m) | 1 |
| Pile spacing to diameter ratio, *s*/*d* | 2.5, 3, 4, 5 |
| Pile length to diameter ratio, *L*/*d* | 10, 15, 20, 25* |

*Standard value if not varied

(*W*avg) is calculated by considering the center settlement (*W*center) and the corner settlement (*W*corner) of the raft and is calculated by Eq. 1 [6]. The proportion of raft and pile load share is measured by the load-sharing ratio ($\alpha$PR), which is defined as the ratio of load carried by the piles in PR to the total applied load on the foundation.

$$W_{\text{avg}} = 1(2W_{\text{center}} + W_{\text{corner}}). \tag{1}$$

## 3.1 Individual Load–Settlement Responses of Raft and Piles in PR

The load–settlement response of PR represents the combined load–settlement responses of both the raft as well as the piles, and therefore, it can be decomposed into the individual components as the raft in PR and piles in PR. The load taken by piles in PR (*P*PR) is obtained by adding the axial loads from the individual pile heads, and the load taken by raft in PR (*R*PR) is determined by subtracting the load taken by *P*PR from the total applied load on the foundation. The individual load versus $W_{\text{avg}}$ responses of the *R*PR and *P*PR are plotted for different foundation configurations, and comparisons are made with those of unpiled raft (UR) and group piles (GP) as shown in Figs. 4, 5, 6, 7, and 8.

Figure 4 shows the individual load versus $W_{\text{avg}}$ responses of the raft in PR and piles in PR compared with UR and GP for different pile spacings, respectively. From Fig. 4a, it can be seen that as compared to UR, the load versus $W_{\text{avg}}$ response of *R*PR is higher for smaller pile spacing (*s*/*d* ratios of 2.5 and 3) (indicating higher load-carrying capacity), whereas for *s*/*d* ratio of 5, the load versus $W_{\text{avg}}$ response of *R*PR is lesser than that of UR. It is observed that as the pile spacing increases, the load-carrying capacity of *R*PR decreases, which means that the proportion of load carried by the piles becomes higher. From Fig. 4b, the load versus $W_{\text{avg}}$ response of *P*PR is higher when compared to group piles for all pile spacing except for *s*/*d* ratio

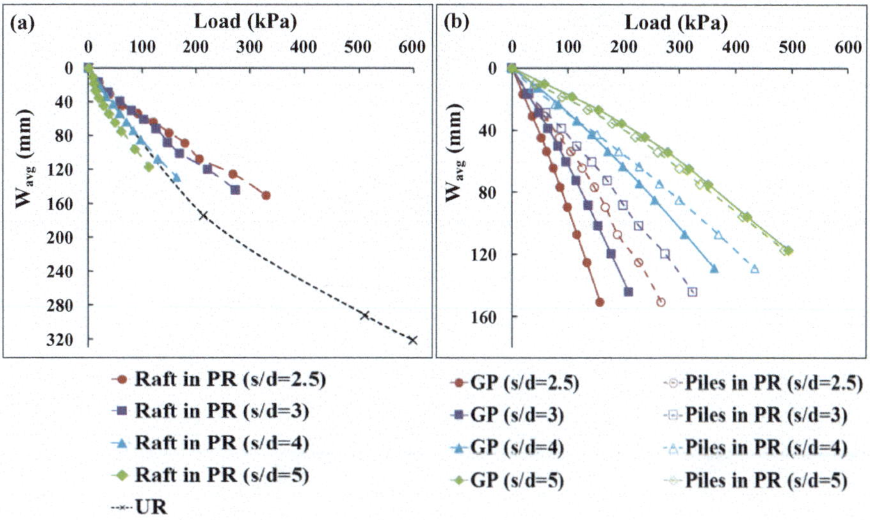


**Fig. 4** Individual load versus $W_{avg}$ responses of **a** raft in PR compared with unpiled raft and **b** piles in PR compared with group piles for different pile spacing

of 5, where the load versus $W_{avg}$ response of *P*PR is slightly lesser than that of GP. As the pile spacing increases, the proportion of load carried by the *P*PR increases.

Figure 5 shows the individual load versus $W_{avg}$ responses of *R*PR compared with UR for different pile lengths at various pile spacings. For all pile spacing, it is observed that the load versus $W_{avg}$ response of *R*PR is higher than UR for all the considered pile lengths (except for the *L*/*d* ratio of 25 at the *s*/*d* ratio of 5). As the pile length increases, there is no significant change in the load versus $W_{avg}$ response of *R*PR in the case of *s*/*d* ratio of 2.5 and 3; however, the load-carrying capacity decreases for pile spacing of *s*/*d* ratio of 4 and 5. The reason for the decrease in the load-carrying capacity of *R*PR when the pile length is increased at larger pile spacing is that there is reduced overlapping of the stress and displacement fields between the piles leading to increased mobilization of pile capacity, which increases as the pile length increases.

Figure 6 shows the individual load versus $W_{avg}$ responses of *P*PR compared with group piles for different pile lengths at various pile spacings. From Fig. 6a and b, at *s*/*d* ratio of 2.5 and 3, it is observed that the load versus $W_{avg}$ response of *P*PR is higher than group piles for all the considered pile lengths. From Fig. 6c, at *s*/*d* ratio of 4, it is noticed that the load versus $W_{avg}$ response of *P*PR is lower than group piles for pile length of *L*/*d* ratio 10 and 15 but for *L*/*d* ratio of 20 and 25, the load versus $W_{avg}$ response of *P*PR is still higher than group piles. From Fig. 6d, at pile spacing of *s*/*d* ratio of 5, the load versus $W_{avg}$ response of *P*PR is significantly lower than group piles for *L*/*d* ratio of 10 and 15 while it is slightly lower for the pile lengths of *L*/*d* ratio 20 and 25. For all *s*/*d* ratios, as the pile length increases (from *L*/*d* ratio of 10

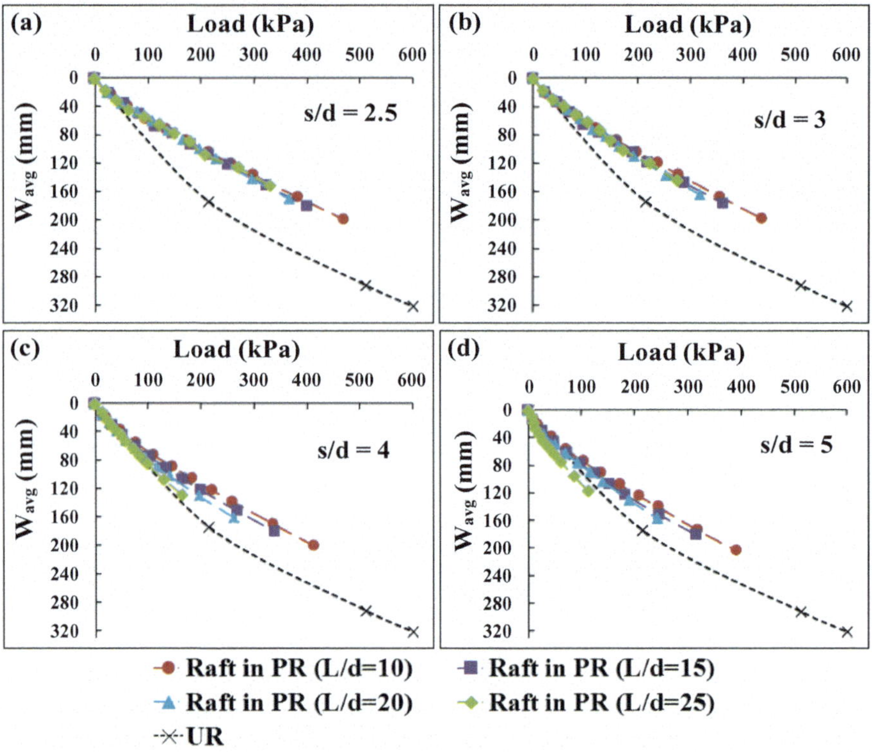


**Fig. 5** Individual load versus $W_{avg}$ responses of raft in PR compared with unpiled raft for different pile lengths

to 25), the load carried by the piles increases, which is found to be more significant for *s*/*d* ratio of 4 and 5.

Figure 7 shows the individual load versus $W_{avg}$ responses of *R*PR compared with UR for different raft thicknesses at various pile spacings. From Fig. 7a and b, at *s*/*d* ratio of 2.5 and 3, it is observed that the load versus $W_{avg}$ response of *R*PR is higher than URs for all raft thicknesses. As the raft thickness increases, the load carried by *R*PR decreases. For *s*/*d* ratio of 4 and 5 (Figs. 7c and d), the load versus $W_{avg}$ response of *R*PR becomes almost equal to that of UR for all raft thicknesses except for raft thickness of 0.7 m (at s/d ratio of 5) where the load versus $W_{avg}$ response of *R*PR becomes less than that of UR. As the raft thickness increases, there is no significant change in the load carried by *R*PR for *s*/*d* ratio of 4; however, for *s*/*d* ratio of 5, the load carried by *R*PR starts increasing as the raft thickness increases. Figure 8 shows the individual load versus $W_{avg}$ responses of *P*PR compared with GP for different raft thicknesses at various pile spacings. From Fig. 8a and b, at *s*/*d* ratio of 2.5 and 3, the load versus $W_{avg}$ response of *P*PR is lower than GP for all raft thicknesses (except for $tr = 0.7$ m). With increasing thickness of raft from 0.7 m to 2 m, the load carried by *P*PR increases but then starts decreasing when the raft

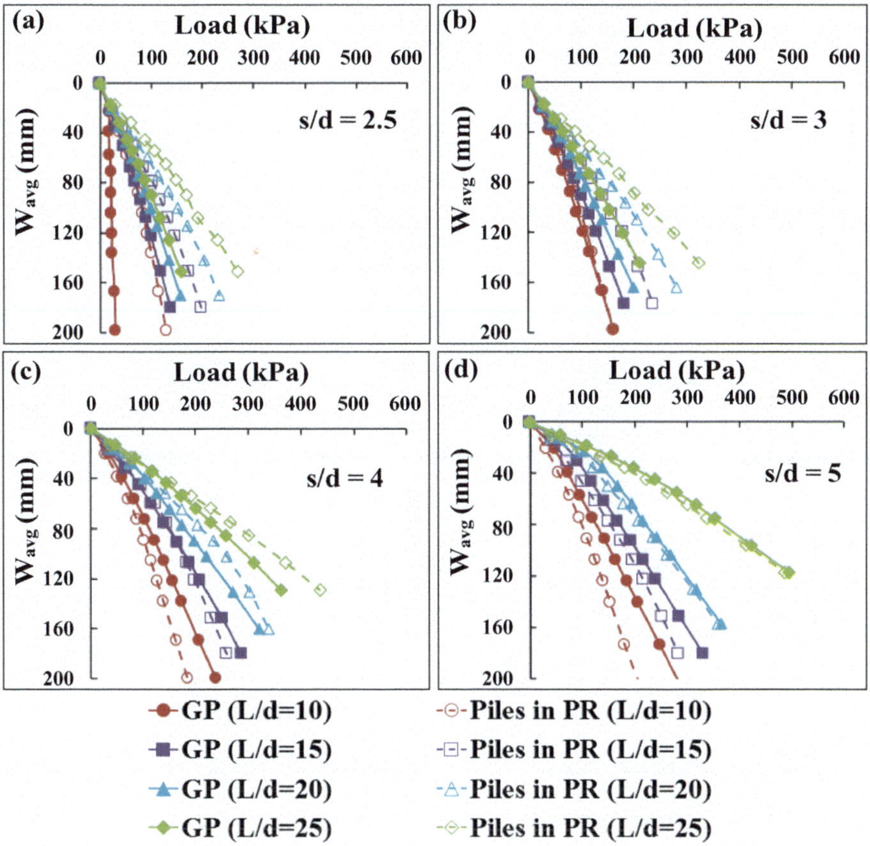


**Fig. 6** Individual load versus $W_{avg}$ responses of piles in PR compared with group piles for different pile lengths

thickness is increased further up to 4.7 m. For *s*/*d* ratio 5 (Fig. 8d), the load versus $W_{avg}$ response of *P*PR also becomes almost equal to that of GP for raft thickness of 0.7 m. As the raft thickness increases, the load carried by *P*PR starts to decrease.

## 3.2 *Variation of Load-Sharing Ratio with Settlement*

In this section, the evaluation of load-sharing ratio of a large PR foundation in dense sand with respect to average settlement by varying pile spacing and length, and raft thickness has been discussed. Figure 9 shows the variation of $\alpha$PR with $W_{avg}$ for different pile spacings. With increasing pile spacing, the $\alpha$PR value increases at all settlement levels. For pile spacing of *s*/*d* ratio 2.5 and 3, the $\alpha$PR value decreases with $W_{avg}$ because the piles are not able to properly mobilize their skin friction

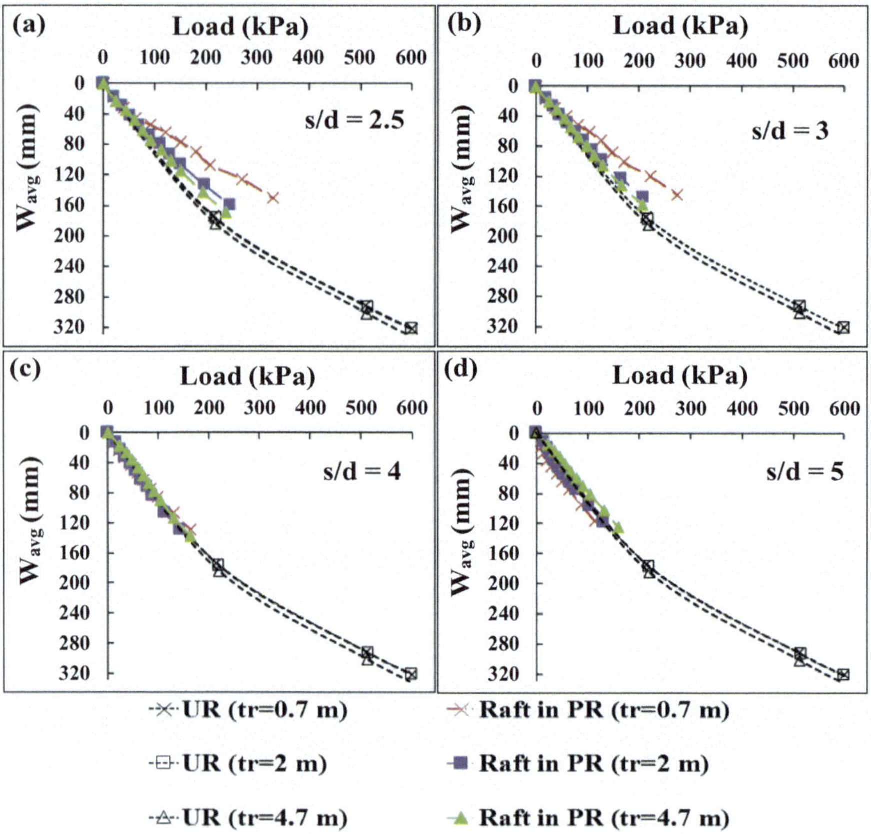


**Fig. 7** Individual load versus $W_{avg}$ responses of raft in PR compared with unpiled raft for different raft thicknesses

at lower pile spacing. For *s*/*d* ratio of 4 and 5, the $\alpha$PR value increases initially at initial settlement and then starts decreasing toward higher settlement because at higher settlement, the pile skin friction got fully mobilized and raft capacity starts to mobilize; hence, the $\alpha$PR value starts decreasing toward higher settlement.

Figure 10 shows the variation of $\alpha$PR with $W_{avg}$ for different pile lengths at various pile spacings. It is observed that for *s*/*d* ratio of 2.5 and 3, the $\alpha$PR value kept decreasing with increasing settlement for all pile lengths; however, for *s*/*d* ratio of 4 and 5, the $\alpha$PR value increases at initial settlement range then decreases toward higher settlement for longer pile lengths of *L*/*d* ratios 20 and 25. As the pile length increases, the $\alpha$PR value increases at all settlement levels which is more significant in case of *s*/*d* ratio of 4 and 5 as compared to *s*/*d* ratio of 2.5 and 3. Figure 11 shows the variation of $\alpha$PR with $W_{avg}$ for different raft thicknesses at various pile spacings. As the raft thickness increases, the $\alpha$PR value increases (for *s*/*d* ratio of 4 and 5) which is significant at initial settlement. The load-sharing ratio increases as

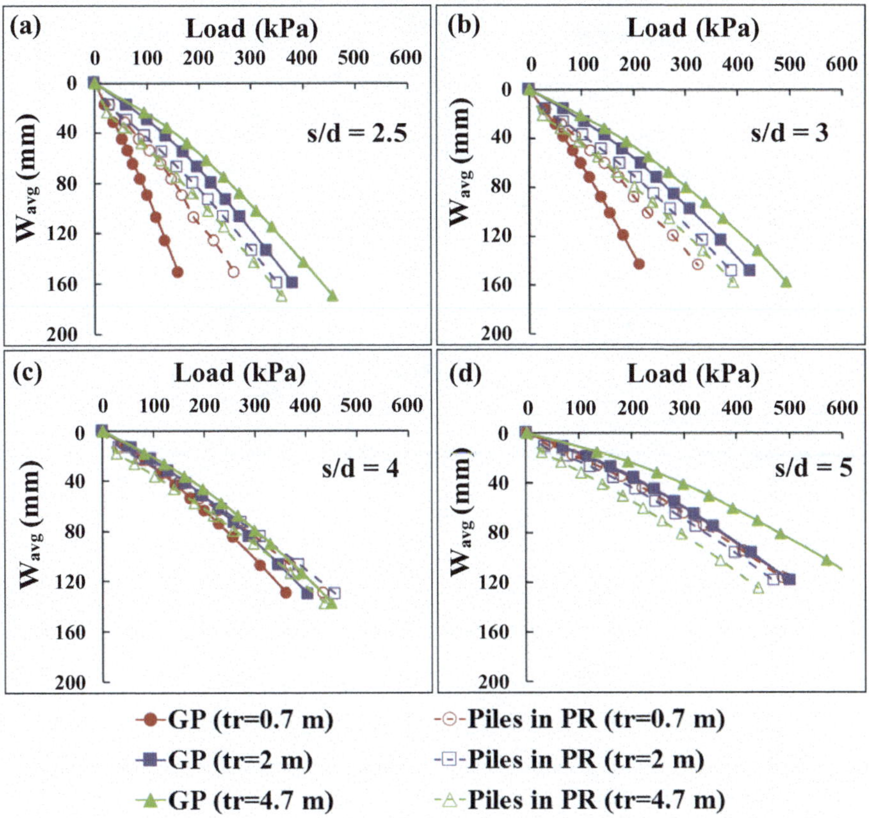


**Fig. 8** Individual load versus $W_{avg}$ responses of piles in PR compared with group piles for different fat thickness

**Fig. 9** Variation of load-sharing ratio with $W_{avg}$ for different pile spacings

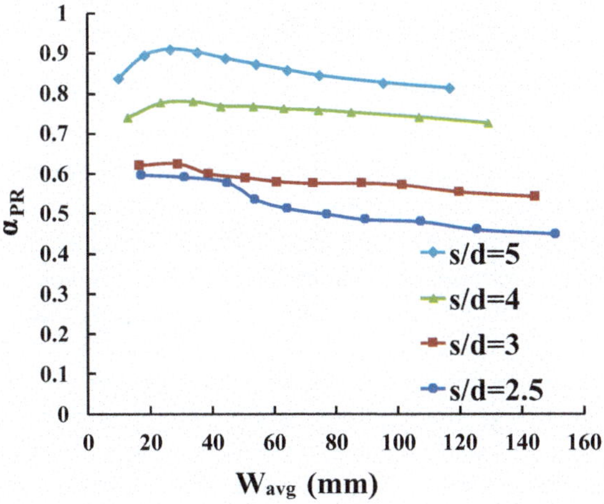

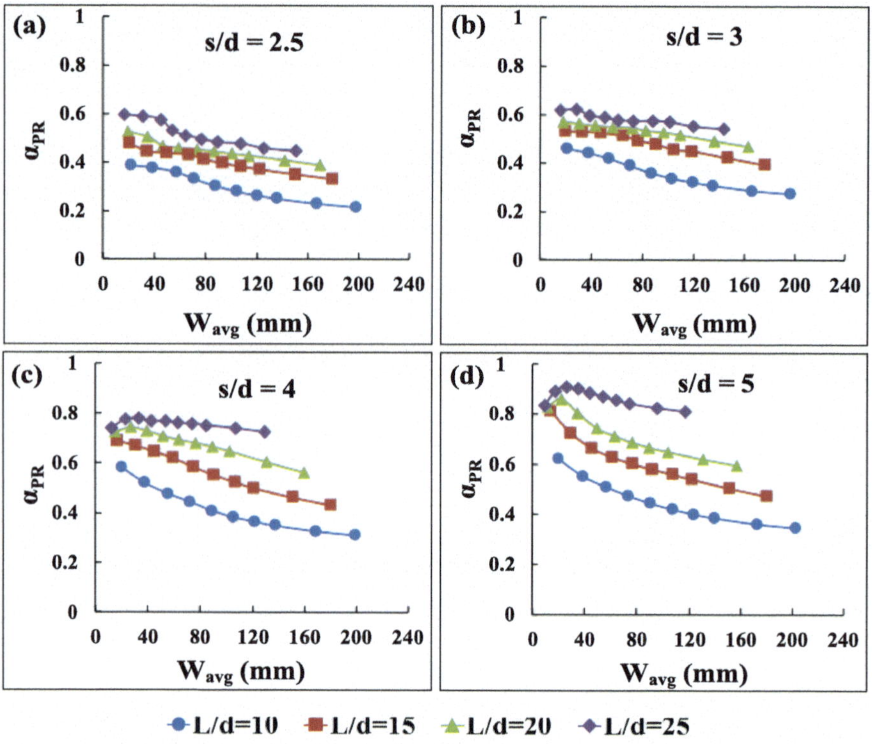


**Fig. 10** Variation of load-sharing ratio with $W_{avg}$ for different pile lengths at various pile spacings

the $W_{avg}$ increases and becomes almost constant toward higher settlement for all raft thicknesses (Fig. 11c and d).

## 4 Conclusions

The settlement-based load-sharing ratio variation for large piled-rafts on dense sand has been investigated through numerical modeling by varying parameters such as pile spacing, pile length, and raft thickness. From the comparison of the individual load vs $W_{avg}$ response of raft and piles in PR with UR and GP respectively, the raft and piles in PR show higher load-carrying capacity than that of UR and GP, respectively, for the PR foundation configuration with raft thickness of 0.7 m, having a smaller pile spacing and longer pile lengths. For the *s*/*d* ratio of 2.5 and 3, the $\alpha$PR value decreases with *W*avg, whereas for the *s*/*d* ratio of 4 and 5, the $\alpha$PR value increases initially at initial settlement and then starts decreasing toward higher settlement. It was observed that the $\alpha$PR value kept decreasing with increasing $W_{avg}$ for all pile lengths (*s*/*d* ratio of 2.5 and 3); however, the $\alpha$PR value increased at the initial settlement range and

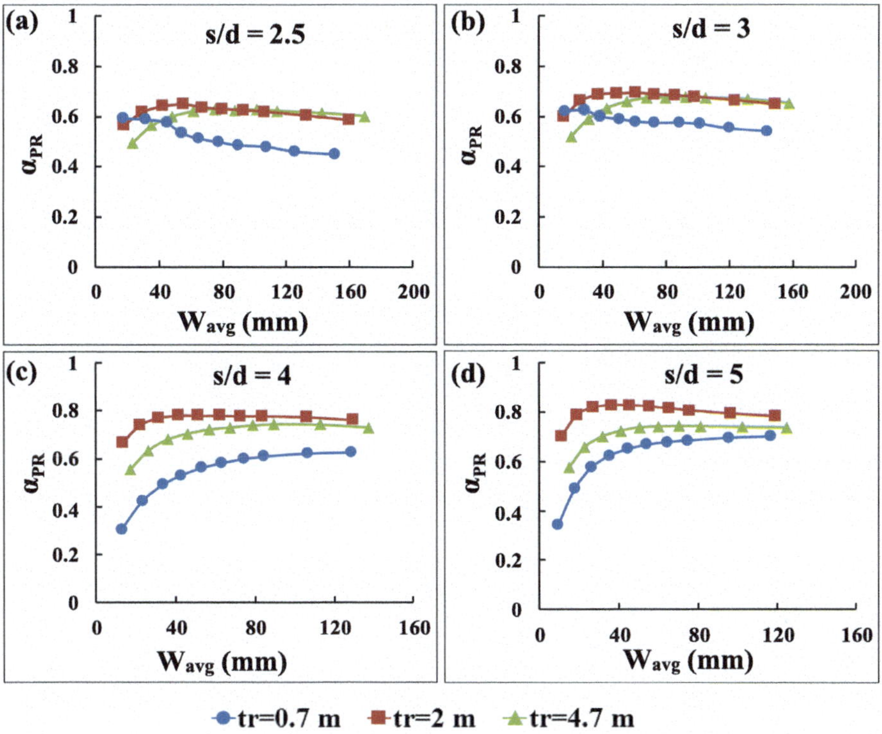


**Fig. 11** Variation of load-sharing ratio with $W_{\text{avg}}$ for different raft thicknesses at various pile spacings

then decreased toward a higher settlement for longer pile lengths (*L*/*d* ratio of 20 and 25) for *s*/*d* ratio of 4 and As the raft thickness increases, the $\alpha$PR value increases (for *s*/*d* ratio of 4 and 5), which is significant at initial settlement.

## References

1. Park D, Lee J (2015) Comparative analysis of various interaction effects for piled rafts in sands using centrifuge tests. J Geotech Geoenviron Eng 141(1):040140821–040140910
2. Lee J, Park D, Park D, Park K (2015) Estimation of load-sharing ratios for piled rafts in sands that includes interaction effects. Comput Geotech 63:306–314
3. Viggiani C, Mandolini A, Russo G (2012) Piles and pile foundation, 1st edn. Spon Press, London & New York
4. Brinkgreve R, Swolfs W, Engin E (2015) PLAXIS user's manual, version 6.1. Balkema, Rotterdam, The Netherlands
5. Nguyen DDC, Jo SB, Kim DS (2013) Design method of piled-raft foundations under vertical load considering interaction effects. Comput Geotech 47:16–27
6. Reul O, Randolph MF (2004) Design strategies for piled rafts subjected to non uniform vertical loading. J Geotech Geoenviron Eng 130(1):1–13

# Numerical Analysis of Reinforced Ring Foundation Subjected to Eccentric-Inclined Loading

**Kishan Kumar, Subhadeep Metya, and Gautam Bhattacharya**

## 1 Introduction

Ring foundation is a type of annular foundation most suitable for a variety of special structures such as cooling towers, silos and oil storages, chimneys, transmission towers, elevated tanks etc. Unlike strip, rectangular, and square footings, circular and ring footings have no weak corners, and hence, no stress concentration occurs in these foundations. In terms of economy, a ring foundation is preferable to a circular foundation (wherever possible) because of less material requirement for its construction as compared with circular foundation. Also, the kern distance is larger for ring foundation as compared to circular foundation. Further, the bearing capacity and settlement analyses for ring foundation, especially the reinforced ring foundation, are not as well established as it is for strip, square, rectangular, and circular foundations.

Based on their research on reinforced ring foundation, Guido et al. [6] and Laman and Yildiz [12] noted that as the tensile strength of the geogrid reinforcement increases, the bearing capacity ratio (BCR) of reinforced ring foundation (defined as the ratio of the bearing capacities of the reinforced and unreinforced foundations) also increases. As the number of reinforcing layers (N) is increased, the BCR continues to increase up to a certain value of N, after which the increase becomes negligible. Laman and Yildiz [11, 12] observed that bearing capacity (BC) increases with increase in stiffness of the reinforcement. They also noted that effective depth obtained in their study was 1.0B for ring foundation. Basudhar et al. [2] noted that rectangular-shaped geogrid is favorable as compared to circular one with respect to material saving. They also noted that B.C. increases substantially for each extra geogrid layer, but the settlement at failure did not change appreciably. Badakhshan and Noorzad [1] observed that as the eccentricity of circular and square

K. Kumar · S. Metya (✉) · G. Bhattacharya
National Institute of Technology Jamshedpur, Jharkhand 831014, India
e-mail: smetya.ce@nitjsr.ac.in



B. T. Jose et al. (eds.), *Proceedings of the Indian Geotechnical Conference 2022 Volume 2*, Lecture Notes in Civil Engineering 477,
https://doi.org/10.1007/978-981-97-1741-5_17

footing with eccentric loading increases, the ultimate load and settlement at ultimate load decrease. Badakhshan and Noorzad [1] also noted that with increase in load eccentricity, reinforcement layers have more influence on circular footing than square footing. Hosamo et al. [7] noted that bearing capacity ratio of ring foundation decreases as the spacing between geogrid layers ($h$) is increased. Further they noted that $n$ ($= r/R$) value between 0.5 and 0.8 gives optimum BC increase in ring footing. In summary, it has been observed that few studies have been reported in the past with concentric or eccentric loading (Boushehrian and Hataf [3], Chakraborty and Kumar [4], Choobbasti et al. [5], Ismael [8], Keshavarz and Kumar [9], Laman and Yildiz [11], Sharma and Kumar [13]). To the best of authors' knowledge, no studies have been reported with inclined-eccentric loading. However, in real life, the applied loading on the foundations is seldom vertical and concentric. Therefore, further studies on ring foundation on reinforced soil are necessary to understand and quantify the improvement in bearing capacity and decrement in settlement.

Based on the literature review presented above, the point-wise objectives of the present study are: (1) to develop a numerical model in PLAXIS 3D using the studies reported in Laman and Yildiz [11, 12] and validate the same; (2) to observe the behavior of a geogrid reinforced ring foundation subjected to inclined and eccentric loading by analyzing the soil displacement pattern and effective stress variation in PLAXIS 3D; and (3) to study the variation of bearing capacity ratio (BCR) of a ring foundation under the influence of eccentrically inclined load with variations in seven parameters, namely, radius ratio of the ring footing (r/R or n), number of geogrid layers below the footing (N), the first layer depth of reinforcement from the footing base (u), the spacing between reinforcing layers (h), the width of reinforcement layer (b), the load eccentricity (e), and the load inclination (i).

## 2 Analysis Methods and Illustrative Example

The example problem considered in the present study concerns a ring foundation with outer diameter (B) of 85 mm and thickness (t) of 20 mm at the surface of a layer of sand. The boundaries of the soil medium are kept equal to the dimensions of the tank designed for experimentation by Laman and Yildiz [11] (i.e., 700 mm × 700 mm × 700 mm). The general layout of geogrid reinforced foundation system considered in the present study is shown in Fig. 1 and has been developed in PLAXIS 3D.

As reported by Laman and Yildiz [11, 12], uniform and fine sand was taken from the Seyhan river bed in Turkey and that was tested in the laboratory. The dry unit weight of the sand was found out to be 17.1 kN/m$^3$. The angle of shearing resistance using direct shear tests was found out to be 41°. For the purpose of modeling the sand, an elasto-plastic hyperbolic model called the hardening soil model (HSM) was selected for the nonlinear sand behavior since it was found out that results obtained by other material models such as linear elastic and Mohr–Coulomb were not in good agreement with the experimental findings. The HSM parameters used to model the

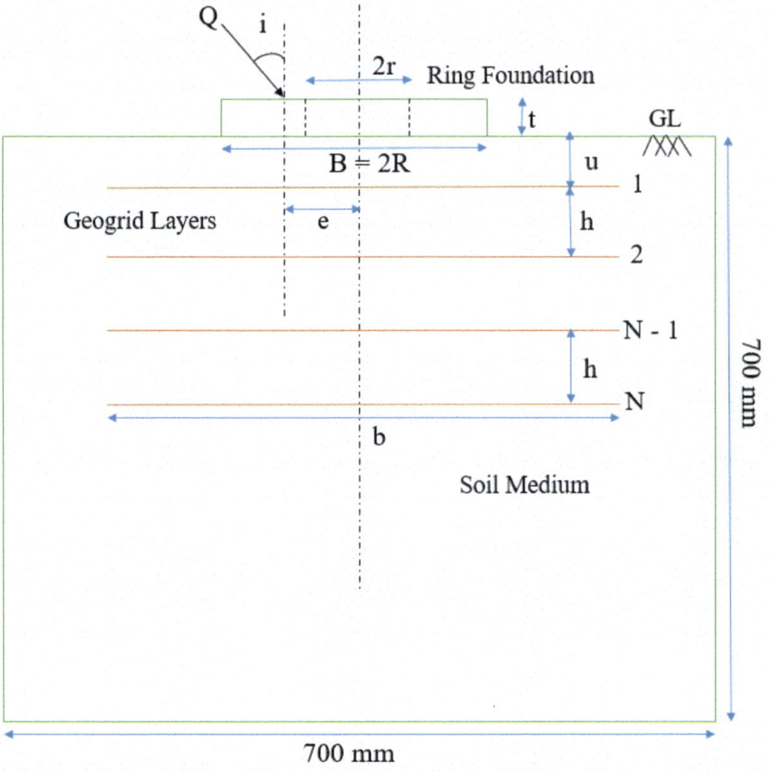


**Fig. 1** Schematic diagram of ring foundation with eccentric-inclined loading

soil are shown in Table 1. The footing and geogrid parameters can be referred to Tables 2 and 3, respectively.

Table 4 summarizes the parameters that are varied in the various parametric studies conducted in this paper while keeping all the other parameters at their mean values.

The validation has been done by comparing the results obtained in PLAXIS 3D and those observed by Laman and Yildiz [12] both for unreinforced and reinforced

**Table 1** Soil parameters used in PLAXIS 3D modeling (Laman and Yildiz [12])

| Sr. No. | Parameter | Value |
|---|---|---|
| 1 | Unit weight of soil $\gamma_s$ (kN/m$^3$) | 17.1 |
| 2 | Secant stiffness $E_{50}$ (kN/m$^2$) | 28,000 |
| 3 | Friction angle $\phi$ (degrees) | 41 |
| 4 | Cohesion c (kN/m$^2$) | 0.3 |
| 5 | Dilatancy angle $\psi$ (degrees) | 11 |
| 6 | Earth pressure at rest ($K_o$) | 0.34 |
| 7 | Poisson's ratio | 0.2 |

**Table 2** Footing parameters used in PLAXIS 3D modeling (Laman and Yildiz [12])

| Sr. No. | Parameters | Value |
|---|---|---|
| 1 | Thickness, $t$ (mm) | 20 |
| 2 | Modulus of elasticity, $E$ (kN/m$^2$) | $207 \times 10^6$ |
| 3 | Unit weight of steel, $\gamma_p$ (kN/m$^3$) | 7850 |
| 4 | Poisson's ratio of steel, $\nu$ | 0.25 |

**Table 3** Geogrid parameters used (Laman and Yildiz [12])

| Sr. No. | Parameters | Values |
|---|---|---|
| 1 | EA (kN/m) | 465 |
| 2 | GA (kN/m) | 232.5 |

**Table 4** Parameters used in different parametric studies

| Studies conducted | Parameter | Mean value | Range considered |
|---|---|---|---|
| Study 1 | r/R | 0.375 | 0–0.75 |
| Study 2 | u/B | 0.6 | 0.2–1 |
| Study 3 | h/B | 0.6 | 0.2–1 |
| Study 4 | N | 3 | 1–5 |
| Study 5 | b/B | 3 | 1–5 |
| Study 6 | e/B | 0.25 | 0–0.5 |

cases. However, because of the limitation of the page number, those figures with further details are not given in this paper and the readers can refer to reference [10].

## 3 Results and Discussion

As already mentioned, in this paper, several parametric studies have been conducted to observe the nature of variations in the bearing capacity of the geogrid reinforced ring foundation subjected to inclined and eccentric loading due to the variation in seven parameters, namely, ratio of internal to external radii of ring foundation (*r*/*R*), ratio of the depth of first layer geogrid to external diameter of footing (*u*/*B*), ratio of inter-layer distance in successive geogrid layers to the external diameter of footing (*h*/*B*), number of geogrid layers (*N*), ratio of width of geogrid layer to the external diameter of footing (*b*/*B*), inclination of loading (*i*), and ratio of eccentricity of loading to the external diameter of footing (*e*/*B*).

### 3.1 *Study 1: Effect of Size of Ring Footing*

In Study 1, to observe the effect of the size of ring foundation on its bearing capacity, the ratio of internal to external radii of ring foundation ($r/R$) has been varied from 0.0 to 0.75 keeping the other parameters constant at their mean values as in Table 4. The variations of bearing capacity ratios (BCR) with the changes in r/R values are shown in Fig. 2.

From Fig. 2, the following points have been observed:

(1) BCR increases slightly from $r/R = 0$ to 0.1875 and then decreases thereafter almost linearly. The decrease in bearing capacity value with increase in size ($n$ value) from 0 to 0.75 has been observed to be from 131 to 63 $kN/m^2$ in unreinforced case and 180–60 $kN/m^2$ in reinforced case with a given eccentric-inclined loading.

(2) Further, it becomes evident that with the eccentric-inclined loading, the design of ring foundation may be permitted with $r/R$ of 0.1875 without compromising the bearing capacity of circular foundation with equal external diameter.

(3) For n value up to 0.5625, the values of the bearing capacity (BC) of reinforced soil are greater than that of unreinforced soil. For $n > 0.5625$, no beneficial effect of geogrid has been observed in increasing the BC of the soil.

(4) From Figs. 3 and 4, it is evident that the immediate settlement value has decreased in unreinforced case from $n = 0$ to $n = 0.1875$. Figures 5 and 6 show that displacement pattern is more uniform in reinforced ring footing with $n = 0.1875$ as compared with pure circular footing.

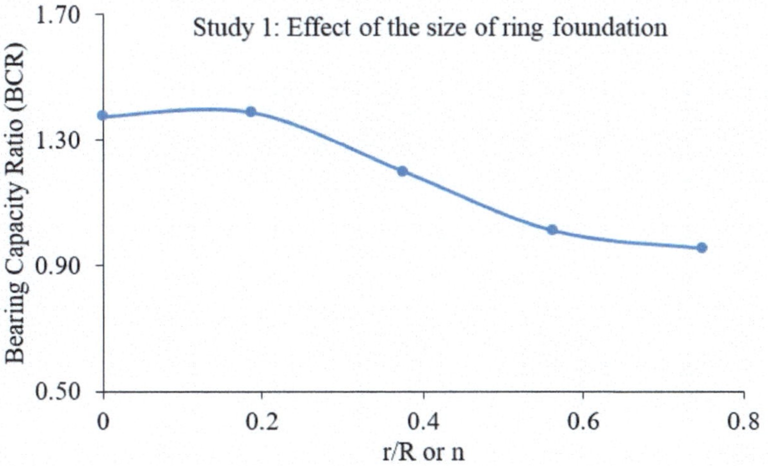


**Fig. 2** Variations of bearing capacity ratios (BCR) with the variation in ring size (r/R or n)

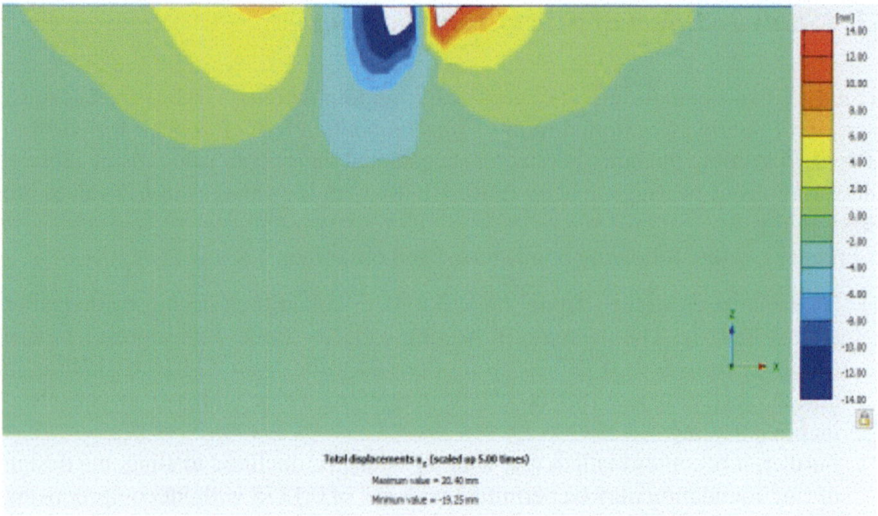


**Fig. 3** Vertical displacement contour (VDC) in unreinforced soil for $n = 0$

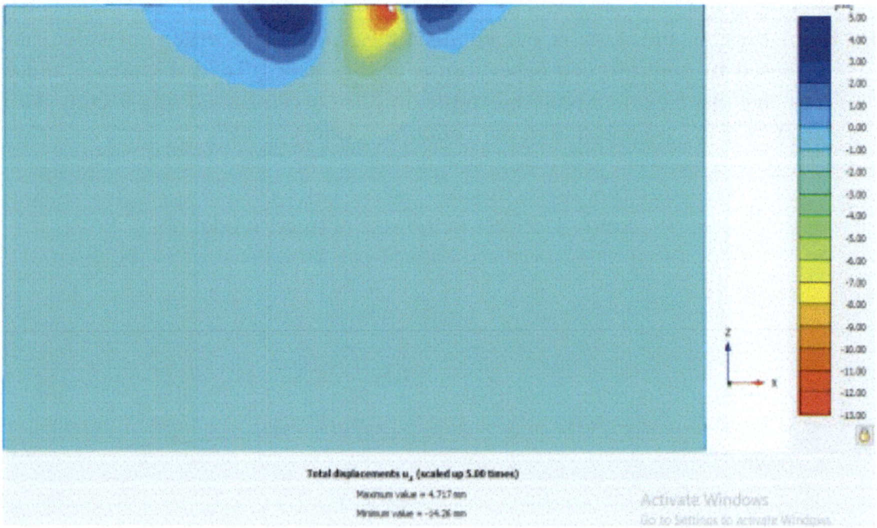


**Fig. 4** VDC in unreinforced soil for $n = 0.1875$

(5) On reinforcing the ring foundation system with $n = 0.5625$, the immediate settlement is decreasing. However, there is not much appreciable change in $z$-direction displacement contour pattern, and hence, it has been observed that BCR value at $n = 0.5625$ is equal to 1 (Figs. 7 and 8).

**Fig. 5** VDC in reinforced soil for $n = 0$

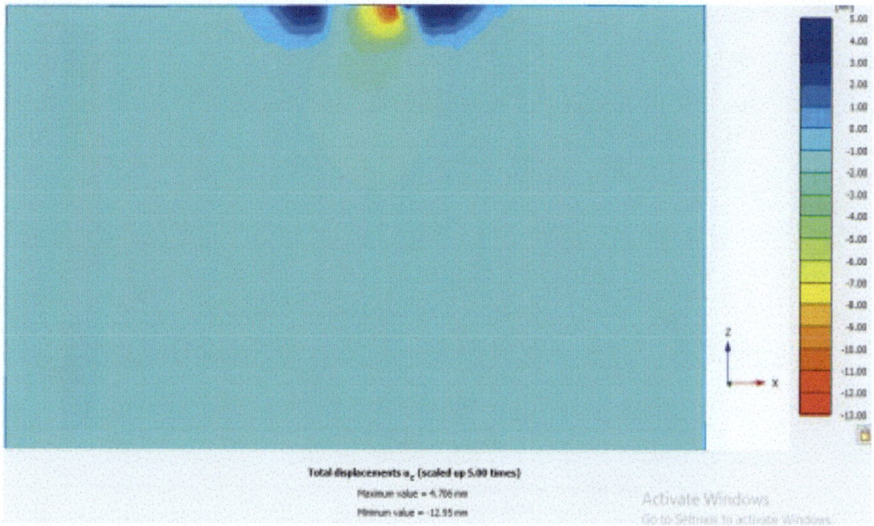


**Fig. 6** VDC in reinforced soil for $n = 0.1875$

## 3.2 *Study 2: Effect of Depth of First Layer of Geogrid*

In Study 2, to observe the effect of depth of first layer of geogrid (from ground level) on its bearing capacity, the ratio of the depth of first layer of geogrid to width of

**Fig. 7** VDC in unreinforced soil for $n = 0.5625$

**Fig. 8** VDC in reinforced soil for $n = 0.5625$

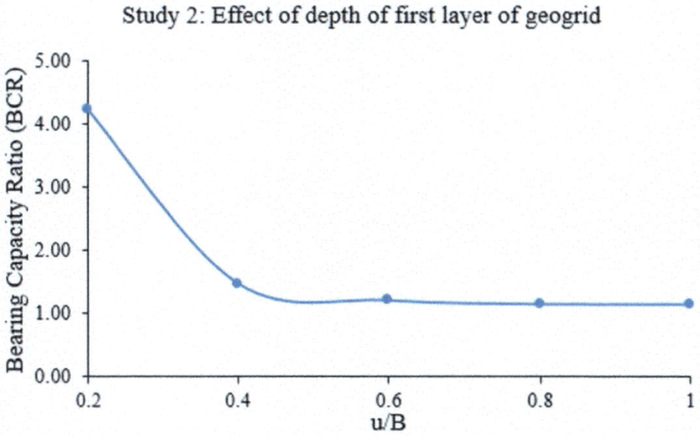


**Fig. 9** Variation of BCR with the variation in depth of first layer of geogrid (*u*/*B*)

footing has been varied from 0.2 to 1.0 keeping the other parameters constant at their mean values as in Table 4. The results of the variation in terms of its BCR are shown in Fig. 9.

From Fig. 9, the following points have been observed:

(1) For *u*/*B* value less than 0.4, an appreciable increase in bearing capacity is observed, while for *u*/*B* value greater than 0.4, an approximate linear graph is obtained which is as good as an unreinforced foundation.
(2) At *u*/*B* equals to 0.2, the bearing capacity ratio was observed to be greater than 4, but with increase in its value to 0.4, the bearing capacity ratio decreases drastically to around 1.5. Further increase in u/B reduces the BCR near about 1. This shows that the effective depth for maximum increase in BCR to observe ($d_{\text{eff}}$) is about 0.2–0.4 times of *B* for the given loading condition.
(3) Figure 9 also shows that as the depth of first layer of geogrid (*u*) increases from *u*/*B* equal to 0.2–0.4, the BC of the ring foundation decreases considerably from 490 to 170 kN/m$^2$. So, as the first geogrid layer is placed nearer to the footing base, the bearing capacity is observed to increase. It has also been observed that after $u/B = 0.6$, the effect of reinforcing the system has negligible effect in increasing the bearing capacity of the foundation.
(4) From Figs. 10 and 11, it is evident that the soil response with $u/B = 0.2$ is better than $u/B = 0.4$. This explains the observation that $u/B = 0.2$ gives higher bearing capacity value than $u/B = 0.4$.

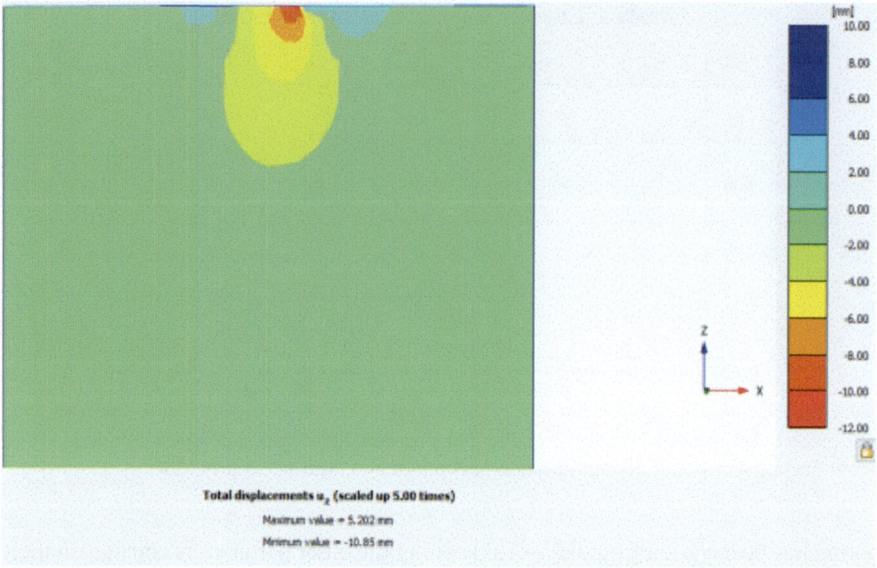


**Fig. 10** VDC in reinforced soil for $u/B = 0.2$

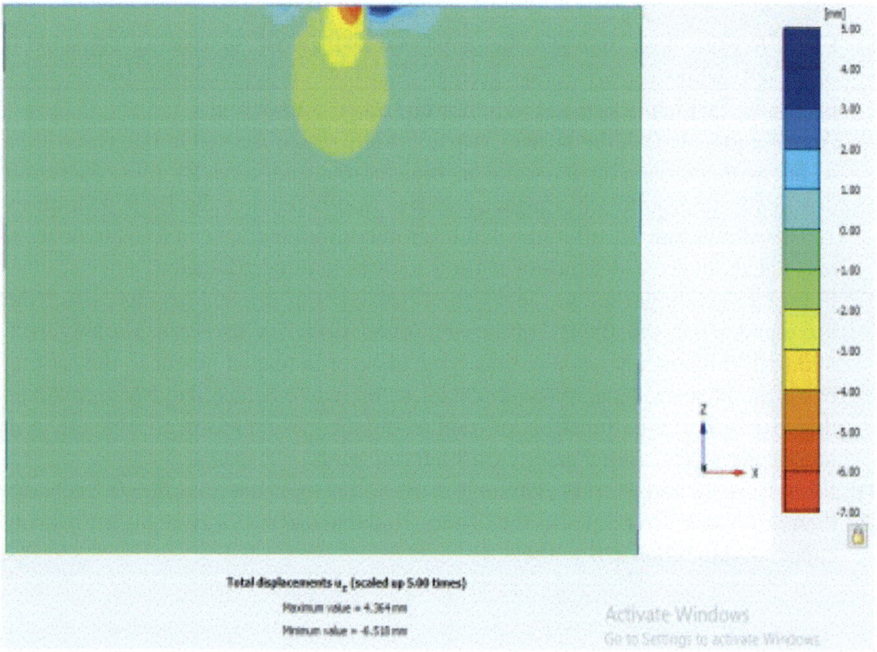


**Fig. 11** VDC in reinforced soil for $u/B = 0.4$

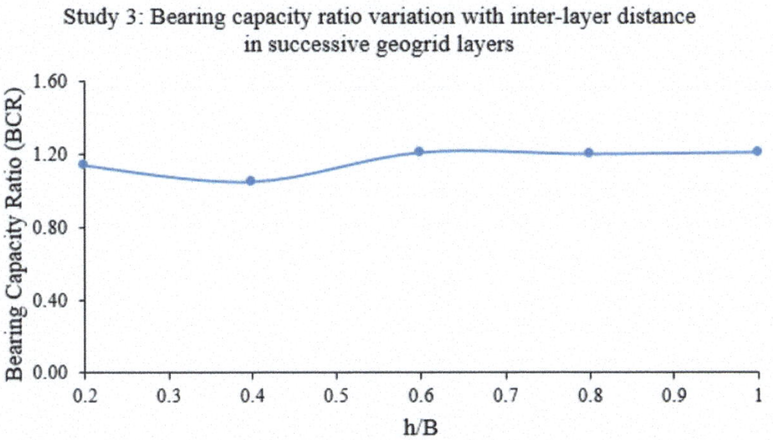


**Fig. 12** Variation of BCR with the variation in inter-layer distance of successive geogrids (*h*/*B*)

## 3.3 Study 3: Effect of Inter-layer Distance in Successive Geogrid Layers

In Study 3, to observe the effect of inter-geogrid layer distance on its bearing capacity, the ratio of inter-geogrid layer distance and width of footing has been varied from 0.2 to 1.0 keeping the other parameters constant at their mean values as in Table 4. The results of the variation in terms of its BCR are shown in Fig. 12. The following points have been observed from Fig. 12:

(1) The BCR value decreases first from 0.2 to 0.4 but increases thereafter and becomes constant after $h/B = 0.6$. The bearing capacity value at first decreases from 132 to 122 kN/m$^2$ and then increases from 122 to 140 kN/m$^2$.
(2) The optimum value of h/B from the observation came out to be 0.6.

## 3.4 Study 4: Effect of Variation of Number of Geogrid Layers

In Study 4, to observe the effect of number of geogrid layers on its bearing capacity, the number of geogrid layers has been varied from 1 to 5 keeping the other parameters constant at their mean values as in Table 4. The results of the variation in terms of its bearing capacity ratio are shown in Fig. 13.

From Fig. 13, the following observations have been made:

(1) As the number of geogrid layers is increased from one to two, there is slight decrease in bearing capacity ratio from 1.19 to 1.17. The corresponding value of bearing capacity decreases from 138 to 136 kN/m$^2$.

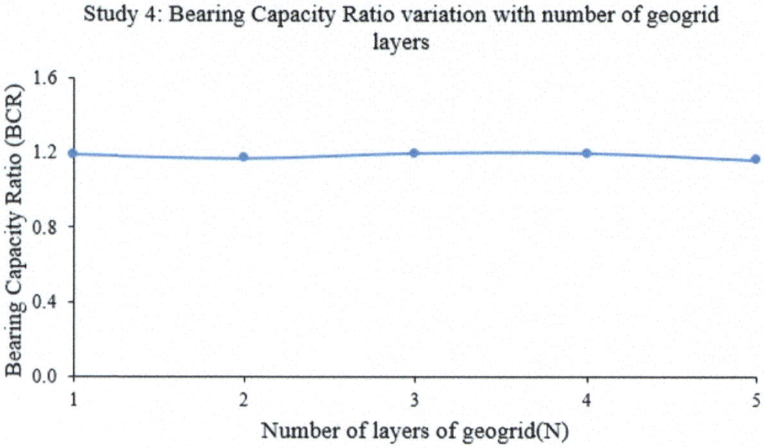


**Fig. 13** Variation of BCR with the variation in number of layers of geogrids (N)

(2) Further increase in number of geogrid layers from two to four results in increasing the bearing capacity ratio from 1.17 to 1.20. The bearing capacity value increases from 136 to 139 kN/m$^2$.
(3) If the value of *N* is increased further from four to five, slight decrease in bearing capacity ratio has been observed. The bearing capacity decreases from 139 to 135 kN/m$^2$.

## 3.5 *Study 5: Effect of Variation of Width of Geogrid Layers*

In Study 5, to observe the effect of the width of geogrid layer on its bearing capacity, the ratio of width of geogrid layers to the width of the footing has been varied from 1 to 5 keeping the other parameters constant at their mean values as in Table 4. The results of the variation in terms of its bearing capacity ratio are shown in Fig. 14.

From Fig. 14, the following observations have been noted:

(1) The bearing capacity ratio decreases from 1.47 to 1.16 as the width of the geogrid layer has been increased from B to 5B. The corresponding decrease in bearing capacity value was from 171 to 134 kN/m$^2$.
(2) The bearing capacity ratio has been observed to be approximately constant after b/B > 3. Therefore, the optimum value of *b*/*B* obtained in the present study is 1.
(3) It has been observed (from Figs. 15 and 16) that the displacement at failure has decreased considerably for b/B = 3 as compared to when the ratio was 2 at the same load. This accounts for the fact that the bearing capacity has decreased from 166 to 139 kN/m$^2$ while increasing *b*/*B* from 2 to 3.

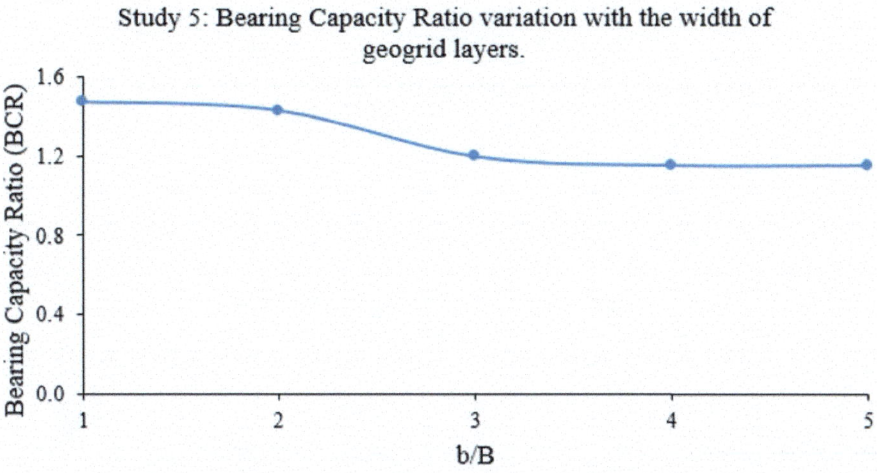


**Fig. 14** Variation of BCR with the variation in width of the geogrid layers (b/B)

**Fig. 15** VDC in reinforced soil for $b/B = 2$

**Fig. 16** VDC in reinforced soil for $b/B = 3$

## 3.6 *Study 6: Effect of Variation of Load Eccentricity and Inclination*

In Study 6, to observe the effect of the load eccentricity and inclination in ring footing on its BC, the ratio of eccentricity and width of footing has been varied from 0.0 to 0.5 keeping the other parameters constant at their mean values as in Table 4. The results of the variation in terms of its bearing capacity ratio are shown in Fig. 17. From Fig. 17, the following observations have been made:

(1) Figure 17 shows the successive decrease in bearing capacity ratio as e/B value increases, and it converges to 1 for $e/B$ equal to 0.5. That is, when eccentricity equals half of the ring foundation width, then there is no effect of reinforcing the foundation system.

(2) The bearing capacity for unreinforced case decreases nonlinearly from 310 kN/m$^2$ for $e/B = 0$ to 36 kN/m$^2$ with $e/B = 0.5$. For the reinforced case, the corresponding decrease is from 484 kN/m$^2$ for $e/B = 0$ to 36 kN/m$^2$ for $e/B = 0.5$.

(3) From the settlement values at failure from Figs. 18, 19, 20, and 21, it can be clearly seen that the soil response pattern is better for e/B = 0 as compared with $e/B = 0.5$. Hence, it has been observed in Fig. 17 that as eccentricity of loading has increased, the bearing capacity ratio has decreased.

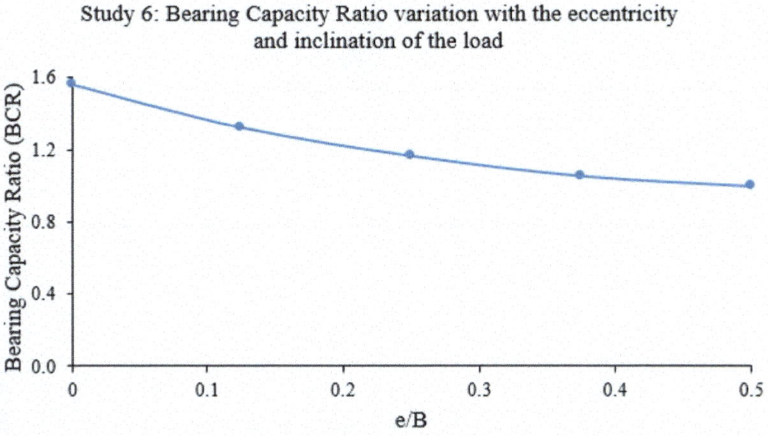


**Fig. 17** Variation of BCR with the variation in load eccentricity and inclination

**Fig.18** VDC in unreinforced soil for $e/B = 0$

(4) It may be noted that the first three values of $e/B$, i.e., 0, 0.125, and 0.25 are inside the kern distance (i.e., 24 mm from the value $(D2 + d2)/4D$) and rest two points are outside the kern distance.

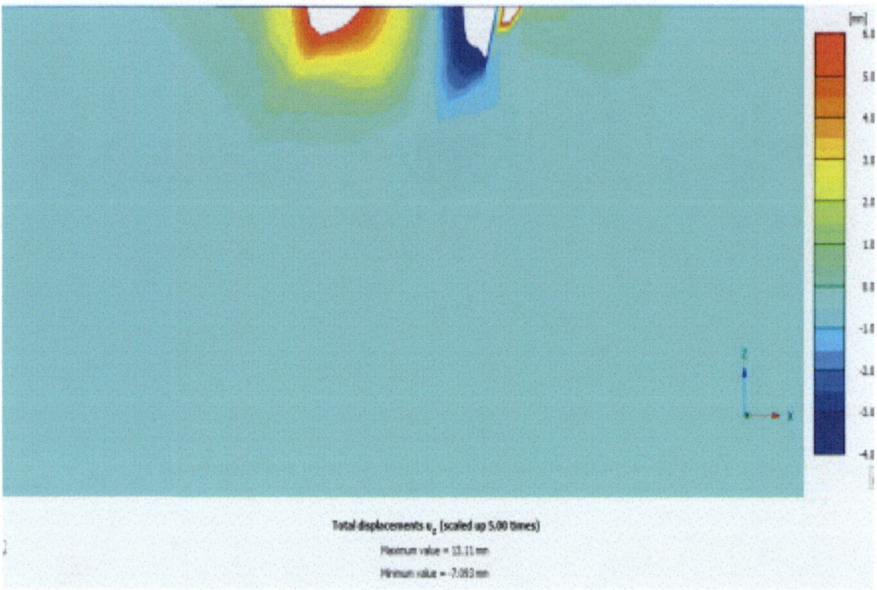


**Fig. 19** VDC in unreinforced soil for $e/B = 0.5$

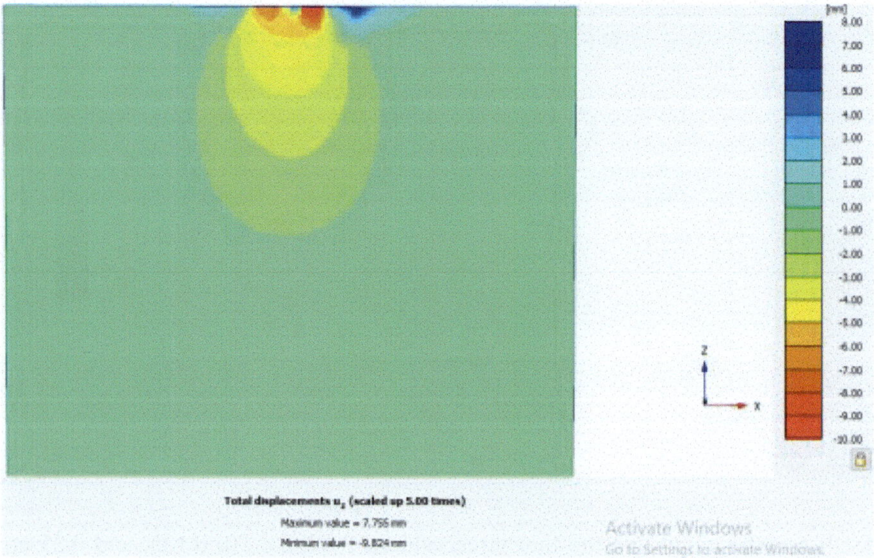


**Fig. 20** VDC in reinforced soil for $e/B = 0$

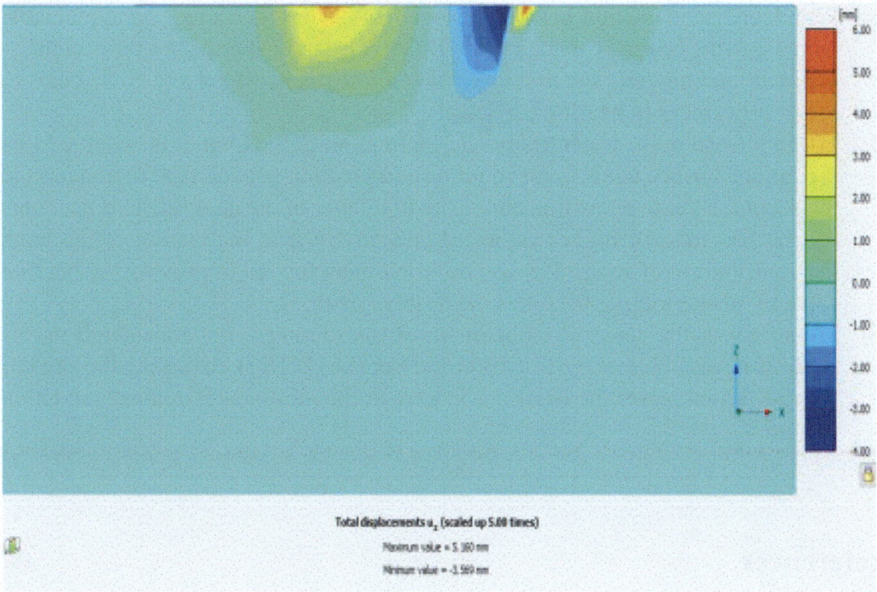


**Fig. 21** VDC in reinforced soil for $e/B = 0.5$

## 4 Conclusions

A comprehensive numerical simulation has been performed to analyze the behavior of a ring footing model over reinforced and unreinforced sand, including identification of the dominant parameters to be considered in design. Based on the observations from the research work undertaken in this paper, the following concluding remarks can be made:

(1) It has been observed that as the internal radius ($r$) of the ring foundation is increased, or in other words, the effective width of the ring is decreased, there is a decrease in the time to failure. This also indicates that the mode of failure changes from general shear failure (GSF) to punching shear failure (PSF). The settlement pattern becomes more uniform with the introduction of geogrid as soil reinforcement.

(2) Further, it has been noted that with the eccentric-inclined loading, the design of ring foundation may be permitted up to the ratio of internal to external radii ($r/R$) of 0.1875 without compromising the bearing capacity of circular foundation with equal external diameter.

(3) The effect of soil reinforcement has been found to be the maximum when the depth of the first layer of reinforcement is kept within the range of (0.2B–0.4B) (where B is the outer diameter of the ring foundation). It has been observed that the shear stress developed due to friction between soil and geogrid is also maximum for this condition.

(4) With increase in the number of reinforcement layer (N), there is no appreciable increase in the bearing capacity ratio (BCR). This indicates that for the eccentric-inclined loading, only the first layer of reinforcement effectively contributes to the improvement in bearing capacity.
(5) With increase in the eccentricity and inclination of the load, it is observed that the failure surface tends to move up from a position between the first layer and the second layer to a position above the first layer of the geogrid. This accounts for the fact that for higher values of eccentricity and inclination of the load, the contribution of geogrid as soil reinforcement toward improving the bearing capacity of the footing becomes negligibly small.
(6) The effect of the depth of embedment of the footing is not considered in this study as it was observed by Sitharam and Sireesh [14] that increasing the embedment depth increases the bearing capacity of the reinforced circular foundation system.

## References

1. Badakhshan E, Noorzad A (2017) Effect of footing shape and load eccentricity on behavior of geosynthetic reinforced sand bed. Geotext Geomembranes 45:58–67. https://doi.org/10.1016/j.geotexmem.2016.11.007
2. Basudhar PK, Saha S, Deb K (2007) Circular footings resting on geotextilereinforced sand bed. Geotext Geomembranes 25:377–384. https://doi.org/10.1016/j.geotexmem.2006.09.003
3. Boushehrian JH, Hataf N (2003) Experimental and numerical investigation of the bearing capacity of model circular and ring footings on reinforced sand. Geotext Geomembranes 21:241–256. https://doi.org/10.1016/S0266-1144(03)00029-3
4. Chakraborty M, Kumar J (2014) Bearing capacity of circular foundations reinforced with geogrid sheets. Soils Found 54:820–832. https://doi.org/10.1016/j.sandf.2014.06.013
5. Choobbasti AJ, Heshami S, Najafi A, Pirzadeh S, Farrokhzad F, Zahmatkesh A (2010) Numerical evaluation of bearing capacity and settlement of ring footing ; case study of Kazeroon cooling towers. Int J Recent Res Appl Stud 263–271
6. Guido VA, Chang DK, Sweeney MA (1986) Comparison of geogrid and geotextile reinforced earth slabs. Can Geotech J 23:435–440. https://doi.org/10.1139/t86-073
7. Hosamo H, Sliteen I, Ding S (2021) Numerical analysis of bearing capacity of a ring footing on geogrid reinforced sand. Buildings 11:1–12. https://doi.org/10.3390/buildings11020068
8. Ismael NF (1996) Loading tests on circular and ring plates in very dense cemented sands. J Geotech Eng 122:281–287. https://doi.org/10.1061/(asce)07339410(1996)122:4(281)
9. Keshavarz A, Kumar J (2017) Bearing capacity computation for a ring foundation using the stress characteristics method. Comput Geotech 89:33–42. https://doi.org/10.1016/j.compgeo.2017.04.006
10. Kumar K (2022) Numerical and sensitivity analysis of reinforced ring foundation subjected to eccentric-inclined loading. National Institute of Technology Jamshedpur, Jharkhand
11. Laman M, Yildiz A (2003) Model studies of ring foundations on geogridreinforced sand. Geosynth Int 10:142–152. https://doi.org/10.1680/gein.2003.10.5.142
12. Laman M, Yildiz A (2007) Numerical studies of ring foundations on geogridreinforced sand. Geosynth Int 14:52–64. https://doi.org/10.1680/gein.2007.14.2.52

13. Sharma V, Kumar A (2017) Influence of relative density of soil on performance of fiber-reinforced soil foundations. Geotext Geomembranes 45:499–507. https://doi.org/10.1016/j.geotexmem.2017.06.004
14. Sitharam TG, Sireesh S (2004) Model studies of embedded circular footing on geogrid-reinforced sand beds. Gr Improv 8:69–75. https://doi.org/10.1680/grim.8.2.69.36369

# Load-Settlement Behavior of Composite Caisson-Pile Foundation (CCPF) in Sand

**Rajan KC, Keshab Sharma, Monu Lal Burnwal, Prishati Raychowdhury, Indra Prasad Acharya, and Jibendra Misra**

## 1 Introduction

Caissons and piles are the two most popular deep water foundation types for bridges spanning rivers and seas. For long-span bridges [1, 2] and in rivers where the depth of alluvial deposits and scour at the base of pier foundations might be rather significant [3, 4], a caisson foundation—also known as a well foundation—is acceptable to meet the criteria for stability and serviceability. Caissons work well in liquefiable soils [5] but are susceptible to strong seismic motion, as shown by recent incidents like the 1995 Kobe Earthquake [6]. However, its construction has been plagued by recurrent obstacles due to difficulties in sinking to the desired depth. For instance, it took more than 10 years to finish the caisson foundations for a five-span, 175 m long bridge in Lumbini, Nepal's strategic road network (Fig. 1) because there were unanticipated hard strata within the foundation's design depth [7]. Numerous bridge projects in Nepal, India, and China had similar issues [2, 8, 9].

Contrarily, pile foundations are more adaptable and favored in situations when the ground's hard strata are significantly deeper and have a much lower bearing capacity than caissons. However, due to its limited ability to withstand vessel impact and difficulty in construction as a result of its extensive length and reduced rigidity, it might not be the best option to use in deep water. To address these issues, the

Rajan KC (✉) · I. P. Acharya
Department of Civil Engineering, Pulchowk Campus, IoE, Tribhuvan University, Lalitpur, Nepal
e-mail: rajankc.ioe@gmail.com

Rajan KC · M. L. Burnwal · P. Raychowdhury
Department of Civil Engineering, Institute of Technology Kanpur, Kanpur, India

K. Sharma
BGC Engineering Inc, Fredericton, NB 3C 0A9, Canada

J. Misra
Department of Roads, Government of Nepal, Kathmandu, Nepal



B. T. Jose et al. (eds.), *Proceedings of the Indian Geotechnical Conference 2022 Volume 2*, Lecture Notes in Civil Engineering 477,
https://doi.org/10.1007/978-981-97-1741-5_18

**Fig. 1** Problematic bridge construction due to caisson sinking issues in Tinau River, Nepal

caisson and piles may be combined to create the cutting-edge composite caisson-pile foundation (CCPF), a hybrid foundation that utilizes the best aspects of both types of foundations while minimizing their drawbacks. Figure 2 shows the schematic diagram of CCPF system.

CCPF provides innovative solutions to complex site conditions in deep water and reduces construction time and cost. Compared to conventional foundations, the CCPF may effectively minimize the length and embedment depth of the caisson and pile foundation, reducing construction complexity and risk [10]. Huang et al. [11] and Zhong and Huang [12] studied the performance of CCPF under vibration loading. The seismic response of the composite foundation was evaluated by Zhong and Huang [12] using centrifuge tests. However, the research on CCPF is still scanty. Moreover, load-settlement behavior of CCPF is still unknown. Due to which, the application of CCPF has been minimal. More research and development initiatives

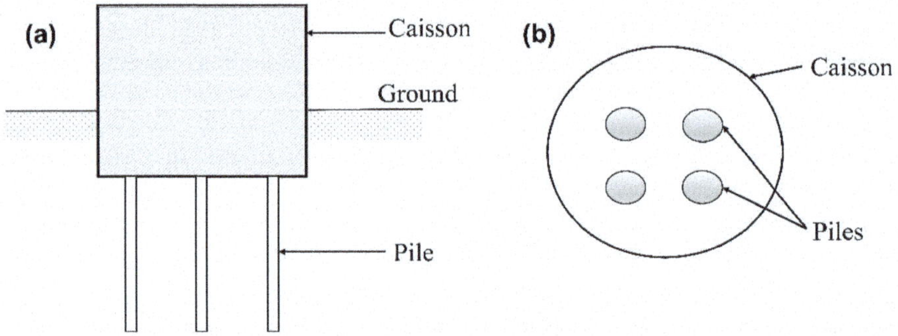


**Fig. 2** Composite caisson-pile foundation (CCPF): a schematic diagram

are desperately needed to give designers more confidence in adopting such innovative foundation systems.

# 2 Material and Methods

## *2.1 Material Characterization*

### Sand

Uniformly graded Ganga sand from Ganga River was used as the soil bed for the model tests at the laboratory setup of the Indian Institute of Technology Kanpur (IITK), Kanpur, India. The physical and mechanical properties of Ganga sand were obtained. The soil type based on USCS classification was SP (poorly graded sand) with sand fraction of 98.19 %. Specific gravity and unit weight of the sand were obtained as 2.67 and 14.07 kN/m$^3$, respectively. Angle of internal friction ($\phi$) was 32.5°, whereas maximum and minimum void ratios were obtained as 0.99 and 0.70, respectively.

### Caisson and Pile Model

To conduct experimental loading tests, the design dimensions of Tinau River Bridge (Lumbini Province, Nepal), designed by the Department of Roads, Nepal, were taken as a prototype. An appropriate similitude philosophy relevant for 1 g load tests given by Iai [13] was used, and a scaling factor of 50 was predicted for a prototype M30 grade RCC caisson and piles. The models were prepared using mild steel with the modulus of elasticity and Poisson's ratio as 210 GPa and 0.3, respectively. The properties of the prototype and model of caisson and piles are shown in Table 1.

**Table 1** Properties of prototype and model of caisson and pile

| Properties | Prototype | Model |
|---|---|---|
| Material | Concrete | Mild steel |
| Modulus of elasticity, E (N/mm$^2$) | 27,386 | 2,10,000 |
| Poisson's ratio, $\upsilon$ | 0.2 | 0.3 |
| External diameter of caisson (mm) | 7000 | 90 |
| Thickness of caisson (mm) | 900 | 18 |
| Caisson depth (mm) | 9000 | 180 |
| Pile diameter (mm) | 800 | 10 |
| Pile depth (mm) | 23,000 | 460 |
| Number of piles | 4 | 4 |

## 2.2 Test Setup and Loading Arrangement

For the model experiments, a 0.6 m sided cubical tank was utilized. To reduce the boundary impact, the side boundaries of the tank are maintained at a distance of more than 2.5 times the diameter of the caisson, and the bottom boundary is maintained at a distance greater than 10 times the diameter of the pile from the pile's top.

The loading tests of CCPF were conducted at structural laboratory of the Indian Institute of Technology (IIT), Kanpur, India. The setup and arrangement of CCPF with piles up to bottom ($CCPF_{bottom}$) and CCPF with piles up to top of caisson ($CCPF_{top}$) are as shown in Fig. 3. In both cases, a 10 kN capacity hydraulic jack was utilized to apply a vertical load on the model foundation. The loading frame consists of four 1.5 m tall vertical columns, two on each side, and two horizontal beams. A 500 kN capacity calibrated load cell was fitted to the jack to measure the load.

In the model setup, the sand bed was prepared to have relative density of 40% (loose) and 70% (dense). The calibration test demonstrated that a 15 cm fall height is optimal for reaching the specified relative density. Likewise, the hammer blow method achieved a relative density of 70%. The total height of the tank (60 cm) was split into equal intervals from the inner side by creating signs every 10 cm in height to facilitate the placement of a specific weight of sand in a defined volume to achieve the desired density through compaction. Figure 4 shows the procedure of preparing the sand bed and installing the test model in the tank.

A total of 12 sensors were used to measure various responses of the foundation and soil as shown in Fig. 3. All sensors were linked to a system for data acquisition (DAQ) system. LabVIEW 2017 software was used to program the system to automate the data collection.

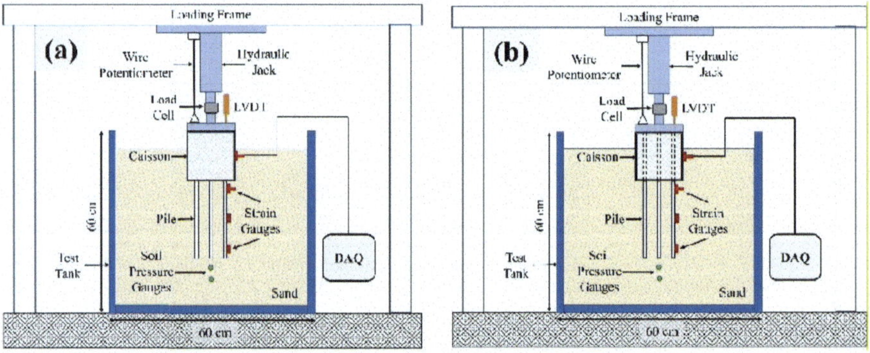


**Fig. 3** Schematic diagram showing the experimental setup for vertical load testing of **a** CCPF with piles up to bottom and **b** CCPF with piles up to top of caisson

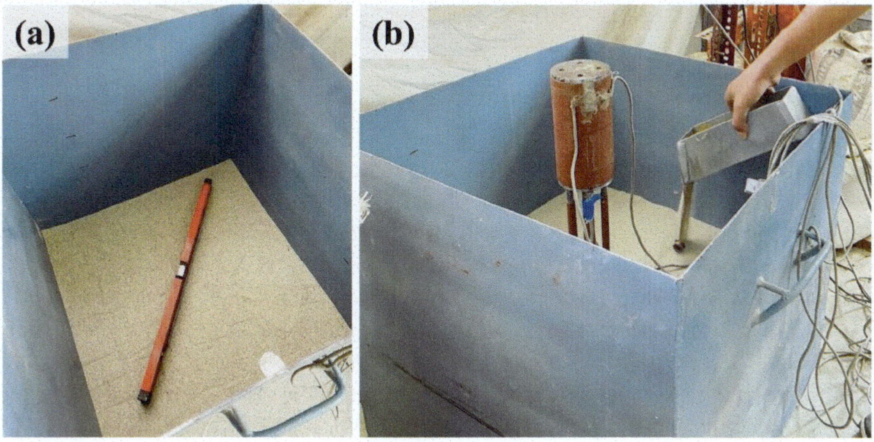


**Fig. 4** Sand bed preparation for CCPF model tests

## 2.3 *Test Procedure and Schemes*

After the installation of model, the load was applied to model through loading plate placed on the caisson. All other sensor instrumentation was conducted, as shown earlier. According to the standard test method suggested by ASTM-D1143 (2013), the failure load can be defined for the settlement exceeding 15% of the caisson diameter for vertical loading. Hence, the vertical loading continued until the settlement reached at least 14 mm. In this way, 6 vertical loading tests were conducted, the details of which are provided in Table 2.

**Table 2** Testing schemes for CCPF

| ID | Test name | Soil condition | Pile length (cm) |
|---|---|---|---|
| 1 | Caisson | Loose | – |
| 2 | Caisson | Dense | – |
| 3 | $CCPF_{bottom}$ | Loose | 28 |
| 4 | $CCPF_{bottom}$ | Dense | 28 |
| 5 | $CCPF_{top}$ | Loose | 46 |
| 6 | $CCPF_{top}$ | Dense | 46 |

## 3 Results and Discussions

### 3.1 Load-Settlement Response

The test findings are reported in terms of load-settlement behavior and load improvement ratio. The load versus settlement plots of single caisson, CCPF with piles up to caisson top ($CCPF_{top}$), and CCPF with piles up to caisson bottom ($CCPF_{bottom}$) in two relative densities (loose and dense) are shown in Fig. 5.

The ultimate load-carrying capacities of single caisson, $CCPF_{bottom}$, and $CCPF_{top}$ measured at 10 mm settlement, i.e., corresponding to 10% of the caisson diameter were found 188, 267, and 334 N, respectively, for loose sand. It can be seen that the $CCPF_{top}$ and caisson alone exhibit the highest (334 N) and lowest (188 N) load-carrying capacities, respectively. Likewise, in dense sand condition, ultimate load-carrying capacities of single caisson, $CCPF_{bottom}$, and $CCPF_{top}$ at 10 mm settlement were observed 904, 1186, and 1213 N, respectively. It is clear that the load-carrying capacity of CCPF increases as the density of the soil bed increases for all studied cases (Fig. 5). The addition of piles is found to be beneficial for improving the load capacity of the caisson as shown in Fig. 5.

As shown in Fig. 5, the load-settlement response of a CCPF system subjected to vertical loading seems to occur in three stages. In the first stage, load-induced settlement increases almost linearly. In the second stage, the load-settlement response becomes more curved, as load-induced settlement increases more than in the first stage. This phase might be referred to as the crucial phase of the foundation. In the third stage, load-settlement behavior returns to being almost linear, although the settlement grows rapidly despite the load not increasing significantly. The results are in close agreement with the results of Wang et al. [13] conducted for CCPF.

### 3.2 Load Improvement Ratio

The enrichment of the external load-carrying capability of the caisson at a particular settlement is defined as the load improvement ratio (LIR). It is a non-dimensional parameter expressed as a fraction of the load conveyed by the CCPF to the caisson alone at a given settlement as given by Eq. 1.

$$\text{LIR} = \frac{P_{ccpf}}{P_c} \tag{1}$$

where $P_{ccpf}$ and $P_c$ are load carried by CCPF and caisson alone at respective settlements, respectively.

In this study, the improvement in the load capacity of the caisson, at 5, 10, 15, 20, 25, and 30 mm settlements, due to the presence of settlement reducing piles is estimated and presented in Fig. 6. It shows the load improvement in different

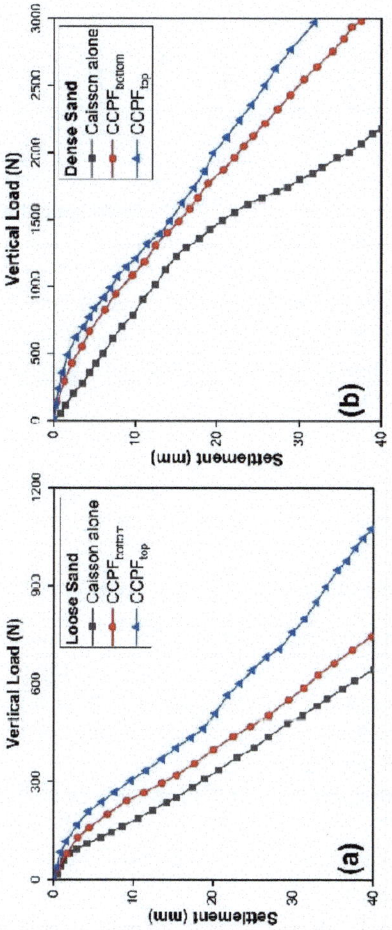


**Fig. 5** Load vs displacement curve for **a** loose (40%) and **b** dense (70%) sand

Fig. 6 Load improvement ratio for loose (40%) and dense (70%) sand

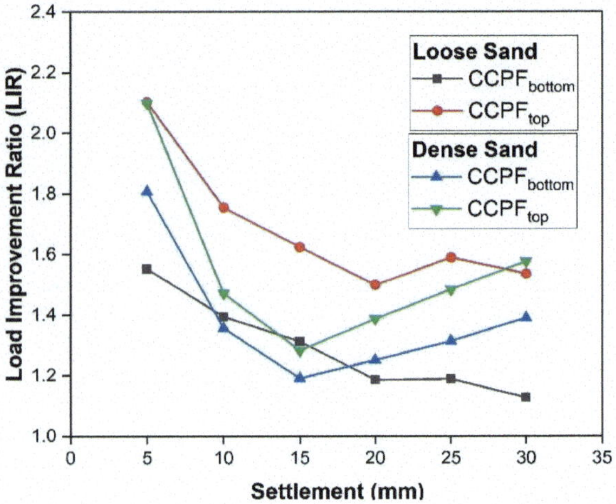


CCPF models in loose and dense sand conditions. It can be seen that, in both sand densities, CCPF with piles up to top of caisson ($CCPF_{top}$) shows higher 'LIR' values than $CCPF_{bottom}$. From the figure, it can be noted that, the 'LIR' varies in almost nonlinear fashion with settlement.

## 4 Conclusions

Following are the conclusions of the experimental study on CCPF:

- The composite caisson-pile foundation (CCPF) maximizes the vertical load-carrying capacity of the soil while successfully controlling the settlement. Generally, the caisson could support the top weight and be utilized as a platform for bridge construction. After bridges are erected, the piles beneath the caisson may be used to regulate the settlement.
- The vertical load-carrying capacity of CCPF increases as the density of the soil bed increases.
- The improvement of vertical load capacity is found to be increasing with increased length of pile.
- Load-settlement response of a CCPF system subjected to vertical loading occurs in three stages: at first increases linearly, then becomes more curved, and finally returns to being almost linear.
- Load improvement ratio (LIR) varies in almost nonlinear fashion with settlement in both vertical as well as lateral loading conditions. $CCPF_{top}$ shows higher LIR values than $CCPF_{bottom}$. In addition, with increase in length of piles, the LIR values have been found increased.

## References

1. Chowdhury I, Tarafdar R, Ghosh A, Dasgupta SP (2017) Dynamic soil structure interaction of bridge piers supported on well foundation. Soil Dyn Earthq Eng 97:251–265. https://doi.org/10.1016/J.SOILDYN.2017.03.005
2. Bag R, Mandal M, Singhal V (2021) Rectification of the tilt and shift of well foundation: a numerical and analytical solution. Indian Geotech J 51(5):1011–1025. https://doi.org/10.1007/s40098-020-00477-5
3. Mukherjee A, Saha D, Harvey CF, Taylor RG, Ahmed KM, Bhanja SN (2015) Groundwater systems of the Indian Sub-Continent. J. Hydrol Reg Stud 4:1–14. https://doi.org/10.1016/J.EJRH.2015.03.005
4. Kothyari UC (2007) Indian practice on estimation of scour around bridge piers a comment. Sadhana Acad Proc Eng Sci 32(3):187–197. https://doi.org/10.1007/s12046007-0017-7
5. Dammala PK, Bhattacharya S, Krishna AM, Kumar SS, Dasgupta K (2017) Scenario based seismic re-qualification of caisson supported major bridges—a case study of Saraighat Bridge. Soil Dyn Earthq Eng 100(June):270–275. https://doi.org/10.1016/j.soildyn.2017.06.005
6. Liang F, Jia Y, Zhang H, Chen, Huang M (2018) Centrifuge shaking table tests on composite caisson-piles foundation. Phys Model Geotech 909–914. https://doi.org/10.1201/9780429438646-22
7. Rajan KC, Sharma K, Raychowdhury P, Acharya IP, Burnwal ML, Misra J (2023) State-of-the-art review of composite caisson-pile foundation (CCPF). Appl Ocean Res 136:103571 (2023). ISSN 0141-1187. https://doi.org/10.1016/j.apor.2023.103571
8. Misra J, Subedi M, Sharma K (2022) Design of composite well-pile foundation (CWPF): a case study of bridge foundation in Nepal
9. Reddi S (2010) Well foundations for bridges are obsolete!. NBM & CW Magazine
10. Zhong R, Huang M (2021) Simplified analytical method for seismic response of composite caisson-piles foundations. Yanshilixue Yu Gongcheng Xuebao/Chinese J Rock Mech Eng 32(5):1009–1019. Accessed May 29 2021. (Online). Available: https://ro.uow.edu.au/eispapers/2202.
11. Huang M, Tu W, Gu X (2018) Time domain nonlinear lateral response of dynamically loaded composite caisson-piles foundations in layered cohesive soils. Soil Dyn Earthq Eng 106:113–130 (2018). ISSN 0267-7261. https://doi.org/10.1016/j.soildyn.2017.12.015
12. Zhong R, Huang M (2014) Winkler model for dynamic response of composite caisson–piles foundations: seismic response. Soil Dyn Earthq Eng 66:241–251. ISSN 0267-7261. https://doi.org/10.1016/j.soildyn.2014.07.005
13. Iai S (1989) Similitude for shaking table tests on soil-structure-fluid model in 1g gravitational field. Soils Found 29(1):105–118

# Behavior of Bridge Abutment Foundation in Mixed Ground Condition

**Chiranjib Sarkar, Sibapriya Mukherjee, and Ankit Som**

## 1 Introduction

Transportation network plays a very vital role in economic growth of fast developing country like India. Railways and roads have always played a very key role in transportation network system [1]. Among transportation systems, bridges are essential components for both rail and road networks [2]. As per conventional approach, a reinforced concrete abutment foundation is to be analyzed by considering the foundation to be rigid [3]. It is fact that this is quite simplify assumption for the design, but it cannot capture the exact representation of the subgrade soil [4]. Therefore, it is better to model the foundation as flexible through its length and to model the subgrade soil by providing closely spaced springs [5]. These springs considered are calculated from the general concept of modulus of subgrade reaction. The different ground soil conditions (medium clay, stiff clay, etc.) provide varying safe bearing capacity, ground settlement, etc. [6]. Soil–structure interaction under different seismic loads is a vastly nonlinear phenomenon. This nonlinear behavior plays very important role in the entire structural response [7].

In this paper, the behavior of bridge abutment and its foundation have been studied in detail under different surrounding subsoil conditions (soft soil, weathered rock, etc.) by varying water table and seismic condition (seismic Zone-III and Zone-IV). As a part of parametric study, various sets of numerical models have been further developed based on Finite Element Method (FEM) to study the nonlinear behavior of bridge abutment with both pile and open foundations. Considering variation in soil and rock layer, six different types of ground conditions have been considered for

C. Sarkar (✉) · S. Mukherjee
Department of Civil Engineering, Jadavpur University, Jadavpur, Kolkata 700032, India
e-mail: chiranjibsarkar.civil@gmail.com

A. Som
AECOM India Pvt. Ltd., Kolkata 700156, India



B. T. Jose et al. (eds.), *Proceedings of the Indian Geotechnical Conference 2022 Volume 2*, Lecture Notes in Civil Engineering 477,
https://doi.org/10.1007/978-981-97-1741-5_19

six different FEM models in this parametric study. The outcomes of model analyses: factor of safety, ground settlement, bending moment, and shear force, have been assessed for all six FEM models.

## 2 Model of the Case Study

In the present paper, different sets of parametric study have been conducted by developing various numbers of models in 2D MIDAS GTS NX software (MIDAS IT Co., Ltd., 2020) for varying subsoil conditions, water tables, and seismicities. Abutment foundation geometrical profile has been further re-modeled for both pile and open foundations. Geometrical profile of embankment cross section along with abutment and soil/rock layers is shown in Fig. 1.

A model view from FEM models of bridge abutment for both pile and open foundations is shown in Fig. 2.

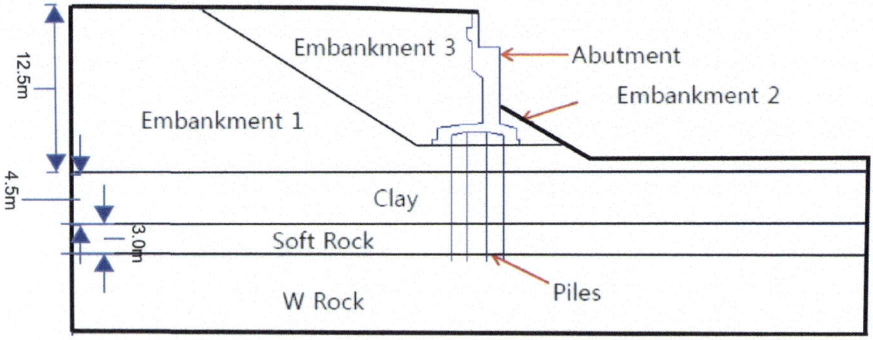


**Fig. 1** Geometrical profile of embankment cross section including abutment, soil, and rock layers

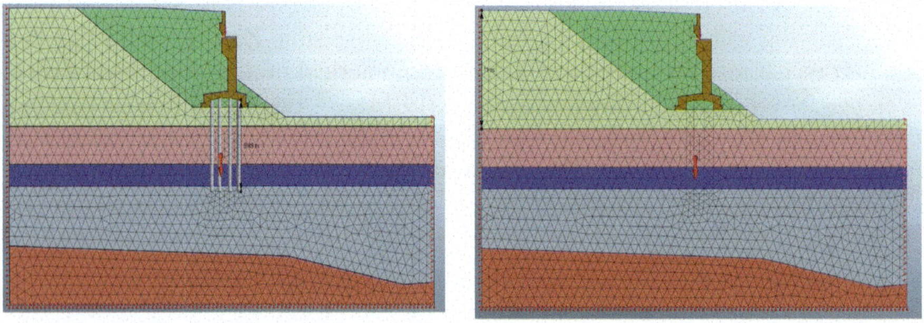

A. Bridge abutment with pile

B. Bridge abutment with open foundation

**Fig. 2** Model views from FEM models of bridge abutment for both pile and open foundations

As per sequence of construction of the bridge abutment, embankment has been subdivided in three zones which are already shown in Fig. 1. Embankment ground conditions with geotechnical parameters are shown in Table 7.

In ground conditions, soil layers followed by two different rock layers have been considered for present parametric study. Considering variation in type and property of soil and rock layer, six different types of ground conditions have been considered for six different FEM models. Ground conditions with geotechnical properties of six different FEM models, i.e., FEM Model-1, FEM Model-2, FEM Model-3, FEM Model-4, FEM Model-5, and FEM Model-6 are tabulated in six tables, i.e., Tables 1, 2, 3, 4, 5, and 6 respectively.

**Table 1** Ground conditions with geotechnical parameters of FEM Model-1

| Ground condition | Soil type | *C* (Avg.) kN/m$^2$ | Φ′ (Avg.) deg | *Y* (Avg.) kN/m$^3$ | *E*′ (Avg.) MPa | *υ*′ (Avg.) |
|---|---|---|---|---|---|---|
| Stratum-I | Medium sandy clayey silt | 50 | 50 | 18.5 | 12.5 | 0.30 |
| Stratum-II | Soft/highly weather rock | 200 | 200 | 22.0 | 150 | 0.25 |
| Stratum-III | Weathered rock | 450 | 35 | 24.0 | 300 | 0.20 |

**Table 2** Ground conditions with geotechnical parameters of FEM Model-2

| Ground condition | Soil type | *C* (Avg.) kN/m$^2$ | Φ′ (Avg.) deg | *Y* (Avg.) kN/m$^3$ | *E*′ (Avg.) MPa | *υ*′ (Avg.) |
|---|---|---|---|---|---|---|
| Stratum-I | Medium sandy clayey silt | 50 | 20 | 18.5 | 12.5 | 0.30 |
| Stratum-II | Soft/highly weather | 200 | 32 | 22.0 | 150 | 0.25 |
| Stratum-III | Weathered Rock | 600 | 35 | 24.0 | 400 | 0.20 |

**Table 3** Ground conditions with geotechnical parameters of FEM Model-3

| Ground condition | Soil type | *C* (Avg.) kN/m$^2$ | Φ′ (Avg.) deg | *Y* (Avg.) kN/m$^3$ | *E*′ (Avg.) MPa | *υ*′ (Avg.) |
|---|---|---|---|---|---|---|
| Stratum-I | Medium sandy clayey silt | 50 | 20 | 18.5 | 12.5 | 0.30 |
| Stratum-II | Soft/highly weather rock | 250 | 32 | 22.0 | 180 | 0.25 |
| Stratum-III | Weathered rock | 450 | 35 | 24.0 | 300 | 0.20 |

**Table 4** Ground conditions with geotechnical parameters of FEM Model-4

| Ground condition | Soil type | $C$ (Avg.) kN/m$^2$ | $\Phi'$ (Avg.) deg | $Y$ (Avg.) kN/m$^3$ | $E'$ (Avg.) MPa | $\upsilon'$ (Avg.) |
|---|---|---|---|---|---|---|
| Stratum-I | Stiff sandy clayey silt | 75 | 20 | 19.5 | 14.5 | 0.30 |
| Stratum-II | Soft/highly weather rock | 200 | 32 | 22.0 | 150 | 0.25 |
| Stratum-III | Weathered rock | 450 | 35 | 24.0 | 300 | 0.20 |

**Table 5** Ground conditions with geotechnical parameters of FEM Model-5

| Ground condition | Soil type | $C$ (Avg.) kN/m$^2$ | $\Phi'$ (Avg.) deg | $Y$ (Avg.) kN/m$^3$ | $E'$ (Avg.) MPa | $\upsilon'$ (Avg.) |
|---|---|---|---|---|---|---|
| Stratum-I | Medium sandy clayey silt | 50 | 20 | 18.5 | 12.5 | 0.30 |
| Stratum-II | Soft/highly weather rock | 250 | 32 | 22.0 | 180 | 0.25 |
| Stratum-III | Weathered rock | 600 | 35 | 24.0 | 400 | 0.20 |

**Table 6** Ground conditions with geotechnical parameters of FEM Model-6

| Ground condition | Soil type | $C$ (Avg.) kN/m$^2$ | $\Phi'$ (Avg.) deg | $Y$ (Avg.) kN/m$^3$ | $E'$ (Avg.) MPa | $\upsilon'$ (Avg.) |
|---|---|---|---|---|---|---|
| Stratum-I | Stiff sandy clayey silt | 75 | 20 | 19.5 | 14.5 | 0.30 |
| Stratum-II | Soft/highly weather rock | 250 | 32 | 22.0 | 180 | 0.25 |
| Stratum-III | Weathered rock | 600 | 35 | 24.0 | 400 | 0.20 |

**Table 7** Embankment ground conditions with geotechnical parameters

| Ground condition | Soil type | $C$ (Avg.) kN/m$^2$ | $\Phi'$ (Avg.) deg | $Y$ (Avg.) kN/m$^3$ | $E'$ (Avg.) MPa | $\upsilon'$ (Avg.) |
|---|---|---|---|---|---|---|
| Embankment | Medium sand | 15 | 29 | 18.0 | 40 | 0.30 |

The basic difference in ground conditions with geotechnical parameters between Model-1 and Model-2 is in the rock parameters of Stratum-III. The difference between Model-1 and Model-3 is in the rock parameters of Stratum-II. The difference between Model-1 and Model-4 is in the soil type/parameters of Stratum-I. In Model-5, there are changes in the rock parameters of Stratum-II and Stratum-III, whereas changes in the soil/rock type and parameters of all three strata are in Model-6.

## 3 Analysis and Presentation of Results

Different sets of numerical models of 1.0 m × 1.0 m mesh size have been developed in 2D MIDAS GTS NX software (MIDAS IT Co., Ltd., 2020) to understand the nonlinear behavior of bridge abutment and its foundation during varying ground conditions. Peak Ground Acceleration (PGA) value of 0.24 and 0.36 for two different seismic zones (Zone-III and Zone-IV as per IS-1893) has been further applied in the present FEM model to understand the seismicity behavior of bridge abutment and its foundation. To notice the effect of water table, existing FEM model has been further developed by considering water profile below the embankment.

As outcomes of present FEM model analysis, factor of safety, ground settlement, bending moment, and shear force have been estimated for all six FEM models. Factor of safety values of the bridge abutment for both pile and open foundations for all six different FEM models has been presented in Tables 8 and 9, respectively.

Ground settlement values of the bridge abutment for both pile and open foundations for all six models have been presented in Tables 10 and 11, respectively.

**Table 8** Factor of safety values of the bridge abutment with pile foundation

| Model-1 | Model-2 | Model-3 | Model-4 | Model-5 | Model-6 |
|---|---|---|---|---|---|
| 3.456 | 3.484 | 3.528 | 3.600 | 3.581 | 3.900 |

**Table 9** Factor of safety values of the bridge abutment with open foundation

| Model-1 | Model-2 | Model-3 | Model-4 | Model-5 | Model-6 |
|---|---|---|---|---|---|
| 1.614 | 1.612 | 1.598 | 1.626 | 1.607 | 1.631 |

**Table 10** Ground settlement values of the bridge abutment with pile foundation

| Model-1 (mm) | Model-2 (mm) | Model-3 (mm) | Model-4 (mm) | Model-5 (mm) | Model-6 (mm) |
|---|---|---|---|---|---|
| 35.946 | 35.54 | 35.77 | 33.923 | 35.37 | 33.137 |

**Table 11** Ground settlement values of the bridge abutment with open foundation

| Model-1 (mm) | Model-2 (mm) | Model-3 (mm) | Model-4 (mm) | Model-5 (mm) | Model-6 (mm) |
|---|---|---|---|---|---|
| 85.536 | 84.812 | 85.184 | 74.694 | 84.465 | 73.618 |

**Table 12** Bending moment values of the bridge abutment with pile foundation

| Model-1 (kN-m) | Model-2 (kN-m) | Model-3 (kN-m) | Model-4 (kN-m) | Model-5 (kN-m) | Model-6 (kN-m) |
|---|---|---|---|---|---|
| 230.65 | 229.12 | 241.81 | 207.69 | 239.98 | 216.57 |

**Table 13** Shear force values of the bridge abutment with pile foundation

| Model-1 (kN-m) | Model-2 (kN-m) | Model-3 (kN-m) | Model-4 (kN-m) | Model-5 (kN-m) | Model-6 (kN-m) |
|---|---|---|---|---|---|
| 68.74 | 60.96 | 68.63 | 83.26 | 60.59 | 75.57 |

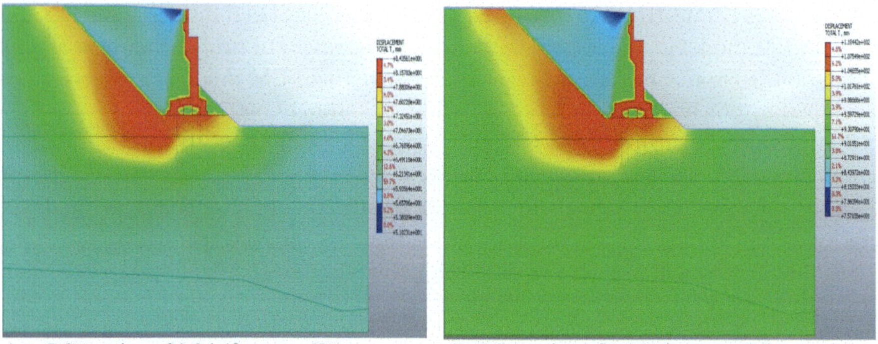

A. PGA value of 0.24 (for zone-III) B. PGA value of 0.36 (for zone-IV)

**Fig. 3** Ground settlements of the bridge abutment with **pile foundation** for PGA value of 0.24 (for Zone-III) and 0.36 (for Zone-IV)

Bending moment and shear force values of the bridge abutment with pile foundation for all six different FEM models have been presented in Tables 12 and 13, respectively.

Ground settlements of the bridge abutment with **pile foundation** for PGA value of 0.24 (for Zone-III) and 0.36 (for Zone-IV) are shown in Fig. 3. Ground settlements of the bridge abutment with **open foundation** for PGA value of 0.24 (for Zone-III) and 0.36 (for Zone-IV) are shown in Fig. 4. Ground settlements and bending moment values of the bridge abutment with pile foundation for water profile below the embankment are shown in Fig. 5.

## 4 Discussions on Results

From the results shown in Table 8, it can be noted that factor of safety values of the bridge abutment with pile foundation gradually increases by 4–12% when ground condition changes from medium to stiff clay followed by weathered rock. From the

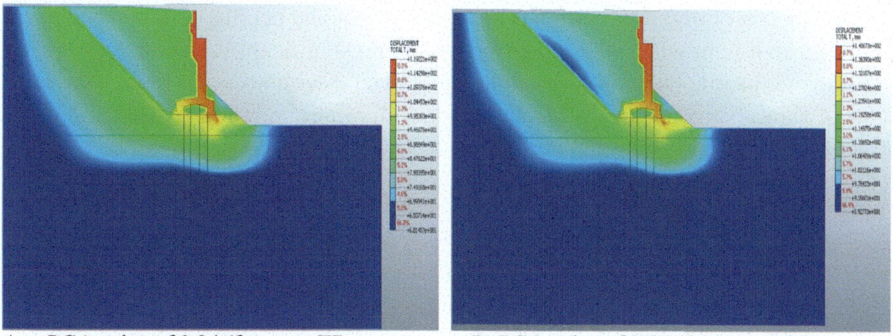

A. PGA value of 0.24 (for zone-III) B. PGA value of 0.36 (for zone-IV)

**Fig. 4** Ground settlements of the bridge abutment with **open foundation** for PGA value of 0.24 (for Zone-III) and 0.36 (for Zone-IV)

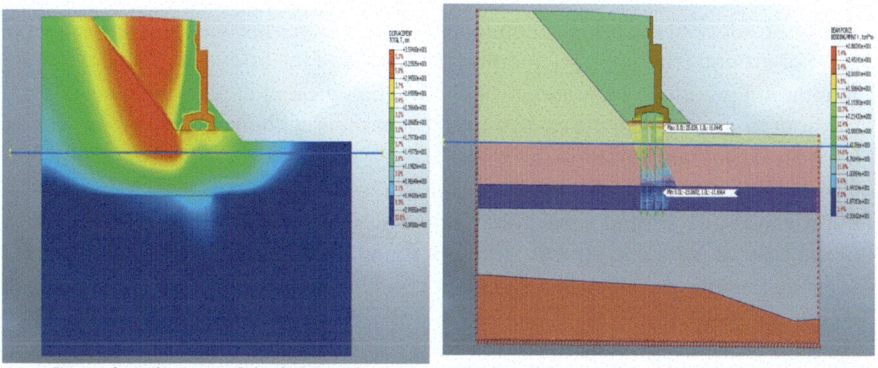

A. Ground settlement of the bridge abutment B. Bending moment of the bridge abutment

**Fig. 5** Ground settlements and bending moment values of the bridge abutment with pile foundation for water profile below the embankment

results shown in Table 9, it is observed that factor of safety values has variation less than 1% for the bridge abutment with open foundation.

From the results shown in Table 10 and Table 11, it is clear that ground settlement value decreases by 6–8% in pile foundation and by 14–16% in open foundation when ground condition changes from medium to stiff clay.

Table 12 shows bending moment value decreases by 6–11% in bridge abutment with pile foundation when ground condition changes from medium to stiff clay, whereas Table 13 shows shear force value increases by 9–17%.

The diagram of Fig. 3 shows that ground settlement of the bridge abutment with pile foundation increases by 30–135% when seismic zone changes from Zone-III to Zone-IV with higher PGA, even for same soil condition, whereas the diagram of Fig. 4 shows that ground settlement of the bridge abutment with open foundation increases by 18–40% when seismic zone changes from Zone-III to Zone-IV.

The diagram of Fig. 5 shows the ground settlement and bending moment values of the bridge abutment with pile foundation having no significant variation between the case of water profile below the embankment and case with no water table. However, factor of safety value of the bridge abutment with pile foundation decreases slightly (around 10%) for the case of water profile below the embankment.

## 5 Conclusions

In the present paper, the behavior of bridge abutment and its foundation has been studied under varying subsoil, water table, and seismic conditions. Different sets of parametric study have been conducted by developing various numbers of FEM models for bridge abutment with both pile and open foundations. Following conclusion can be drawn from the current study's analysis and results:

- Factor of safety of the bridge abutment gradually increases with gradual improvement of ground condition. Factor of safety values of the bridge abutment with pile foundation gradually increases by 4–12% when ground condition changes from medium to stiff clay. In the case of bridge abutment with open foundation, values have no significant variation (around 1%).
- Ground settlement of the bridge abutment gradually decreases with gradual improvement of ground condition. Ground settlement value decreases by 6–8% in the case of bridge abutment with pile foundation and by 14–16% in the case of bridge abutment with open foundation when ground condition changes from medium to stiff clay.
- With the change of ground condition from medium to stiff clay, bending moment value decreases by 6–11% in bridge abutment with pile foundation, whereas shear force value increased by 9–17%.
- With the change of seismic zone from Zone-III to Zone-IV, ground settlement of the bridge abutment with pile foundation increases by 30–135% even for same soil condition, whereas ground settlement increases by 18–40% in the case of bridge abutment with open foundation.
- Factor of safety of the bridge abutment with pile foundation decreased by 10% for the case of water profile below the embankment. However, ground settlement and bending moment values of the bridge abutment with pile foundation having no significant variation between the case of water profile below the embankment and case with no water table.

Outcomes of the present case study may be useful to the practicing engineers as a reference for similar situations.

## References

1. Indian Road Congress (1998) Hill road manual (1998 ed., vol SP:48). Indian Road Congress, New Delhi, India
2. Najia M, Firoozib AA, Firoozic AA (2020) A review: study of integral abutment bridge with consideration of soil-structure interaction. https://doi.org/10.1590/1679-78255869
3. Rao NK (2010) Foundation design: theory and practice. Wiley
4. Jardine RJ, Potts DM, Fourie AB, Burland JB (1986) Studies of the influence of non-linear stress–strain characteristics in soil–structure interaction. Geotechnique 36(3):377–396
5. Winkler E (1867) Die Lehre yvon der Elastizitat und Festigkeit. Prague: H. Dominicus 141:182–184
6. Balla B, Manandhar S (2019) Analysis of abutment foundation using conventional method and FEM model. In: 2nd International conference on earthquake engineering and post disaster reconstruction planning (ICEE-PDRP 2019)
7. Spyrakos CC (1992) Seismic behaviour of bridge piers including soil-structure interaction. Comput Struct 4(2):373–384

# Pile Capacity Estimation Considering Variability in Soil Adhesion Factor

**C. Supritha Reddy, Goushya Begum, and B. Hanumantha Rao**

## 1 Introduction

Any high-rise building constructed on weak-bearing soils needs a robust, deep foundation system to prevent significant settlements over time. According to Wrana [14], a pile is considered to be a structural element that utilizes its end point and skin friction to transfer the axial load of the superstructure to the soil. However, the soil–pile interaction has a significant impact on a pile's skin friction capability. Large pile settlements may occur if the soil strata are composed of soft soils.

Skin friction of pile in clayey soil is calculated taking into consideration, predominantly, of soil–pile interface parameter, i.e., adhesion factor, which in clay is indeed depended on the soil parameters like plasticity index, pile geometry, undrained shear strength, overburden stress, etc. [4]. Arifin [2] has analyzed the large bored pile through numerical approach (PLAXIS 2D) and confirmed the adhesion factor of 0.97 and 0.94 for very soft and soft clayey layer soils, while 0.56 and 0.48 for stiff and very stiff clayey layer soils. Kulhaway and Jackson [11] Coduto [5], in their studies, have asserted that the obtained adhesion factors for various undrained shear strength of soil are in good agreement with adhesion distribution. As per Banerjee [3], it was found that required piled group-raft area ratio for minimizing the differential settlement of a raft in a layered soil should be within a range of 0.4–0.6.

C. Supritha Reddy · G. Begum
School of Infrastructure, IIT Bhubaneswar, Argul, Bhubaneswar, Odisha 752050, India
e-mail: csr10@iitbbs.ac.in

G. Begum
e-mail: gb18@iitbbs.ac.in

B. Hanumantha Rao (✉)
Department of Civil Engineering, IIT Bhubaneswar, Argul, Bhubaneswar, Odisha 752050, India
e-mail: bhrao@iitbbs.ac.in

 
B. T. Jose et al. (eds.), *Proceedings of the Indian Geotechnical Conference 2022 Volume 2*, Lecture Notes in Civil Engineering 477,
https://doi.org/10.1007/978-981-97-1741-5_20

A squared pile raft foundation was modeled using PLAXIS-3D for different pile lengths, spacing between piles, and a number of piles to investigate the pile behavior in terms of pile load capacity and load settlement curve under uniform vertical load. At a specific load, the settlement of the pile increases as the pile spacing increases.

Generally, the ultimate bearing capacity of a pile is determined by pile load tests. The obtained results from pile load tests can further be employed to ascertain the quality of numerical models. As per Ezzat et al. [6], three different numerical models, i.e., Modified Mohr–Coulomb (MMC), Mohr–Coulomb (MC), and soft soil models, can be used to simulate the field load tests. As we know that MC model is elastic perfectly plastic model, the other two models have a nonlinear relation between axial strain and deviatoric stress. The primary difference to note in these models is that stiffness in MC model is constant, whereas in SS and MMC model, stiffness is stress dependent. It was observed that very good agreement was found between field settlements and the calculated settlements using MMC model.

The objective of the present study is to calculate the skin friction capacity of a pile under axial loading by numerical approach (PLAXIS-3D) and compare it with the pile load test of the field. The numerical results are used to assess the accuracy of the empirical equations of adhesion factor distribution suggested by different authors.

## 2 Methodology

The skin friction of a pile for cohesive and granular soils can be determined by the analytical equation given in IS 2911 (Part 1) [7]. As the adhesion factor is dependent on various parameters, it is estimated using empirical equations suggested by them. Consequently, the skin friction of the pile is calculated by the analytical equation suggested by the IS 2911 code provision. In the present study, a bored pile of 1200 mm diameter and 25 m length is modeled using PLAXIS-3D software with an application of axial point load. The soil–pile interaction factor (adhesion factor) suggested by different authors is given as input data to find out the variation in skin friction capacity obtained from the load settlement curve, as interpreted in IS 2911 (Part 4) [8]. The load settlement curve obtained for different adhesion factor values is compared with the field load settlement curve to assess the accuracy of the adhesion factor determined as per the empirical equations.

## 2.1 Analytical Equations for Estimating Adhesion Factor Values

Skin friction for cohesive soils can be estimated as per IS 2911 (Part 1) [7], given as:

$$Q_s = \sum_{i=0}^{n} \alpha_i * c_i * A_s, \tag{1}$$

where $Q_s$ = ultimate skin friction; $\alpha_i$ = adhesion factor of the *i*th layer of soil strata; $c_i$ = undrained shear strength of *i*th layer strata; $A_s$ = surface area of the pile.

The values of the adhesion factor, in Eq. (1), are determined based on the empirical approaches proposed by different researchers. Please note that each empirical approach accounts for different soil properties. Thus, there is a chance to understand and signify which soil parameter has major impact on the skin friction capacity of a pile installed in cohesive soils. The various empirical approaches are given herein.

As per the API [1], the adhesion factor is a function of undrained shear strength alone.

$$\begin{aligned} &\alpha = 1-(c_u - 25)/90 - \text{when } 25\text{ kPa} < c_u < 70\text{ kPa},\\ &\alpha = 1 - \text{when } c_u <= 25\text{ kPa},\\ &\alpha = 0.5 - \text{when } c_u > 70\text{ kPa}. \end{aligned} \tag{2}$$

The adhesion factor equation proposed by Sladen [13] is dependent on two soil parameters such as effective stress and undrained shear strength (Fig. 1).

$$\alpha = 0.5 * (\sigma'_v/c_u)^{0.45}. \tag{3}$$

According to Karlsud et al. [9], the adhesion factor varies with the plasticity index of $\alpha$ clayey soils. The correlation of $\sigma_u/c'_v$ and Plasticity Index ($I_p$) with $\alpha$ is introduced in the form of trend lines described in the below graph:

The adhesion factor estimated by Kulhaway and Jackson [11] is dependent on effective stress and undrained shear strength parameters of soil:

$$\begin{aligned} &\alpha = 0.21 + 0.26 * (c_u/P_a)\\ &\text{for } c_u/P_a <= 3 \text{ and } \alpha <= 1. \end{aligned} \tag{4}$$

Variation of $\alpha$ with $c_u/P_a$ and slenderness ratio (L/D) as proposed by Kolk and Van der Velde [10] is given as:

$$\alpha = 0.5 * (D/L)^{0.2} * (c_u/\sigma'_v)^{0.3}. \tag{5}$$

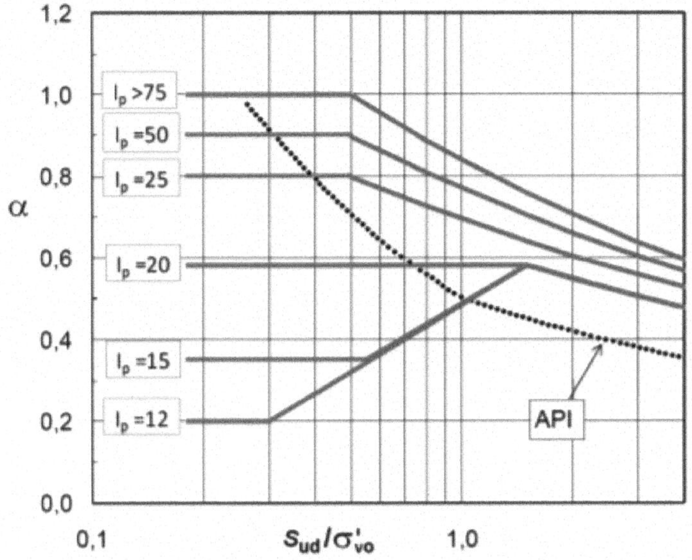


**Fig. 1** Variation of $\alpha$ with $\sigma_u/c_v'$ and Plasticity Index ($I_p$)

## 2.2 Finite Element Modeling

The pile was modeled using PLAXIS-3D in layered cohesive soil whose parameters are given in Table 1. PLAXIS 3D is a three-dimensional program for deformation, stability, and flow analyses for different types of geotechnical applications. The program uses a comfortable graphical user interface to quickly create a geometry model and a finite element mesh. It provides different models to simulate the soil behavior which are linear elastic, Mohr–Coulomb, strain hardening, and soft soil creep models. The soil behavior is assumed to follow the Mohr–Coulomb failure criterion in the model. In PLAXIS, to simulate the Mohr–Coulomb model, five parameters of soil are required. Young's modulus ($E$) and Poisson's ratio ($\nu$) are inputted as stiffness parameters, and cohesion ($c$), friction angle ($\varphi$), and Dilatancy Angle ($\psi$) are used as strength parameters. The value of these parameters are listed in Table 1.

As regards to boundary conditions, the bottom of soil body is fixed in all three directions. The side faces of the soil body are fixed in $X$- and $Y$-directions but free to move along the $Z$-direction.

The analyzed pile is cylindrical and isolated element subjected to axial loading, allowing for three-dimensional embedded pile simulation. Soil properties vary considerably from layer to layer. The embedded pile model can be used to model the pile and parameters used in PLAXIS-3D are given in Table 2. The embedded pile is connected to the adjacent soils by special interfaces named skin interfaces and foot

**Table 1** Undrained parameters of Paradip soil used in PLAXIS-3D software

| Parameter | Sand | Clay | Sand |
|---|---|---|---|
| Range of depth (m) | 0–13 | 13–18 | 18–40 |
| Material model | Mohr–Coulomb | Mohr–Coulomb | Mohr–Coulomb |
| Material behavior | Undrained (A) | Undrained (B) | Undrained (A) |
| $S_u$ (KN/m$^2$) | 0 | 22 | 0 |
| Young's modulus (KN/m$^2$) | 8000 | 8000 | 12,000 |
| Poisson's ratio | 0.33 | 0.2 | 0.33 |
| Angle of internal friction | 33 | 0 | 34 |
| Strength reduction factor | 1 | Variable—0.27, 0.52, 1 | 1 |

**Table 2** Parameters of the pile in Paradip soil

| Pile data | Value |
|---|---|
| Pile length | 25 m |
| Pile diameter | 1.2 m |
| *E* (MPa) | 29,580 |
| Model of pile | Embedded beam |
| Material behavior | Linear elastic |
| Poisson's ratio | 0.2 |

interfaces. A point load is applied over the pile with increments and the maximum load applied is 3800 kN.

Medium mesh is selected as the optimum mesh generation element considering excessive time consumption of fine, and very fine meshes and the deformed soil body mesh are shown in Figs. 2 and 3.

## *2.3 Interface Modeling*

The interface between soil and pile is modeled by input parameter called interface reduction factor ($R_{\text{inter}}$) [12]. If $R_{\text{inter}}$ is 1, the soil is fully bonded to the surface of pile. Interface properties indirectly depend on the soil strength. Undrained shear strength of interface can be determined by Eq. 6, and adhesion of soil–pile interface is given by Eq. 7.

$$s_{u,i} = R_{\text{inter}} s_{u,\text{soil}}, \tag{6}$$

$$s_{a,i} = \alpha s_{u,\text{soil}}, \tag{7}$$

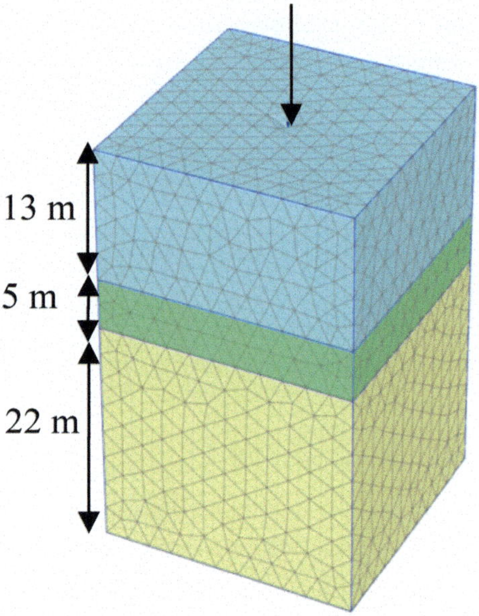


**Fig. 2** Deformed mesh of soil body

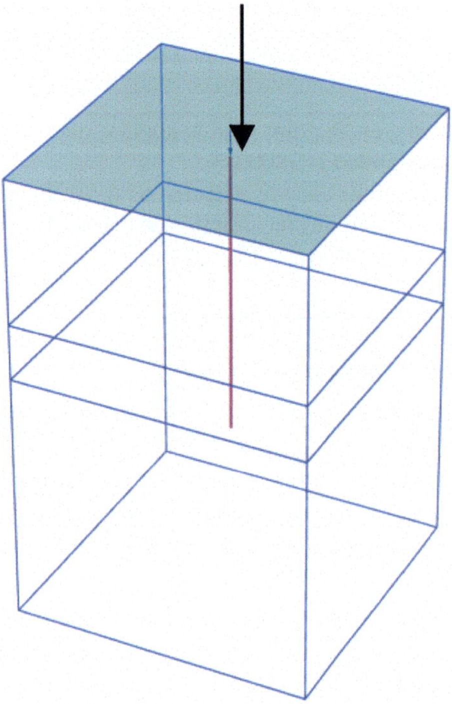

**Fig. 3** Pile centered in the soil

where $s_{u,i}$ is undrained shear strength of interface, $s_{u,\text{soil}}$ is undrained shear strength of soil, $s_{a,I}$ is the adhesion of soil, and $\alpha$ is the adhesion factor of clay.

Equation 6 can be compared to Eq. 7 when the adhesion of soil is equal to adhesion factor multiplied by the undrained shear strength of soil. So, the $R_{\text{inter}}$ resembles the adhesion factor of soil and $s_{u,i}$ resembles the adhesion of interface (mobilized shear strength). So, $R_{\text{inter}}$ which is the adhesion factor of clay is given as input in the PLAXIS and varied for the clayey soil located at a depth of 13 m. Adhesion factor of clayey soil of thickness 5 m suggested by various authors is given in Table 4. So, the $R_{\text{inter}}$ parameter is varied and the differences in the plot of load displacement are observed for each value of adhesion factor ($R_{\text{inter}}$).

## 2.4 Pile Load Test Data

The pile load test conducted at Paradip port which is a natural deep-water port on the East coast of India in Odisha was used for the study purpose. The static load test (SLT) was performed on a cylindrical concrete pile having a length of 25 m and diameter of 1200 mm. The soil parameters based on the geotechnical investigations are given in Table 3.

**Table 3** Bore log details of Paradip

| Depth of soil (m) | Group of soil | SPT | Cohesion (kPa) | Plasticity index (*I*p) |
|---|---|---|---|---|
| 0.5 | SM | – | – | – |
| 1.5 | SM | 6 | – | – |
| 3 | SC | 19 | – | 19 |
| 4.5 | SP-SM | 22 | – | – |
| 6 | SP-SM | 25 | – | – |
| 7.5 | SP | 33 | – | – |
| 9 | SP | 40 | – | – |
| 10.5 | SP | 33 | – | – |
| 12 | SP | 18 | – | – |
| 13.5 | CI | 2 | – | 21 |
| 15 | CI | – | 22 | 22 |

**Table 4** Adhesion factor estimated by various methods for Paradip soil

| Depth (m) | Adhesion factor, $\alpha$ | | | | |
|---|---|---|---|---|---|
| | API [1] | Sladen [13] | Karlsud [9] | Kulhaway and Jackson [11] | Kolk and vander Velde [10] |
| 0–2 | – | – | – | – | – |
| 2–13 | – | – | – | – | – |
| 13–18 | 1 | 1.16 | 1 | 0.27 | 0.52 |
| 18–27 | – | – | – | – | – |

# 3 Results

## 3.1 Analytical Results of Adhesion Factor Using Various Methods

The adhesion factor of clay located at 13 m depth estimated by different empirical approaches is given in Table 4. The adhesion factors suggested by different authors are in range of 0.27–1.16. As various parameters are taken into consideration while estimating adhesion factor, differences in suggested values of α are also high. The adhesion factor for other layers is ignored as the remaining layers are granular soils (non-cohesive soils) for which the skin friction capacity of pile is calculated by the effective stress approach as specified by IS 2911 Section 1.

On the other hand, the skin friction of the soil sample is interpreted from load displacement curve of the plate load test as per the IS 2911 code (Fig. 4). Furthermore, skin friction is calculated as per adhesion factor (Table 4), and the values of it are presented in Fig. 5. The load capacity of pile predicted using the empirical equations is overestimated when compared to the skin friction obtained from pile load test.

## 3.2 Results of Numerical Analysis

In addition to estimation of skin friction capacity, the load displacement curve is predicted using PLAXIS-3D for static axial load applied on the top of pile. The variation in the load displacement curve obtained for $R_{\text{inter}}$ (adhesion factor) suggested by different authors and field load displacement curve is plotted in Fig. 6.

As the $R_{\text{inter}}$ is increased to 1, the load displacement curve is also approached toward the field load displacement curve. The skin friction for the load displacement curve calculated as per IS code 2911 is shown in Fig. 7. The variation in skin friction capacities is very low as adhesion factor of clay is changed and also the skin friction of pile obtained by numerical model by varying the adhesion factor is underestimated wile compared to that obtained from plate load test. The skin friction calculated as per adhesion factor values suggested by API and Karlsud is close to that of field load

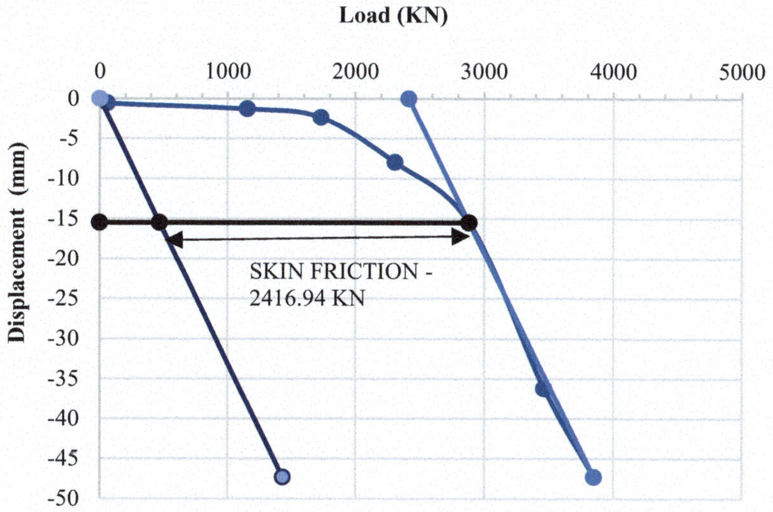


**Fig. 4** Interpretation of skin friction as per IS method for Paradip Soil

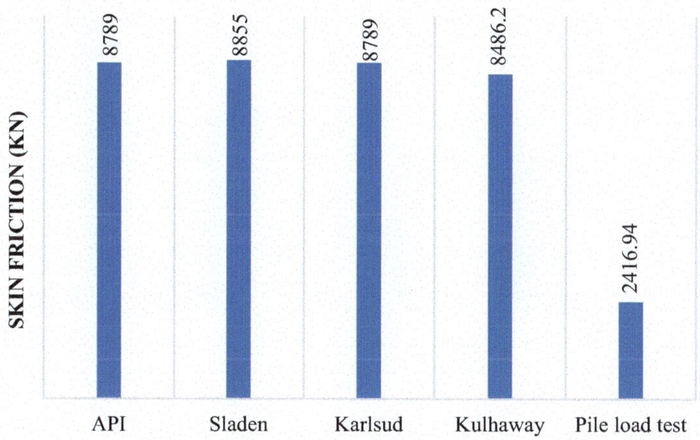


**Fig. 5** Skin friction as per the adhesion factor suggested by different authors

test. So, the adhesion factor estimated by API and Karlsud as 1 can be used for the clayey soil in Paradip situated at a depth of 13.5 m. The settlement of pile in the soil body was comparatively high with those of field settlements as shown in Fig. 8.

Fig. 6 Load displacement curve obtained from numerical modeling and pile load test

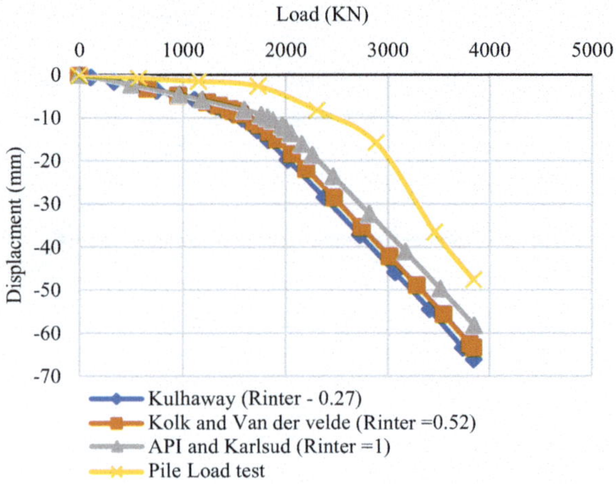


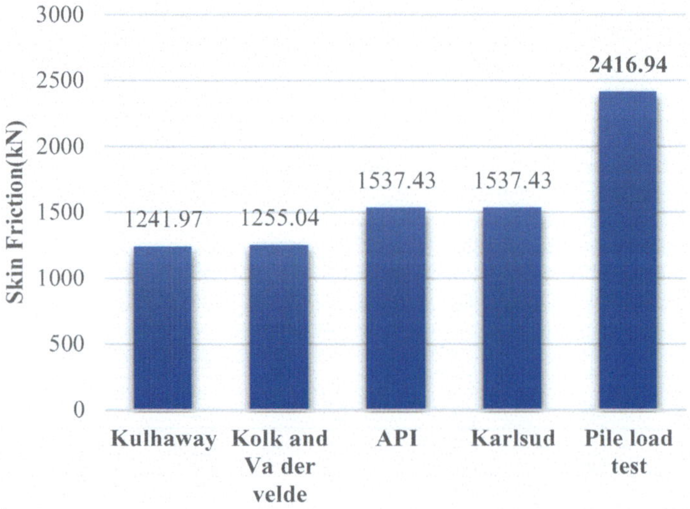


Fig. 7 Skin friction capacity from load displacement curves obtained from numerical modeling as per IS 2911 code

## 4 Conclusions

Based on the numerical studies and analysis of experimental data, arrived at the following conclusions given below:

- The variation in skin friction capacity is noticed to be very low with the change in the adhesion factor values.

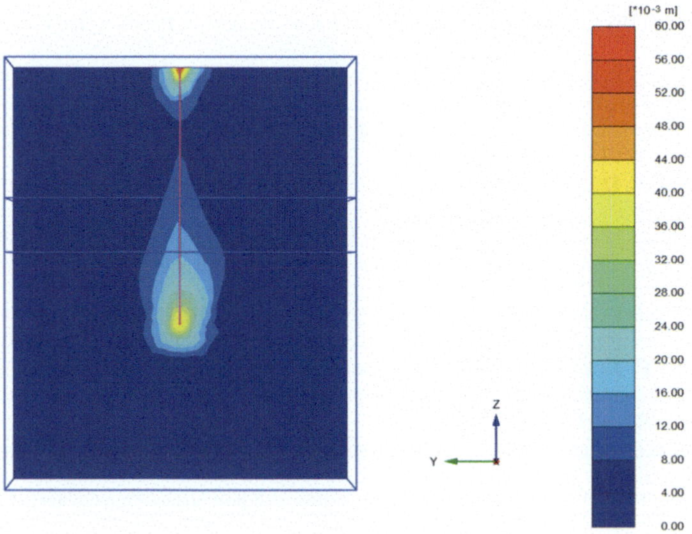


**Fig. 8** Contour of settlement of pile in soil strata

- Settlements obtained by the numerical analysis are high when compared with that of field data.
- It is found that the load settlement curve obtained when adhesion factor of soft clay of unity is close to that of the field load settlement curve.
- Adhesion factors estimated by empirical methods are excessively simplified and cannot validate the reason for the change in adhesion factor for all kinds of soils.
- It is necessary to check the wide range of experimental data to know the reason behind the change in adhesion factor for different soil types and its effect on the calculation of skin friction.
- The skin friction predicted by the numerical approach is lesser when compared with that of experimental test. This may account for the constant stiffness of soil assumed in model and stiffness does not as the soil undergoes settlement.

## References

1. American Petroleum Institute (1984) API recommended practice for planning, designing and constructing fixed off-shore platforms. API, Washington, DC
2. Arifin YF (2022) Uji Plagiasi (Turnitin) Adhesion factor of large diameter bored Piles in soft to stiff clay in the Sei Alalak bridge replacement project–South Kalimantan, Indonesia
3. Banerjee R, Bandyopadhyay S, Sengupta A, Reddy GR (2022) Settlement behaviour of a pile raft subjected to vertical loadings in multilayered soil. Geomech Geoengineering 17(1):282–296
4. Cherubini C, Vessia G (2007) Reliability approach for the side resistance of piles by means of the total stress analysis (α method). Can Geotech J 44(11):1378–1390

5. Coduto DP (1989) Foundation design, principles, and practices. Prentice Hall Inc., Eaglewood Cliffs, N.Y
6. Ezzat M, Zaghloul Y, Sorour T, Hefny A, Eid M (2019) Numerical simulation of axially loaded to failure large diameter bored pile. Int J Geotech Geol Eng 13(5):283–297
7. IS 2911 (Part 1/Sec1): 2010 Indian Standard design and construction of pile foundations—code of practice part 1 concrete piles
8. IS 2911 (Part 4): 2013 Design and construction of pile foundations—code of practice part 4 load test on piles
9. Karlsrud K, Clausen CJF, Aas PM (2005) Bearing capacity of driven piles in clay, the NGI Approach. In: Proceedings of international symposium on frontiers in offshore geotechnics, vol 1. Perth, pp 775–782
10. Kolk HJ, Van der Velde E (1996) A reliable method to determine friction capacity of pile driven into clays. In: Proceedings of the 28th offshore technology conference, pp 337–346
11. Kulhawy FH, Jackson CS (1989) Some observations on un-drained side resistance of drilled shafts. In: Foundation engineering: current principles and practices congress Evanston Illinois, vol 2, pp 1011–1025
12. PLAXIS 3D—version 9.02 (2018) Reference manual, Delft University of Technology, Delft, Netherlands
13. Sladen JA (1992) The adhesion factor: applications and limitations. Can Geotech J 29(2):322–326
14. Wrana B (2015) Pile load capacity–calculation methods. Stud Geotech Mech 37(4):83–93

# V–H Capacity Envelopes of Strip Footings on Cohesionless Soil Overlying Soft Rock Mass

S. K. Kumawat, D. Raj, and S. Dangayach

## 1 Introduction

Shallow foundations are widely used in practice, where the shear strength of upper strata is high and capable enough to support the associated structures. Many literature [1–3] and standards [4, 5] are available to estimate the vertical bearing capacity of shallow foundations resting on flat homogeneous ground. However, in the case of offshore structures subjected to wind or seismic loading, the foundations are often subjected to interactive vertical/axial force (V) and shear force (H). The effect of the combined loading can be dealt by including the effect of horizontal shear as the load inclination factor and moment as the effective width of the foundation [4, 5]. To deal with the shallow foundation subjected to the combined interactive loading, an alternate approach, the capacity envelope is used primarily for offshore structures. In the past, several researchers have estimated the normalized V–H capacity envelope for shallow foundation, using analytical solutions and developed empirical equations, resting either on homogeneous cohesive soil [6–14] or homogeneous cohesionless soil [15–24] or cohesive-frictional soil [25, 26]. From the literature review, it has been found that the maximum shear capacity of foundation was achieved at vertical load, varying between 0.4 and 0.6 times of vertical capacity. In some cases, the foundation is constructed on a layered soil system with varying strengths along the depth, supported by an overlying rock mass [27]. An ample of literature is available on estimation of the bearing capacity of shallow foundation resting on rock mass [28–38], using analytical and numerical methods. Recently, Das and Chakraborty [27] estimated the bearing capacity of strip footing resting on cohesionless soil overlying rock mass. In the author's capacity, the estimation of V–H capacity envelope for strip

S. K. Kumawat · D. Raj (✉) · S. Dangayach
Department of Civil Engineering, MNIT Jaipur, Jaipur, Rajasthan 302017, India
e-mail: dhiraj.ce@mnit.ac.in

 
B. T. Jose et al. (eds.), *Proceedings of the Indian Geotechnical Conference 2022 Volume 2*, Lecture Notes in Civil Engineering 477,
https://doi.org/10.1007/978-981-97-1741-5_21

footing resting on cohesionless soil overlying soft rock mass has not been studied yet.

In this article, a numerical study has been conducted for the estimation of capacity envelope of the strip footing, located on the surface of cohesionless soil overlying soft rock, subjected to V–H combined loading using lower bound finite-element limit analysis (LB-FELA). The maximum vertical bearing capacity ($V_0$) of the considered strip footing located on the surface of cohesionless soil overlying soft rock mass has been estimated. The results are presented in the form of V–H capacity envelope varying with soil strength parameters and soil-rock thickness ratios ($T_s/B$, ratio of thickness of the cohesionless soil layer to the width of strip foundation.). The variation in bearing capacity ratio ($B_{cr}$, ratio of bearing capacity of cohesionless soil to the bearing capacity of cohesionless soil overlying rock mass) with $T_s/B$ has also been presented, herein.

## 2 Problem Statement

To develop V–H capacity envelope, a rough and rigid strip footing with width, $B$, kept on the top surface of the homogeneous cohesionless soil with varying internal friction angle ($\phi$) has been considered in this study, as shown in Fig. 1.

In Fig. 1, the thickness of top soil layer consists of cohesionless soil following Mohr–Coulomb yield criteria, which has been considered as $T_s$, whereas the thickness of bottom layer consists of soft rock mass having Hoek–Brown parameters

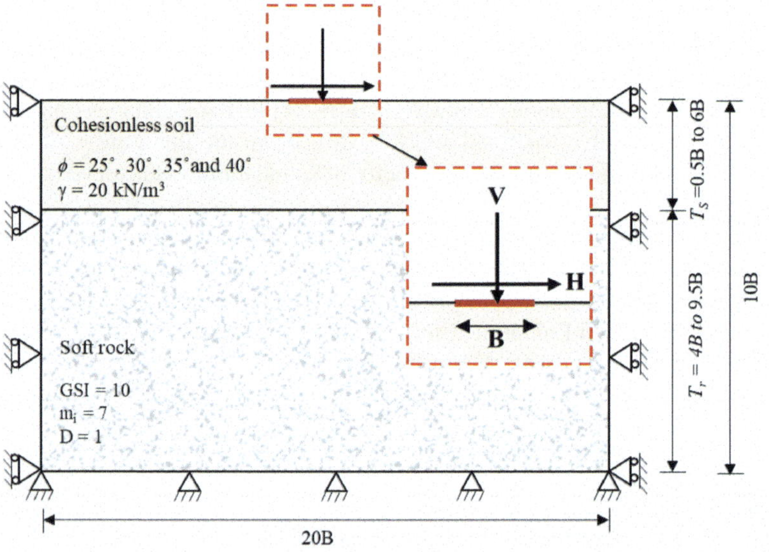


**Fig. 1** Problem geometry of strip footing placed on cohesionless soil overlying soft rock

**Table 1** Material properties

| Properties | Cohesionless soil | Rock |
|---|---|---|
| Unit weight ($\gamma$ kN/m$^3$) | 20 | 0 |
| Poisson's ratio ($\nu$) | 0.3 | 0.3 |
| Angle of internal friction ($\phi$) | 25°, 30°, 35°, 40° | – |
| Geological strength index (GSI) | – | 10 |
| $m_i$ | – | 7 |
| $D$ | – | 1 |

such as geological strength index (GSI), Hoek–Brown yield parameter ($m_i$), and disturbance factor ($D$), which has been considered as $T_r$. Table 1 shows the material properties used in the present study.

Based on the sensitivity analysis, the width and depth of the FE model have been kept 20B and 10B, respectively.

The effect of the varying soil-rock thickness ratios, $T_s/B$ ($T_s/B = 0.5$, 1, 1.5, 2, 3, 4, 5, and 6), on the vertical bearing capacity and V–H capacity envelope has been explored here.

## 3 FE Modeling and Analysis

In the present study, lower bound finite-element limit analysis (LB-FELA) has been employed to evaluate the maximum vertical bearing capacity and V–H capacity envelope of strip footing using OptumG2 [39] software. A 2D plain strain FE model has been developed for a strip foundation lying on top of flat two-layered soil-rock system. Among these two layers, the top layer consists cohesionless soil obeying Mohr–Coulomb failure criterion with associated flow rule, whereas the bottom layer consists of soft rock following Generalized Hoek–Brown failure criterion [40] with associated flow rule. The strip footing has been considered as weightless and modeled using rigid plate. Based on the sensitivity analysis, the width and depth of the FE model have been kept 20B and 10B, respectively. In this study, the value of $T_s$ and $T_r$ has been considered varying from 0.5B to 6B and from 4B to 9.5B, respectively, as considered in past study [27]. In this study, as a special case $T_s = 10\text{B}$ or $T_r = 0$ represents the pure cohesionless soil.

The interface element (with reduction factor, $R = 1$) has been used between the foundation and soil layer. Fan meshes have been used at both ends of the footing to handle stress concentration or singularity issue, The FE model consists of soil and rock mass which has been discretized with three-noded triangular plane-strain elements. The number of elements utilized in each analysis has been varied from 8000 to 10,000 with three adaptive iterations. The force-based swipe analysis [25, 41] has been performed to obtain the limit load and to develop entire V–H capacity envelope, where V and H were applied at the mid-point of the strip footing in a fixed

(H/V) ratio varying between 0.90 and 40. All the movements were restricted at the base of FE model, whereas only horizontal displacement was restricted at the lateral boundaries.

The normalized vertical load, $\overset{\bullet}{V}$, and horizontal load, $\overset{\bullet}{H}$, have been obtained as:

$$\overset{\bullet}{V} = \frac{V}{V_o}, \tag{1}$$

$$\overset{\bullet}{H} = \frac{H}{V_o}. \tag{2}$$

## 4 Results and Discussions

In this section, the results of the present study have been discussed in terms of, (a) variation in bearing capacity ratio ($B_{cr}$) with $T_s/B$ and (b) the capacity envelope varying with soil strength and $T_s/B$. Figure 2 shows the variation in bearing capacity ratio ($B_{cr}$) with $T_s/B$. It is interesting to note that with the increase in the value of $T_s/B$ for a particular soil having constant value of $\phi$, the value of $B_{cr}$ increases and becomes constant (= 1) at higher value of $T_s/B$. It can also be observed that, at a particular value of $T_s/B$, with the increase in the value of $\phi$, the value of $B_{cr}$ decreases.

Figures 3a–h present the variation in the shape of normalized capacity envelope ($H/V_0$ with $V/V_0$) with varying $\phi$ and $T_s/B$ values. It can be observed from Fig. 3(a)

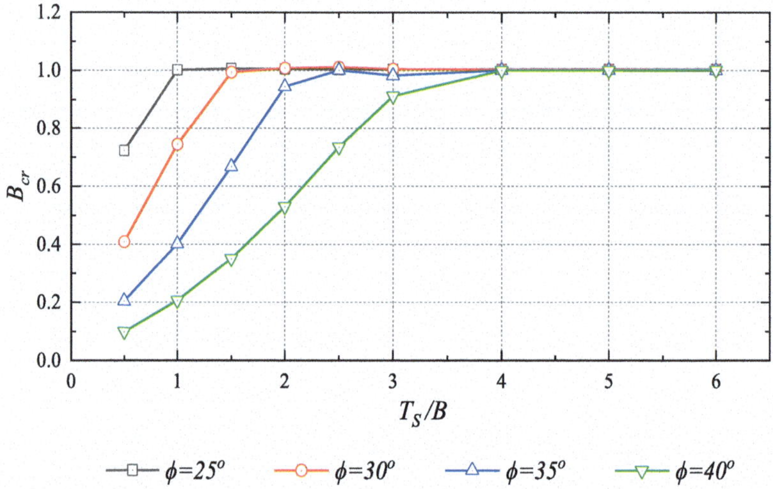


**Fig. 2** Variation in bearing capacity ratio ($B_{cr}$) with soil-rock thickness ratios ($T_s/B$)

that the effect of varying $\phi$ value has no significant influence on the normalized capacity envelope for strip footing placed at top of pure cohesionless soil [21].

It is interesting to note that, the $T_s/B$ value (ranging from 0.5 to 3.0) has significant influence on the normalized capacity envelope for a particular soil with higher value of $\phi$. Further, the normalized capacity envelope was found unaffected by $T_s/B$ value beyond 4.0, for a soil with any value of $\phi$. It can also be observed that the maximum value of normalized shear capacity ($H/V_0$) for soil-rock system was found 1.7 times that of the pure cohesionless soil with $\phi = 40°$, at $T_s/B = 1.0$.

Figures 4a–j show the failure surface represented by shear dissipation profile for foundation resting on soil (with $\phi = 30°$ and 40°) overlying soft rock system with different values of $T_s/B$ and subjected to combined V–H loading. It can be noted that the failure surface is always asymmetric or one-sided for all the considered cases. It is also found that the shape and depth of the failure surface are dependent on the both $T_s/B$ and $\phi$ values.

## 5 Conclusions

The V–H capacity envelope for strip footing placed on cohesionless soil overlying soft rock has been developed by performing LB-FELA using OptumG2. The effect of $T_s/B$ and $\phi$ on bearing capacity ratio ($B_{cr}$), shape of capacity envelope, and corresponding failure surface of soil-rock system has been explored here in details. It was noted that, with the increase in $T_s/B$ value for a particular soil, the value of $B_{cr}$ increases and then becomes constant (= 1) at higher value of $T_s/B$. It was also observed that at a particular $T_s/B$ value, the $B_{cr}$ decreases with the increase in $\phi$. For strip footing place on top of two-layered soil-rock system, a significant variation in capacity envelope has been seen with increasing $\phi$ value and $T_s/B$ ranging between 0.5 and 3.0, whereas for $T_s/B$ greater than 4.0, the variation in capacity envelope is insignificant and became independent of both $T_s/B$ and $\phi$ values. The present analysis has been performed by assuming the soil and rock which consist of homogenous material and limit loads are obtained using LB-FELA; hence, the results are restricted to the investigated case only. For more realistic understanding, a comprehensive study considering material shear strength varying spatially is required.

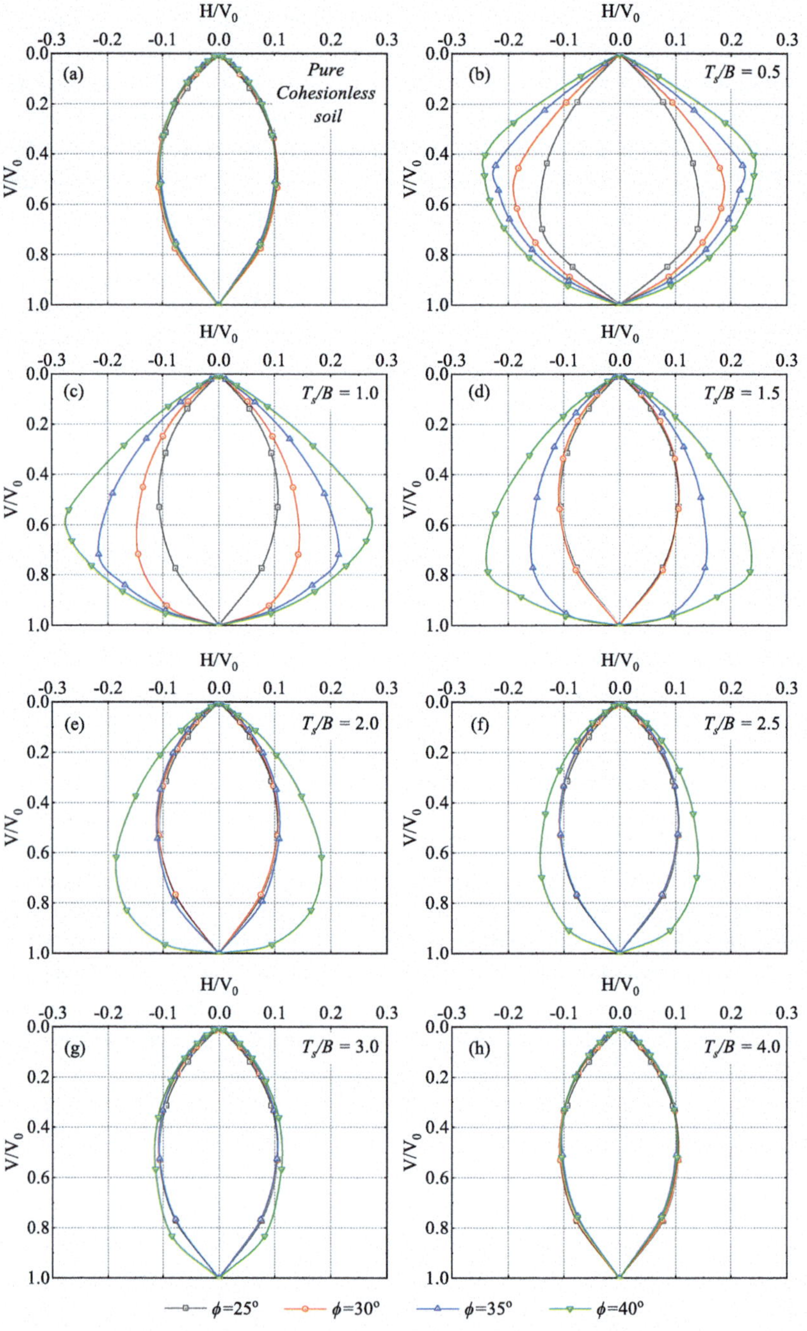


**Fig. 3** Comparison of V–H capacity envelope for strip footing placed on cohesionless soil having different $\phi$ values and overlying soft rock mass with: **a** $T_s/B = 10$ (pure cohesionless soil); **b** $T_s/B = 0.5$; **c** $T_s/B = 1.0$; **d** $T_s/B = 1.5$; **e** $T_s/B = 2.0$; **f** $T_s/B = 2.5$; **g** $T_s/B = 3.0$; and **h** $T_s/B = 4.0$

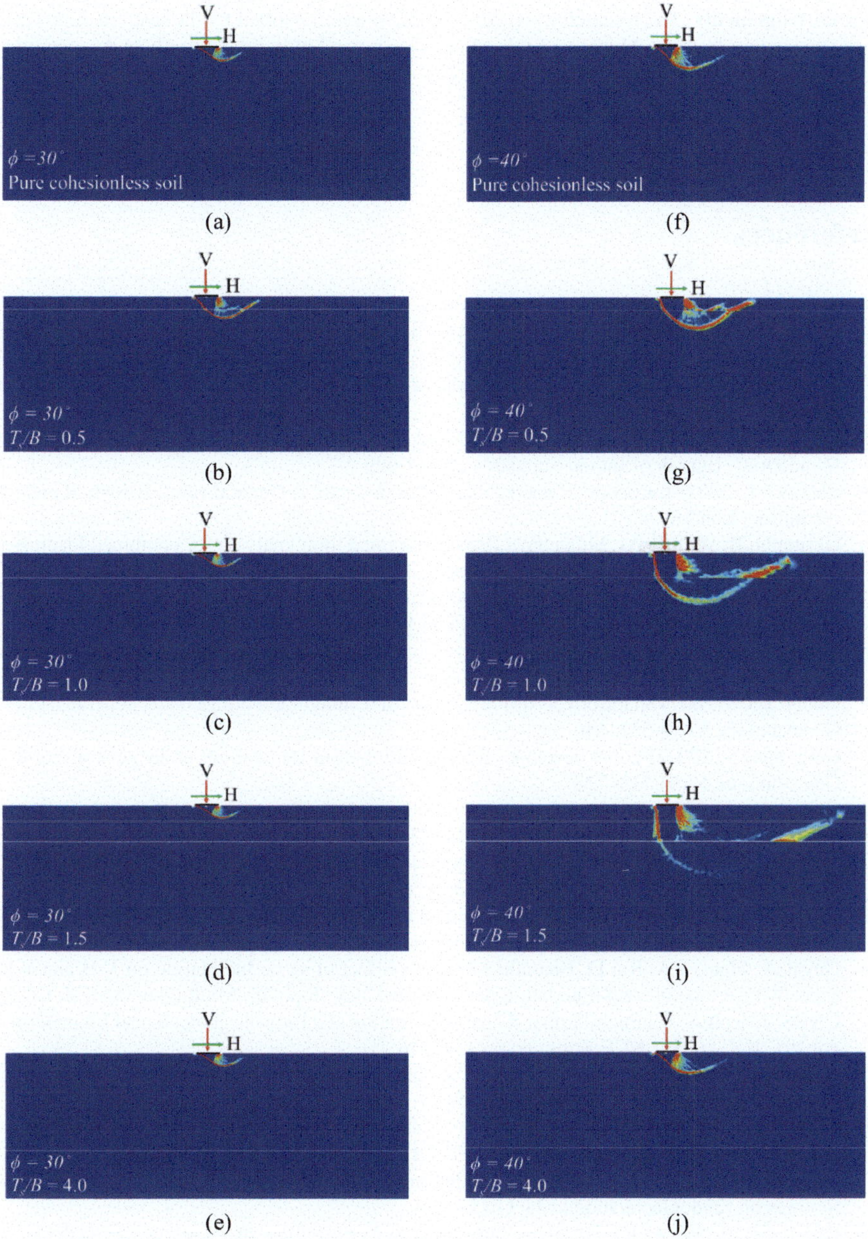


**Fig. 4** Failure surface represented by shear dissipation profile for the strip footing subjected to interactive V–H loading and located on cohesionless soil overlying soft rock with: **a** $\phi = 30°$, $T_s/B = 10$; **b** $\phi = 30°$, $T_s/B = 0.5$; **c** $\phi = 30°$, $T_s/B = 1.0$; **d** $\phi = 30°$, $T_s/B = 1.5$; **e** $\phi = 30°$, $T_s/B = 4.0$; **f** $\phi = 40°$, $T_s/B = 10$; **g** $\phi = 40°$, $T_s/B = 0.5$; **h** $\phi = 40°$, $T_s/B = 1.0$; **i** $\phi = 40°$, $T_s/B = 1.5$; and **j** $\phi = 40°$, $T_s/B = 4.0$

**Acknowledgements** The research work presented here was supported by the Institute fellowship to the first author from the Ministry of Education, Government of India. The authors are grateful to 'Optum Computational Engineering' (OptumCE) for providing free academic license of OptumG2 software.

## References

1. Terzaghi K (1943) Theoretical soil mechanics. Wiley, New York, USA
2. Meyerhof GG (1963) Some recent research on the bearing capacity of foundations. Can Geotech J 1(1):16–26
3. Vesić AS (1973) Analysis of ultimate loads of shallow foundations. J Soil Mech Found Div 99(1):45–73
4. BIS (2002) IS6403: Code of practice for determination of bearing capacity of shallow foundations. Bureau of Indian Standards, New Delhi
5. EN1997-1 (2004) Eurocode 7: geotechnical design—part 1: General rules. British Standards Institution, London
6. Ukritchon B, Whittle AJ, Sloan SW (1998) Undrained limit analyses for combined loading of strip footings on clay. J Geotech Geoenviron Eng 124(3):265–276
7. Bransby MF (2001) Failure envelopes and plastic potentials for eccentrically loaded surface footings on undrained soil. Int J Numer Anal Methods Geomech 25(4):329–346
8. Yun G, Bransby MF (2007) The horizontal-moment capacity of embedded foundations in undrained soil. Can Geotech J 44(4):409–424
9. Gourvenec S (2007) Shape effects on the capacity of rectangular footings under general loading. Géotechnique 57(8):637–646
10. Gourvenec S (2007) Failure envelopes for offshore shallow foundations under general loading. Géotechnique 57(9):715–728
11. Gourvenec S (2008) Effect of embedment on the undrained capacity of shallow foundations under general loading. Géotechnique 58(3):177–185
12. Shen Z, Feng X, Gourvenec S (2016) Undrained capacity of surface foundations with zero-tension interface under planar V-H-M loading. Comput Geotech 73:47–57
13. Xiao Z, Tian Y, Gourvenec S (2016) A practical method to evaluate failure envelopes of shallow foundations considering soil strain softening and rate effects. Appl Ocean Res 59:395–407
14. Goyal A, Maurya A, Raj D, Bharathi M (2022) Effect of slope Inclination on V-H and V-M capacity envelope of strip foundation on undrained clay slope. In: Muthukkumaran K, Reddy CNVS, Joseph A, Senthamilkumar S (eds) Foundation and forensic geotechnical Engineering, vol 295. IGC 2021. LNCE Springer, pp 139–146
15. Gottardi G, Butterfield R (1993) On the bearing capacity of surface footings on sand under general planar loads. Soils Found 33(3):68–79
16. Gottardi G, Butterfield R (1995) The displacement of a model rigid surface footing on dense sand under general planar loading. Soils Found 35(3):71–82
17. Butterfield R, Gottardi G (1994) A complete three-dimensional failure envelope for shallow footings on sand. Géotechnique 44(1):181–184
18. Paolucci R, Pecker A (1997) Seismic bearing capacity of shallow strip foundations on dry soils. Soils Found 37(3):95–105
19. Gottardi G, Houlsby GT, Butterfield R (1999) Plastic response of circular footings on sand under general planar loading. Géotechnique 49(4):453–469
20. Houlsby GT, Cassidy MJ (2002) A plasticity model for the behaviour of footings on sand under combined loading. Géotechnique 55(2):117–129
21. Loukidis D, Chakraborty T, Salgado R (2008) Bearing capacity of strip footings on purely frictional soil under eccentric and inclined loads. Can Geotech J 45(6):768–787

22. Krabbenhoft S, Damkilde L, Krabbenhoft K (2012) Lower-bound calculations of the bearing capacity of eccentrically loaded footings in cohesionless soil. Can Geotech J 49(3):298–310
23. Kim D, Youn J, Jee S, Choi J, Lee J, Choo Y (2014) Numerical studies on combined VH loading and inclination factor of circular footings on sand. J Korean Geotechical Soc 30(3):29–46
24. Tang C, Phoon K-K, Toh K-C (2015) Effect of footing width on Nγ and failure envelope of eccentrically and obliquely loaded strip footings on sand. Can Geotech J 52(6):694–707
25. Raj D, Singh Y, Kaynia AM (2019) Behavior and critical failure modes of strip foundations on slopes under seismic and structural loading. Int J Geomech 19(6):04019047
26. Raj D, Singh Y, Kaynia AM (2019) V-H–M seismic capacity envelopes of strip foundations on slopes for capacity design of structure-foundation system. Bull Earthq Eng 17(6):2963–2987
27. Das S, Chakraborty D (2022) Effect of soil and rock interface friction on the bearing capacity of strip footing placed on soil overlying Hoek-Brown rock Mass. Int J Geomech 22(1):04021257
28. Kulhawy FH, Carter JP (1992) Settlement and bearing capacity of foundations on rock masses. In: Engineering in rock masses. Elsevier, pp 231–245
29. Yang XL (2009) Seismic bearing capacity of a strip footing on rock slopes. Can Geotech J 46(8):943–954
30. Chakraborty M, Kumar J (2015) Bearing capacity of circular footings over rock mass by using axisymmetric quasi lower bound finite element limit analysis. Comput Geotech 70:138–149
31. Saada Z, Maghous S, Garnier D (2008) Bearing capacity of shallow foundations on rocks obeying a modified Hoek-Brown failure criterion. Comput Geotech 35(2):144–154
32. Beygi M, Keshavarz A, Abbaspour M, Vali R, Saberian M, Li J (2022) Finite element limit analysis of the seismic bearing capacity of strip footing adjacent to excavation in c-φ soil. Geomech Geoengineering 17(1):246–259
33. Lyamin AV, Sloan SW (2002) Lower bound limit analysis using non-linear programming. Int J Numer Meth Eng 55(5):573–611
34. Merifield RS, Lyamin AV, Sloan SW (2006) Limit analysis solutions for the bearing capacity of rock masses using the generalised Hoek-Brown criterion. Int J Rock Mech Min Sci 43(6):920–937
35. Jaiswal S, Chauhan VB (2021) Ultimate bearing capacity of strip footing resting on rock mass using adaptive finite element method. J King Saud Univ Eng Sci. https://doi.org/10.1016/j.jksues.2021.09.004
36. Rahaman O, Kumar J (2022) Seismic bearing capacity of a strip footing on rock media. J Rock Mech Geotech Eng 14(2):560–575
37. Prakash S, Raj D, Singh Y (2021) Foundation bearing capacity estimation on rock-mass using Hoek–Brown failure criterion and equivalent Mohr–Coulomb parameters. In: Patel S, Solanki CH, Reddy KR, Shukla SK (eds) Proceedings of the Indian geotechnical conference 2019, vol 133. LNCE Springer, pp 213–224
38. Yang X-L, Yin J-H (2005) Upper bound solution for ultimate bearing capacity with a modified Hoek-Brown failure criterion. Int J Rock Mech Min Sci 42(4):550–560
39. Krabbenhoft K, Lyamin A, Krabbenhoft J (2016) OptumG2: theory. Optum computational engineering. Available on: https://optumce.com/wp-content/uploads/2016/05/Theory.pdf
40. Hoek E, Brown ET (1980) Empirical strength criterion for rock masses. J Geotech Eng Div 106(9):1013–1035
41. Raj D, Singh Y, Shukla SK (2018) Seismic bearing capacity of strip foundation embedded in c-ϕ soil slope. Int J Geomech 18(7):04018076

# Numerical Analysis of Skirted Foundation on Sand for Load–Displacement Behaviour

**Suresh Prasad Singh, Sumit Kumar, and Sumit Sarkar**

## 1 Introduction

Skirted foundations are shallow foundations in which the footing is upheld through vertical plates or skirts. It's one of the more innovative ways to improve the performance of shallow foundations by adding a skirt. The soil around the foundation performs a crucial role in the behaviour of the foundation. The skirted foundation is a foundation type where a vertical or slanted fence encompasses one or more borders of the soil below the footing. The skirts provide a walled-in area where the soil is monitored and filled in as a unit to transfer superstructure load to the soil at the skirt tip, resulting in an increase in bearing limit and a decrease in building settlement. Geotechnical experts are looking for an alternative technique to increase the bearing limit and reduce foundation settlement. While many methodologies of soil adjustment are highly recognised, they can be expensive and are restricted for few site conditions. In few circumstances it is difficult to adapt those methods[1]. For those cases, a skirt foundation might be an effective solution for increasing the bearing limit and reducing foundation settlement. Footing laid on the ground having skirt walls at the boundaries to restrict the soil below it. This detention of soil additionally controls the tilt and accordingly expands the bearing limit of the foundation. In this way, a substitute methodology is expected to enhance the bearing limit of the footing by utilising a structural skirt fixed at the boundary. This improvement method does not require soil excavation and is not limited by the presence of a high water table.

S. P. Singh (✉) · S. Kumar · S. Sarkar
Department of Civil Engineering, National Institute of Technology, Rourkela 769008, India
e-mail: spsingh@nitrkl.ac.in

S. Kumar
e-mail: 220ce1014@nitrkl.ac.in

S. Sarkar
e-mail: 221ce1184@nitrkl.ac.in

 
B. T. Jose et al. (eds.), *Proceedings of the Indian Geotechnical Conference 2022 Volume 2*, Lecture Notes in Civil Engineering 477,
https://doi.org/10.1007/978-981-97-1741-5_22

The vertical and horizontal load capacity of footing is improved by confinement and restricted to sliding. Subsequently, it could transform into a substitute approach for enhancing the bearing capacity of footing. This numerical analysis has been done to sort out the approach to know the foundation's performance with two sides, vertical and inclined skirt.

Skirted foundations can be a low-cost option for structures built on granular soils with high water tables, such as those found in many waterfront and nearby shore developments. Despite this, research into the performance of skirted shallow foundations resting on soils and carrying vertical loads is quite limited. This study aims to determine the impacts of lateral sand confinement by comparing the performance of skirted foundations to those of surface foundations. As a result, bearing capacity and settlement extents were used to illustrate such consequences. This should assist in forecasting how skirted foundations will behave under structural loading.

This work aims to innovate a strategy to increase the bearing limit and diminish the settlement of the foundation laying on the soil by skirt establishment.

## 2 Literature Review

Bell [2] makes sense that shallow seaward establishments gain their strength by having the establishment situated on the seabed. They can be idealised as huge round establishments exposed to vertical, horizontal, and transient loads. Bransby and Randolph [3] demonstrated that soil footing geometry and strength profiles affect longitudinal and vertical limits using finite element and plastic analysis. Yun and Bransby [4] compared the load–displacement reaction from centrifuge test data to the finite element result of a circular skirt footing with varying roughness and skirt profundity up to five times the diameter of the base. It was discovered that compared to a typical foundation, a skirted foundation increases the horizontal burden limit by roughly 3–4 times. Alaghbari and Mohamedzein [5] organised several foundation model experiments to investigate the factors that influence skirted foundation behaviour. Elements incorporate foundation friction, skirt profundity, skirt lateral roughness, skirt solidness, and soil compaction. Because of these boundaries, they propose a condition for skirted nails. Alaghbari [6] led a progression of tests on the circular establishment with and without skirt structure in an enormous tank to concentrate on the settlement of round establishment in the sand. These trial results show that primary layers diminish subfloor settlement as a function of applied pressure and skirt profundity and change in foundation settlement behaviour. Yun and Bransby [7] used analytical and experimental methods to demonstrate the vertical bearing limit of skirted foundations on typically consolidated undrained soils. They show that the boundary of the skirted foundation in the influence of vertical load normally conveys as though it were a strong groundwork with an embedment profundity equivalent to the depth of the skirt. Kudsk et al. [8] used ordinary and mathematical finite element soil association to show the world's largest jack-up, skirted footing, and changing stresses on stratified soil. An envelope for offshore foundations with similar soil characteristics was provided based on the results. Yun and Bransby [9] demonstrated how deformity of the soil between the external skirts could result in an essentially lower bearing limit of the establishment than a comparable inserted strong groundwork. The particular geometry of the establishment should be considered during the design. Wakil [10] conducted 12 burden tests on small-sized rounded baseboards subjected to horizontal loads. From lab testing, the skirts were viewed as ready to change the failure method of the shallow pins from sliding to a turning system. Skirts appended to the foundation of the nail essentially expanded the transverse capacity of the shallow nail. Eid [11] used a small-scale physical model to conduct experiments. The bearing limit and settling of the foundation were evaluated with footing size, sand shear strength, and skirt depth. The findings reveal that the wall footing has a similar load-bearing limit and settling to, but is not identical to, a pile foundation of equal width and profundity. The findings show that the consolidation reaction is influenced by the roughness of the skirt-soil contact and the embedment ratio. The results of Pusadkar and Bhatkar [12] reveal that using a skirt surface for the foundation significantly impacts working on the bearing limit. This improvement increases as the skirt depth increase with varied raft sizes. Mana et al. [13] discussed the failure mechanism of skirted foundations subjected to uplift and compression loadings.

## 3 Methodology

Although several well-known and well-developed soil stabilisation methods may be prohibitively expensive and constrained by site factors. In some circumstances, applying them to existing foundations can be difficult. This method of enhancing bearing capacity requires no soil excavation and is not restricted by a high groundwater table. The numerical analysis is used to investigate the load-carrying carrying capacity of foundations with different skirt ratios. The vertical skirts are supposed to increase foundation capacity by trapping soil beneath the raft and between the skirts and transferring applied weight to the ground at the skirt points. The shear failure of the soil beneath the shallow footings is the leading cause of failure. Such failures can be avoided by enclosing the earth beneath the footing in some form of enclosure. Skirts with foundations operate as a soil plug to transfer super-structural weight to the earth and form an enclosure where the soil is tightly confined. In this analysis, two parameters were evaluated. The bearing capacity for a constant settlement comes first, followed by the settlement for a constant bearing pressure. The performance of the skirted raft footing is investigated using Mohr–Coulomb models for various skirt depths and footing sizes. The bearing capacities of foundations are determined through numerical analysis of skirted foundations utilising numerous skirts in this work.

## 4 Numerical Analysis

A comprehensive 2D finite element analysis is used to study the effects of skirts on the behaviour of axially strained surface foundations resting on the sand. Steel sheets with a thickness of 50 mm were used to simulate the skirts in the analysis. This thickness was chosen to match the rigidity (moment of inertia) of commercially available steel sheet-pile sections used for foundation skirting. Fifteen-node planar strain elements are adopted to represent sand, footing, and skirts. To lessen the boundary effect, the distance between the finite element mesh boundary and the foundation edges was set to 15B (where B is the width of the footing). Footing and skirts are made of linear elastic fabrics. Sand behaviour is represented using the Mohr–Coulomb model. The stress-settlement linkages have no evident break-point or peak. As a result, the ultimate bearing capacity was determined in this numerical study as the stress corresponding to a settlement. The numerical research revealed considerable differences in the foundation types' behaviour when comparing the stress-settlement correlations. These adjustments were mirrored in the overall displacement diagrams generated from the numerical investigation by applying the same applied stress to the surface and skirting foundations. The saturated unit weight taken for soil is 20 kN/m$^3$, while for $Y_{unsat}$ is 17 kN/m$^3$. While the value of cohesion is zero, the friction angle has the value of 35° for the analysis. Using comprehensive finite element analysis, the impacts of skirts on the behaviour of a uniformly loaded surface foundation on the sand are explored. The geometry of the finite element soil model employed in this study is 15B × 20B, with B = 10 unit to 20 unit raft sizes and Ds (skirt depth) of 0.5B, 1B, 1.5B, 2B, 2.5B, and 3B skirt depths. The soil employed for the analysis was essentially sandy cohesionless soil. Fifteen-node triangular elements are used to depict the behaviour of sand in the Mohr–Coulomb model. The skirts and the footing were both considered linear elastic materials (Fig. 1).

The total stress variation as the load for specific settlement to the structure is shown in Fig. 2 above. The figure shows the variation in stress near the provided skirt. It shows the variation of stress across the boundary and along the depth. The change in settlement and load-carrying capacity can be seen in the above image. The variation for skirt ratio from 0 to 3 is utilised to build the model in the same manner

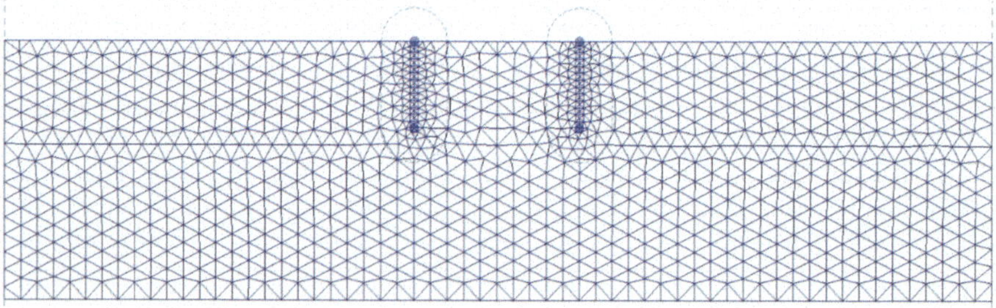

**Fig. 1** Generated mesh of the model

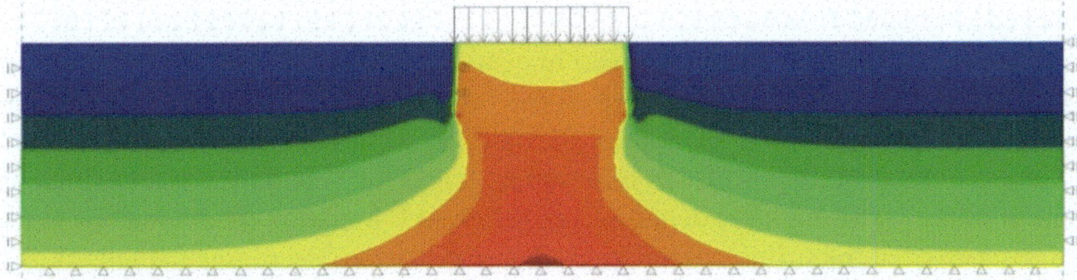

**Fig. 2** Represents the variation of stress across the skirted footing

for all the models, and the value of the respective settlement is noted down. The variation in settlement with various skirt ratios and the corresponding differential change is noticed. The results are analysed, and the discussion is processed based on the change that is seen for different skirt ratios and varied skirt orientations.

## 5 Results and Discussions

The foundations of 10-unit, 15-unit, and 20-unit widths with different skirt depths for the vertical skirt are analysed. This study generated six numerical models by altering the skirt's depth below the ground while keeping the foundation width constant. The skirt ratio is adjusted in 0.5 increments from zero to three. As the depth of the skirt increases, there is a change in settling and bearing capacity (Fig. 3).

Table 1 shows how the length of the skirt beneath the soil affects the settlement of a foundation with a skirt. The breadth of the foundation is kept constant, while the depth of the skirt varies. Models (a) to (g) depict seven different foundation models with skirts. Based on the findings, the footing accomplishes less settlement by increasing the length of the skirt in comparison with the unskirted surface foundation. By keeping the foundation width constant and adjusting the length of the skirt that has been given around the foundation edges in a vertical direction, it has been observed that by increasing the length of the skirt, a lower value of the settlement is

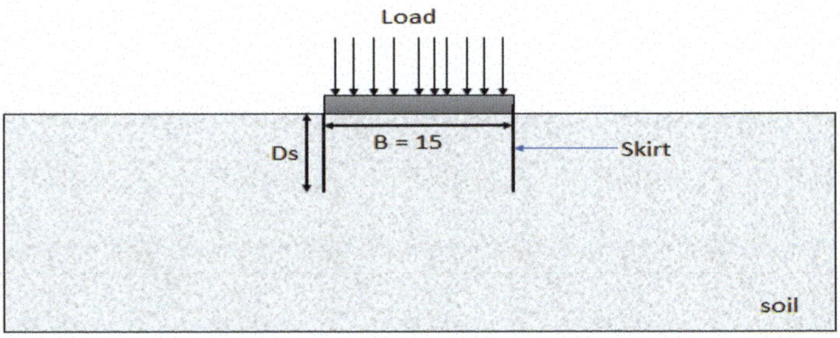


**Fig. 3** Geometry model of foundation with vertical skirt having the foundation of 15 m width

achieved compared to a surface foundation. As the skirt ratio varies from 0.5 to 3 at 0.5 intervals, a percentage improvement in bearing capacity of roughly 12 to 50% is noticed compared to the surface foundation. The load-carrying capacity of a foundation with a skirt varies as the length of the skirt beneath the soil increases, as given in Table 2. Only the depth of the skirt varies, while the foundation width remains fixed. Models (a) through (g) denote seven different foundation models with a skirt. It is observed that the surface foundation's bearing capacity increases with the skirt's length. The bar chart in Fig. 4 depicts the percentage improvement in foundation-bearing capacity due to the addition of the skirt. By gradually increasing the skirt ratio, it is seen that there is a significant improvement in bearing capacity. In this analysis, the variation for different permissible settlements is also observed, which results in a near-identical percentage difference in foundation carrying capacity. By increasing the skirt ratio from 0 to 3, the bearing capacity is increased by 50%. It means that when the length of the skirt increases, the bearing capacity also enhances. The number of times bearing capacity increased over the unskirted surface foundation (bearing capacity ratio) was calculated using the preceding graph plotted between *Qsk*/*Qsu* Vs skirt ratio (Fig. 5). It is discovered that the bearing capacity is improved by approximately 1.5 times. When a similar analysis is performed for $B = 10$ unit, 15 unit and 20 unit, it is observed that keeping the skirt away from the centre results in a drop in bearing capacity. As a result, when the value of B is low, that is, low footing width, the value of bearing capacity improves significantly more than in other cases. The graph in Fig. 6 illustrates this by showing the variety in percentage increases in bearing capacity for *B* equal to 10, 15, and 20.

**Table 1** Settlement of foundation of 15 unit width with different skirt depth for the vertical skirt

| Model | Ds/B | Settlement for 25 kN/m² (mm) | Settlement for 50 kN/m² (mm) | Settlement for 100 kN/m² (mm) | Settlement for 150 kN/m² (mm) | Settlement for 200 kN/m² (mm) | Settlement for 250 kN/m² (mm) |
|---|---|---|---|---|---|---|---|
| Model a | 0 | 20.2 | 40.5 | 81.1 | 122.2 | 163.1 | 204.4 |
| Model b | 0.5 | 20.2 | 40.4 | 81.0 | 122.0 | 163.0 | 204.0 |
| Model c | 1 | 19.8 | 39.6 | 79.4 | 120.3 | 160.3 | 200.9 |
| Model d | 1.5 | 19.1 | 38.2 | 76.8 | 116.3 | 155.3 | 195.2 |
| Model e | 2 | 18.4 | 36.8 | 74.1 | 112.5 | 150.2 | 189.4 |
| Model f | 2.5 | 17.8 | 35.6 | 71.7 | 107.9 | 146.4 | 183.0 |
| Model g | 3 | 17.3 | 34.6 | 69.7 | 105.1 | 142.0 | 178.1 |

**Table 2** Bearing capacity of foundation for 15 m width ($B = 15$ m) with different skirt depths for different allowable settlement

| Model | *Ds*/*B* | Bearing capacity for 25mm settlement (kN/m$^2$) | Bearing capacity for 40mm settlement (kN/m$^2$) | Bearing capacity for 50mm settlement (kN/m$^2$) | Bearing capacity for 75mm settlement (kN/m$^2$) | Bearing capacity for 100mm settlement (kN/m$^2$) |
|---|---|---|---|---|---|---|
| Model a | 0 | 33.10 | 52.79 | 65.84 | 98.27 | 130.47 |
| Model b | 0.5 | 36.88 | 58.81 | 73.4 | 109.53 | 145.47 |
| Model c | 1 | 40.09 | 63.99 | 79.87 | 119.4 | 158.87 |
| Model d | 1.5 | 42.75 | 68.27 | 85.27 | 127.53 | 169.73 |
| Model e | 2 | 45.11 | 72.07 | 89.93 | 134.60 | 179.2 |
| Model f | 2.5 | 47.23 | 75.47 | 94.20 | 141.00 | 187.67 |
| Model g | 3 | 49.14 | 78.53 | 98.00 | 146.73 | 195.33 |

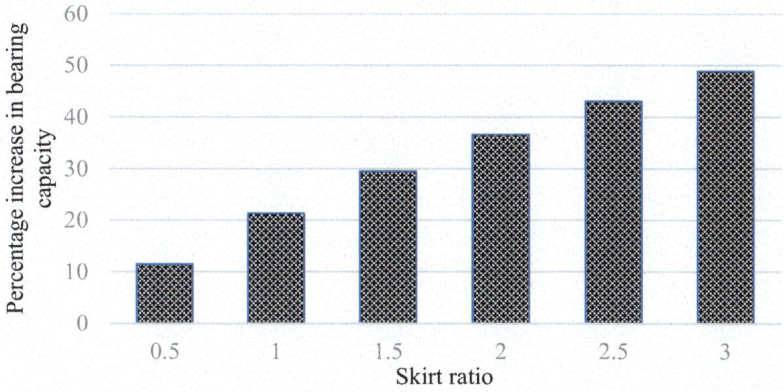


**Fig. 4** Percentage increase in bearing capacity for $B = 15$

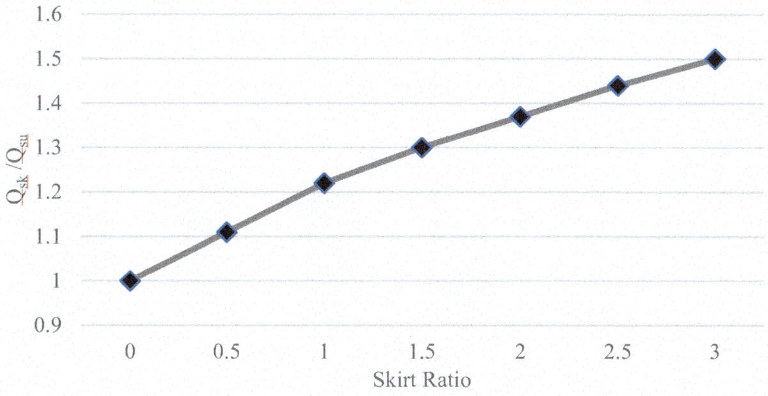


**Fig. 5** Variation of bearing capacity ratio with the skirt ratio

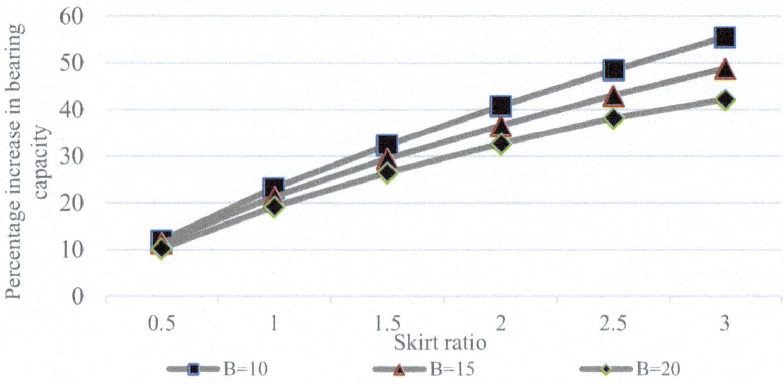


**Fig. 6** Percentage increase in bearing capacity for *B* equals 10, 15, and 20 units

When comparing the bearing capacity and settlement for one-side and two-side skirts of 15-unit width, it is clear that two-sided skirts are always preferable to one-sided skirts. The bar chart (Fig. 7) shows that varying the skirt ratio from 0.5 to 3 increases the bearing capacity for two-sided skirts over one-sided skirts. For similar conditions, the percentage decrease in settlement for two-sided skirts over one-way skirts is around 1.5 times. Six numerical models are generated by altering the depth of the skirt below the ground while keeping the foundation's width constant. In this study, the inclined skirt is modelled at a 5-degree to 30-degree angle from vertical. The skirt ratio is adjustable in 0.5 increments from zero to three. With increasing skirt depth, there is a change in settlement and bearing capacity (Table 3).

Table 4 shows how the length of the skirt beneath the soil affects the settlement of a foundation with a skirt. The breadth of the foundation is kept constant, while the depth of the skirt varies. Models (Ia) to (Ig) represent seven different foundation models with skirts. Based on the findings, one can accomplish less settlement by increasing the length of the skirt in comparison with that of the surface foundation. Table 5 shows how the bearing capacity of a foundation with a skirt varies as the

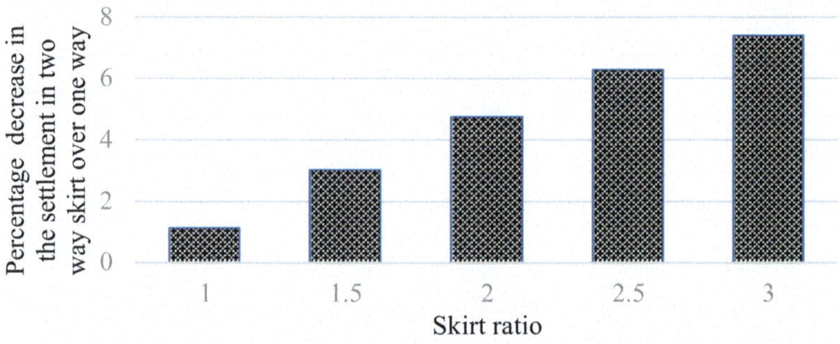


**Fig. 7** Percentage decrease in the settlement in two-side skirts over one-side skirts

**Table 3** Comparison of one-side and two-side vertical skirts

| *Ds/B* | Settlement for 100 kN/m$^2$ (mm) for one-side vertical Skirt | Settlement for 100 kN/m$^2$ (mm) for two-side vertical Skirt |
|---|---|---|
| 0 | 81.0 | 81.0 |
| 0.5 | 81.1 | 81.0 |
| 1 | 80.3 | 79.4 |
| 1.5 | 79.2 | 76.8 |
| 2 | 77.8 | 74.1 |
| 2.5 | 76.5 | 71.7 |
| 3 | 75.3 | 69.7 |

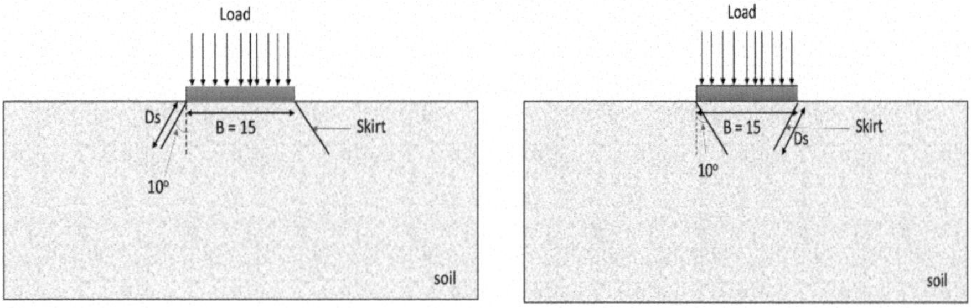


**Fig. 8** Geometry model of foundation with two-sided skirt. (a) 10° positive and (b) 10° negative skirt

length of the skirt beneath the soil increases. The breadth of the foundation is kept constant, while the depth of the skirt varies. Models (Ia) to (Ig) represent seven different foundation models with skirts. According to the analysis, increasing the length of the skirt can result in a higher bearing capacity than what we currently have for the surface foundation.

When a similar analysis is performed for two-sided 10° interior (negative) inclined skirt and compared with vertical and 10° positive inclined skirts (Fig. 8), while examining the graph shown in Fig. 9, it is observed that the fluctuation of the percentage decrease in settling for inclined skirts against vertical skirts, from the analysis it is found that the inclined skirt with a 10° inclination outperforms the vertical skirt with the same attribute. A 10° inclined skirt away from the foundation and a 10° negative inclined skirt are also checked. The 10° negative inclined skirt has a lesser bearing capacity improvement than the 10° positive inclined skirt, according to the findings of this experiment. Comparing vertical and inclined skirts for various authorised settlements, inclined skirts perform better than vertical skirts. Further investigation reveals that, up to a point, an inclined skirt outperforms a vertical skirt in terms of increasing load-carrying capability. It means that while comparing the inclined skirt for different angles, it is discovered that as the inclination increases beyond a certain

**Table 4** Settlement of foundation of 15 m width ($B$ = 15 m) with different skirt depth for 10 degrees inclined skirt

| Model | $Ds/B$ | Settlement for 25 kN/m$^2$ (mm) | Settlement for 50 kN/m$^2$ (mm) | Settlement for 100 kN/m$^2$ (mm) | Settlement for 150 kN/m$^2$ (mm) | Settlement for 200 kN/m$^2$ (mm) | Settlement for 250 kN/m$^2$ (mm) |
|---|---|---|---|---|---|---|---|
| Model Ia | 0 | 20.2 | 40.5 | 81.1 | 121.9 | 162.7 | 203.6 |
| Model Ib | 0.5 | 20.0 | 40.0 | 80.3 | 120.7 | 161.3 | 202.0 |
| Model Ic | 1 | 19.7 | 39.4 | 79.2 | 119.3 | 159.8 | 200.5 |
| Model Id | 1.5 | 18.9 | 38.1 | 76.7 | 115.9 | 155.5 | 195.6 |
| Model Ie | 2 | 18.3 | 36.8 | 74.3 | 112.3 | 150.9 | 190.0 |
| Model If | 2.5 | 17.8 | 35.7 | 72.0 | 108.9 | 146.5 | 184.8 |
| Model Ig | 3 | 17.3 | 34.7 | 70.0 | 106.1 | 142.8 | 180.2 |

**Table 5** Bearing capacity of foundation for 15 m width ($B$ = 15 m) with different skirt depths for various allowable settlements for 10 degrees inclined skirt

| Model | $Ds/B$ | Bearing capacity for 25 mm settlement (kN/m$^2$) | Bearing capacity for 40 mm settlement (kN/m$^2$) | Bearing capacity for 50 mm settlement (kN/m$^2$) | Bearing capacity for 75 mm settlement (kN/m$^2$) | Bearing capacity for 100 mm settlement (kN/m$^2$) |
|---|---|---|---|---|---|---|
| Model Ia | 0 | 33.1 | 52.79 | 65.84 | 98.27 | 130.47 |
| Model Ib | 0.5 | 38.18 | 61.06 | 76.27 | 114.27 | 151.47 |
| Model Ic | 1 | 41.10 | 65.63 | 81.93 | 122.67 | 163.33 |
| Model Id | 1.5 | 44.09 | 70.4 | 87.93 | 131.73 | 175.4 |
| Model Ie | 2 | 46.67 | 74.53 | 93.13 | 139.47 | 185.73 |
| Model If | 2.5 | 48.97 | 78.27 | 97.73 | 146.4 | 195 |
| Model Ig | 3 | 51.03 | 81.53 | 101.87 | 152.6 | 203.27 |

**Table 6** Comparison of settlement for 10 degrees positive and negative skirt

| *Ds*/*B* | Settlement for 25 kN/m$^2$ (mm) When B = 15 for the vertical skirt | Settlement for 25 kN/m$^2$ (mm) for 10 negative skirt | Settlement for 25 kN/m$^2$ (mm) for 10 degrees positive skirt |
|---|---|---|---|
| 0 | 20.2 | 20.2 | 20.2 |
| 0.5 | 20.1 | 20.0 | 19.9 |
| 1 | 19.8 | 19.8 | 19.7 |
| 1.5 | 19.1 | 19.2 | 18.9 |
| 2 | 18.4 | 18.5 | 18.3 |
| 2.5 | 17.8 | 18.0 | 17.8 |
| 3 | 17.3 | 17.6 | 17.3 |

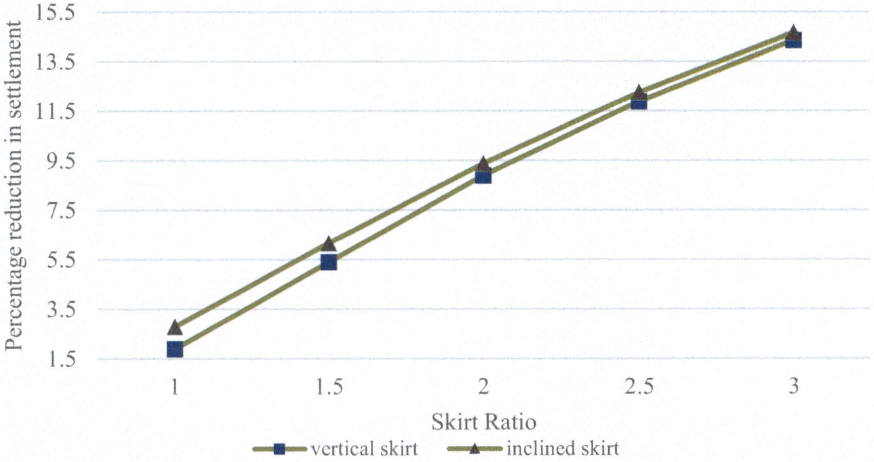


**Fig. 9** Percentage reduction in settlement for the vertical and inclined skirt

point, i.e. 12 degrees in our analysis, the settlement increases (Table 6) and the load-carrying capacity (Table 7) decreases because it is unable to bind soil between the footing's edges.

## 6 Conclusions

Some of the significant conclusions that can be drawn from the analysis are as follows:

- It is inferred from the analysis that a reduction in settlement of about 15% is observed on the installation of the skirt.
- One-side skirt's functioning capacity is also much lower than that of both side skirts.

Table 7 Comparison of bearing capacity for 10 degrees positive and negative skirt

| *Ds*/*B* | Bearing capacity for 25 mm settlement (kN/m$^2$) for a 10-degree interior inclined skirt | Bearing capacity for 25 mm settlement (kN/m$^2$) for a 10-degree exterior inclined skirt |
|---|---|---|
| 0 | 33.1 | 33.1 |
| 0.5 | 34.28 | 38.18 |
| 1 | 38.78 | 41.10 |
| 1.5 | 41.01 | 44.09 |
| 2 | 42.93 | 46.67 |
| 2.5 | 44.58 | 48.97 |
| 3 | 46.25 | 51.03 |

- Based on the findings, it is deduced that varying the inclination of the skirt initially yields a favourable result, i.e. the settlement decreases. There is a decline in bearing capacity after the inclination exceeds 12 degrees.
- A variation in the width of 10 to 20 units at a 5-unit interval reveals that extending the skirt beyond the centre causes an increase in settlement and a reduction in bearing capacity owing to independent action.
- When comparing the results obtained from the vertical and inclined skirts for inclination up to 12 degrees, it is observed that the result obtained from the inclined skirt is quite promising.
- The investigation also revealed that a skirt angled towards the foundation (negative skirt) has a detrimental effect, but a skirt angled away from the foundation (positive skirt) results in higher bearing capacity.

## References


1. Naik S, Dasaka S, Chauhan V (2015) Numerical study of offshore skirted foundations subjected to combined loading. Conference paper Dec 2015. https://doi.org/10.3850/978-981-09-7520-3_137
2. Bell RW (1991) The analysis of offshore foundations subjected to combined loading. Master of Science Hilary Term
3. Bransby MF, Randolph MF (1998) Combined loading of skirted foundations. Géotechnique 48(5):637–655
4. Yun GJ, Bransby MF (2003) Centrifuge modelling of the horizontal capacity of skirted foundations on drained loose sand. In: Proceedings of the international conference organised by British geotechnical association, Dundee, Scotland on 2–5th September 2003. Thomas Telford Publishing, pp 975–984
5. Al-Aghbari MY, Mohamedzein Y (2004) Bearing capacity of strip foundations with structural skirts. Geotech Geol Eng 22(1):43–57
6. Al-Aghbari MY (2007) Settlement of shallow circular foundations with structural skirts resting on the sand. J Eng Res [TJER] 4(1):11–16
7. Yun G, Bransby MF (2007) The undrained vertical bearing capacity of skirted foundations. Japan Geotech Soc Soils Found 47(3):493–505

8. Kellezi L, Kudsk G, Hofstede H (June, 2008) Skirted footings capacity for combined loads and layered soil conditions. In: Proceedings of the BGA international conference on foundations. Dundee, Scotland, pp 24–27
9. Bransby MF, Yun GJ (2009) The undrained capacity of skirted strip foundations under combined loading. Géotechnique 59(2):115–125
10. El Wakil AZ (2010) Horizontal capacity of skirted circular shallow footings on sand. Alex Eng J 49(4):379–385
11. Eid HT (2013) Bearing capacity and settlement of skirted shallow foundations on sand. Int J Geomech 13(5):645–652
12. Pusadkar SS, Bhatkar T (2013) Behaviour of raft foundation with inclined skirt using PLAXIS 2D. In: Indian geotechnical conference, vol 3, no 6
13. Mana DS, Gourvenec SM, Randolph MF, Hossain MS (2012) Failure mechanisms of skirted foundations in uplift and compression. Int J Phys Modell Geotech 12(2):47–62

# Evaluation of Nonlinear Response of Block Foundation Subjected to Vertical Vibration

**Kavita Tandon, Bappaditya Manna, and G. V. Ramana**

## 1 Introduction

Developing effective and economical designs for footings subjected to dynamic loads has become more critical in recent years. The detrimental effects of vibrations emanating from larger machines and industrial installations have been the primary concern for industrialists and geotechnical engineers. In this regard, various studies have been reported on the dynamic response of machine foundations for the last 60–70 years. However, several researchers conducted analytical and experimental investigations over the previous few decades. The vibration of soils and foundations has led to the development of fundamental methods for formulating soil-foundation systems. The conceptual machine-foundation design limits the vibrational amplitudes so that the satisfactory operations of the machines and the people working in the immediate vicinity remain unaffected by the impact of vibrations. The vibration studies are essential as the dynamic properties of the soil play a crucial part in the overall response of foundations (shallow and deep). Various researchers have made significant contributions to geotechnical engineering, considering the dynamic soil properties of foundations subjected to machine-induced vibrations [1, 2, 4, 7] that showed the nonlinear behavior in the dynamic response of the block foundations. The small-scale field tests help to examine the behavior of block foundations subjected to dynamic loads. The field tests are also conducted to avoid spurious reflection of waves often present in the laboratory tests in the tank due to the boundary zone effect. Among various researchers and experimental studies reported in the literature, Novak and Beredugo [6] conducted field tests on two blocks at different embedment

K. Tandon (✉)
Department of Civil Engineering, HBTU, Kanpur 208002, India
e-mail: ktandon@hbtu.ac.in

B. Manna · G. V. Ramana
Department of Civil Engineering, Indian Institute of Technology, Delhi 110016, India



B. T. Jose et al. (eds.), *Proceedings of the Indian Geotechnical Conference 2022 Volume 2*, Lecture Notes in Civil Engineering 477,
https://doi.org/10.1007/978-981-97-1741-5_23

**Table 1** Geotechnical characterization of soil with depth

| Depth (m) | Dry unit weight (kN/m$^3$) | Particle size distribution (%) | In situ SPT-N | *Vs* (m/s) |
|---|---|---|---|---|
| 1 | 15.23 | | 11 | 113.36 |
| 2 | 15.34 | Sand = 80, | 14 | 113.36 |
| 3 | 16.27 | Silt = 18, | 15 | 113.36 |
| 4 | 17.08 | Clay = 2 | 18 | 128.21 |
| 5 | 17.20 | | 18 | 128.21 |

ratios and densities of the backfill soil under vertical vibration. Beredugo (1971) observed strong nonlinearity in the soil-foundation system. The experimental results were compared with Baranov's analytical approach, and it was reported that the analytical approach overestimates the resonant amplitude. In the current research work, an attempt has been made to analyze and investigate the dynamic nonlinear response of the small-scale surface block foundation subjected to machine-induced vertical vibrations. The nonlinear approach proposed by Novak [5] was used to evaluate the effective mass with dynamic impedances, i.e., stiffness and damping from the experimental response curves. The dynamic frequency–amplitude curves were back-calculated by the theoretical approach and compared with the field test results.

## 2 Geotechnical Characterization of Site Soil

The test location identified between Block II and Block III at IIT Delhi, New Delhi, India, was characterized with the help of various field and laboratory tests. SPT and MASW tests have obtained the geotechnical properties of the site, and it was found that the subsurface consists of sands with fines (SM—silty sands and silt mixtures). The soil properties explored to the depth of 5.0 m are presented in Table 1.

## 3 Experimental Investigations

The forced vibration tests subjected to vertical loading were conducted on a square (*L/B*) block foundation under a static load of (*WS*) 6.6 kN. The dynamic tests were carried out for four excitation intensities ($W.e$ = 0.221 Nm, 0.868 Nm, 1.450 Nm, and 1.944 Nm). The dynamic forces were induced on the block foundation using a Lazan-type mechanical oscillator connected with a flexible shaft to the DC motor. The vibration responses were recorded using the data acquisition system connected with the unidirectional accelerometers for frequency intervals of 5 Hz. The vertical vibration test setup for the surface block foundation is shown in Fig. 1.

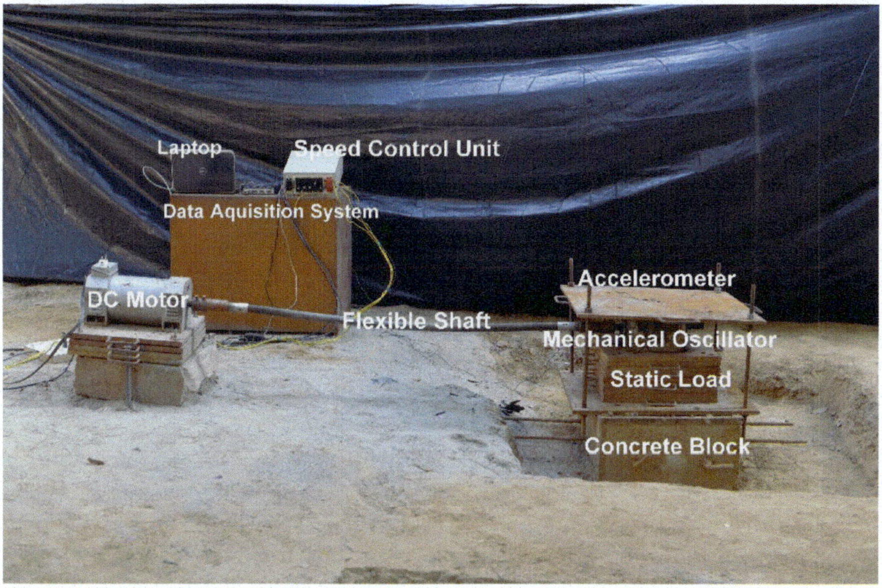


**Fig. 1** Experimental test setup for vertical vibration

## 4 Theoretical Investigations

A typical feature of nonlinear vibration is the resonant frequency and amplitude change with different excitation intensities. The experimental frequency–amplitude responses determine the effective mass and dynamic impedances. The nonlinear dynamic responses were back-calculated using Novak's theory of nonlinear vibration from experimental response curves. It can be established from experimentally obtained frequency response curves that the undamped natural frequency of the soil-foundation oscillator system reduces with the increasing amplitudes of vibration. The undamped natural frequency was calculated from restoring force and the damping force. The restoring force was assumed to be nonlinear, whereas the damping force was linearly viscous. The restoring force was calculated using Eq. 1

$$\Omega\Omega = \sqrt{\omega_1\omega_2}, \tag{1}$$

where $\omega_1$ and $\omega_2$ are the undamped frequencies corresponding to the interaction points between the experimental response curve and a line passing through the origin. The present study considers the nonlinear frequency–amplitude response obtained from dynamic field tests ($L/B = 1.0$ and $h/H = 0$) for a static weight of 6.6 kN. The backbone curve $\Omega(A)$ for each frequency–amplitude curve is constructed by intersecting the response curves by a trace of lines, as shown in Fig. 2. It is observed from each response curve that the impedance of the machine-foundation soil-oscillator system varies with the degree of eccentricity and its respective excitation moment. The results reported are based on the equivalent linearization of both damping and restoring forces, known as the first approximation accuracy. The steady-state harmonic force excitation amplitude proportional to the square of the frequency was considered, and the vertical stiffness was expressed for the translational amplitudes *A*. The vertical damping and effective mass were determined [5] using the geometrical properties of the nonlinear response curves. The effective mass (*m*) was found to be greater than the mass of the foundation loading system ($m = Wst/g$; *Wst* is the total mass of the foundation, including steel plates, ingots, and the exciter) as the surrounding soil participates in the dynamic analysis of the foundation. The apparent additional mass in terms of the mass coefficient is expressed as

$$\xi = \frac{m_{eff} - m_s}{m_s}. \tag{2}$$

The restoring force characteristics and effective mass were also calculated from the backbone curve $\Omega$ for each response curve. The typical nonlinear force–displacement characteristic curve is shown in Fig. 3. The effective mass, stiffness, and damping for different eccentric moments determined from the backbone curves are given in Table 2. It can be observed from Table 2 that the vertical stiffness of the soil-foundation oscillator system decreases with an increase in the excitation force. It can also be seen from Table 2 that with the increase in the eccentric moment, the vertical damping values increased; however, the vertical stiffness values and the effective mass reduced with higher eccentric moments. The frequency–amplitude curves are back-calculated using the nonlinear theoretical approach. The curves are obtained

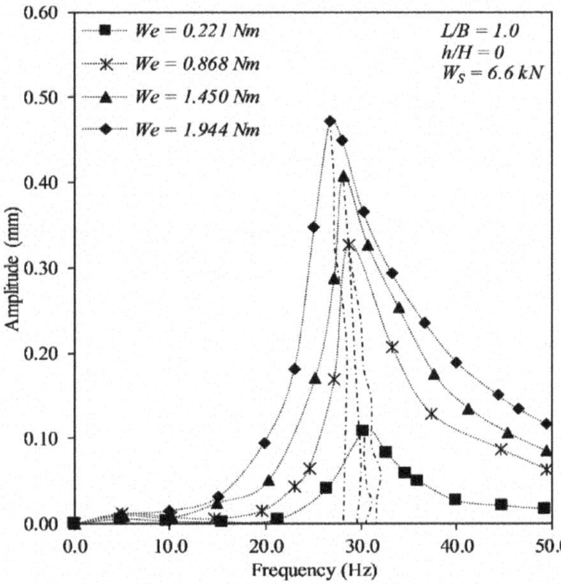


**Fig. 2** Experimental and back-calculated response curves of block foundation ($L/B = 1.0$) subjected to vertical vibration

from the calculated soil-foundation oscillator system mass, vertical damping, and restoring force characteristics. The calculated response curves using the inverse problem are then compared with the dynamic field test results, as shown in Fig. 2. It can be noticed that the theoretical back-calculated response curves match well with the test data obtained from the vertical vibration test. So, it can be concluded that the nonlinear parameters obtained from the theoretical analysis for foundations resting on the ground surface and partially or fully embedded [5] are accurate enough to be considered in the response analysis.

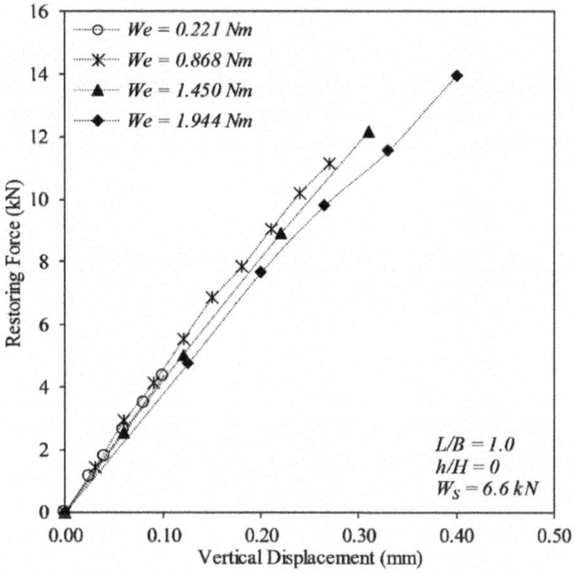


**Fig. 3** Nonlinear force characteristic curve for square block foundation resting on the ground surface ($WS = 6.6$ kN)

**Table 2** Nonlinear parameters of block foundation soil-oscillator system under vertical vibration

| Exciting moment (Nm) | Mass of the soil-foundation system, *ms* (kg) | Mass *meff* (kg) | Mass coefficient, ξ | Damping | Stiffness (kN/m) |
|---|---|---|---|---|---|
| 0.221 | 1008 | 1182.559 | 0.17317 | 0.147 | 4.33E + 07 |
| 0.868 | 1008 | 1173.645 | 0.16433 | 0.135 | 4.00E + 07 |
| 1.450 | 1008 | 1165.511 | 0.15626 | 0.136 | 3.97E + 07 |
| 1.944 | 1008 | 1124.959 | 0.11603 | 0.138 | 3.42E + 07 |

## 5 Conclusions

The present study measured the dynamic responses, and the frequency–amplitude response characteristics were evaluated for machine-induced vertical vibration. Based on the vertical vibration test results on the surface block foundation, the following interpretations were drawn:

1. The experimental response curves agree with the theoretically obtained response curves.
2. The force characteristics' curve represents the nonlinear behavior of the foundations as the decrease in resonant amplitudes and increases in resonant frequency are observed with the increase in excitation level.
3. The impedance characteristics, i.e., dynamic stiffness and damping parameters obtained from the frequency–amplitude curves, also indicate the nonlinear behavior of the foundation.

## References

1. Baidya DK (2000) Natural frequency of vibrating foundation on layered soil—an experimental investigation. In: Proceedings of international conference of GeoEng-2000. Melbourne, Australia
2. Baidya DK, Rathi A (2004) Dynamic response of footing on a sand layer of finite thickness. J Geotech Geoenviron Eng 130(6):651–655
3. Gazetas G, Stokoe KH (1991) Free vibration of embedded foundations: theory versus experiment. J Geotech Eng 117(9):1382–1401
4. Gazetas G (1983) Analysis of machine foundation vibration: state of the art. Struct Dynam Earthquake Eng 2(1):1–42
5. Novak M (1971) Data reduction from nonlinear response curves. J Eng Mechan ASCE 97(EM4):1187–1204
6. Novak M, Beredugo YO (1972) Vertical vibration of embedded footings. J Soil Mech Found Div ASCE 98(SM12):1291–1310
7. Novak M (1987) State-of-the-art in analysis and design of machine foundations. Dev Geotech Eng 43:171–192

# Lateral Behavior of a Long Pile with Load Application Above Ground: A Case Study

**B. Govind Raj, Anurag Chafale, Madan Kumar Annam, and K. V. Babu**

## 1 Introduction

Pile foundations are typically used for providing stability for the structures subjected to high vertical and lateral loads. The lateral loads can be due to wind and seismic cases and the understanding of the load deflection behavior is an essential requirement for the design of the structures.

40 m deep piles, 1m diameter was necessitated for an elevated tank in the West Coast of India. The project site is located in active Seismic Zone, subjected to high lateral forces. Behavior of the piles is studied by Finite Element and Finite Difference approaches. Lateral pile load test was performed at 2 m above ground level.

This paper discusses theoretical and actual behavior of laterally loaded piles tested above ground level. The P–Y method proposed by Reese and Van Impe [1] was used for the analysis. The governing differential equation is solved by the Finite Difference Techniques [2]. The results obtained are compared with 3D Finite Element Models and the load test results. In addition, the diameter effect which needs to be considered while using P–Y curve method is also discussed.

B. G. Raj (✉) · A. Chafale
Keller Ground Engineering India Pvt. Ltd., Mumbai 400059, India
e-mail: govind@kellerindia.com

M. K. Annam
Keller Ground Engineering India Pvt. Ltd., Chennai 600024, India

K. V. Babu
L&T Hydrocarbon Engineering Limited, Vadodara-19, India

 
B. T. Jose et al. (eds.), *Proceedings of the Indian Geotechnical Conference 2022 Volume 2*, Lecture Notes in Civil Engineering 477,
https://doi.org/10.1007/978-981-97-1741-5_24

**Table 1** Subsoil stratification at the job site

| Sl. No. | Type | Elevation, m | N | cu kPa | $\phi'(^\circ)$ |
|---|---|---|---|---|---|
| 1 | Silty sand | 0.0–3.0 | 8 | – | 31 |
| 2 | Stiff clay | 3.0–7.5 | 15 | 68 | – |
| 3 | Silty sand | 7.5–10.5 | 46 | – | 34 |
| 4 | Stiff clay | 10.5–18.0 | 27 | 122 | – |
| 5 | Stiff clay | 18.0–25.5 | 38 | 171 | – |
| 6 | Stiff clay | 25.5–31.5 | 26 | 117 | – |
| s7 | Stiff clay | 31.5–36.0 | 39 | 176 | – |
| 8 | Silty sand | 36.0–50.0 | 133 | – | 34 |

## 2 Site and Subsoil Conditions

An elevated tank was proposed in the West Coast of India. Project specifications demand high lateral load requirement for the piles. The piles are 1m diameter, 40 m deep, and cut-off level is 2 m above existing ground level.

Two boreholes were explored to 50 m deep below the existing ground level (EGL). The subsoil consists of mixed soil strata (layers of sand and stiff to hard layers). Ground water table was observed to be 2 below EGL at the time of testing. The stratification along with the subsoil data is presented in Table 1.

## 3 Numerical Analysis

Figure 1 shows a typical arrangement of lateral pile load test. The pile is modeled by P–Y Method and 3D Finite Element Method. P–Y curve analysis is performed using the method proposed by Reese and Matlock. The parameters used to develop P–Y curves in clay are undrained shear strength and the normal strain at 50% of the ultimate deviator stress in a triaxial test. The parameters used in sand are the angle of friction and the horizontal modulus of subgrade reaction. The governing differential equation is solved by the Finite Difference Techniques.

Embedded pile/beam option is considered for 3D Finite Element Method for pile and Mohr–Coulomb model parameters are used for surrounding soil. The clay layers are modeled as undrained soil, whereas sand is modeled as drained type (Fig. 2).

Influence of mesh coarseness in Finite Element Method and diameter effect in the P–Y curve method is studied.

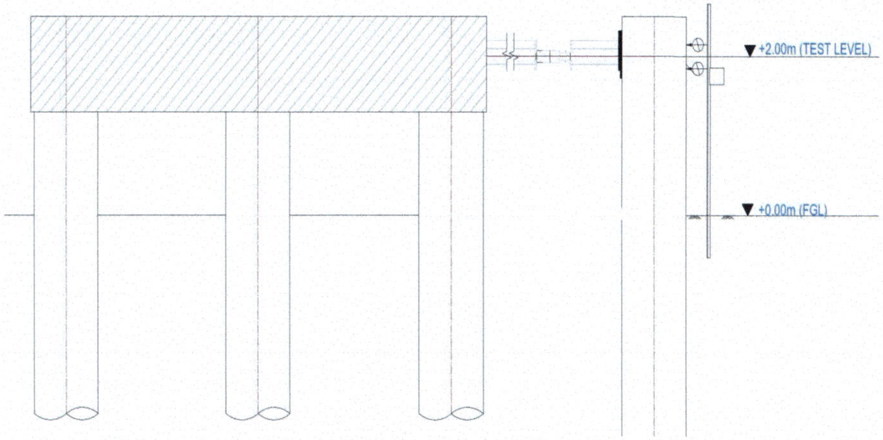


**Fig. 1** Typical arrangement illustrating lateral pile load test

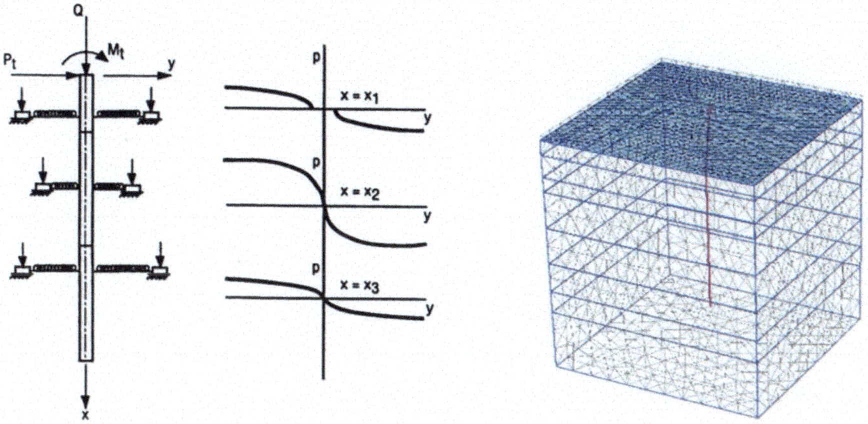


**Fig. 2** P–Y method and FEM method

## *3.1 Effect of Mesh Coarseness on Deflection Pattern (FEM Analysis)*

Four mesh types (fine, medium, coarse, and very coarse) are considered to identify the influence on the deflection patterns in a Finite Element model. A typical load case of 920 kN is considered in the analysis (Fig. 3).

It is observed that load settlement behavior of an embedded pile is mesh dependent. Very coarse and coarse meshes under predicted the deflection patterns, while the medium and fine meshes yielded comparable results.

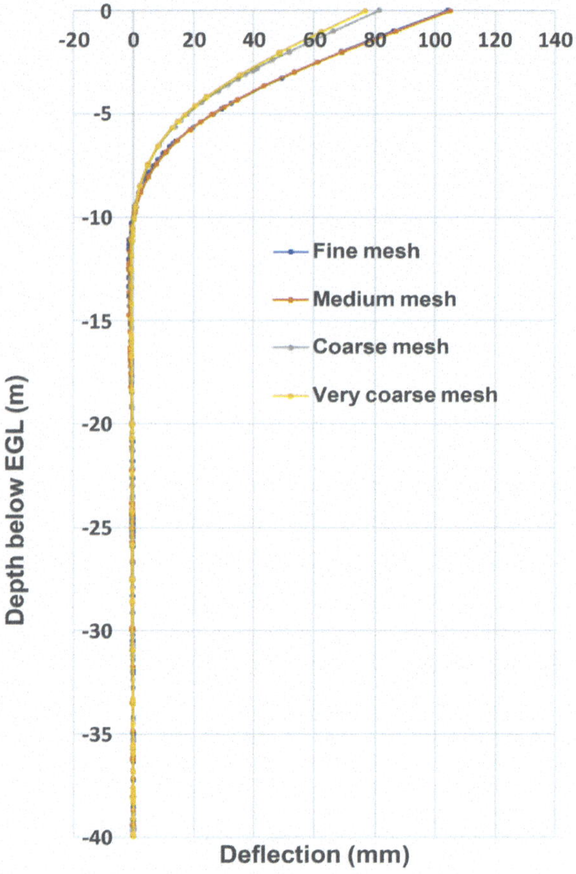


**Fig. 3** Influence of mesh on the deflection profiles (1 m diameter)

## *3.2 Diameter Effects (P–Y Method of Analysis)*

P–Y curves are originally developed based on the lateral load tests on the piles ranging between 0.30 and 1.20 m diameters. An exercise is made to assess deflection of 1.0 m, 1.5 m, and 2.0 m dia. piles for varying lateral loads in both P–Y and 3D FEM. The results are tabulated (Table 2).

The present analysis shows differences in the deflection patterns for large diameter piles at high loads; however, detailed studies are to be done on the same.

Data from experiments across the globe are insufficient to make a conclusion on the diameter influence and further extensive studies are required for the same. Hence, it is opinioned that the assessment of big diameter piles through P–Y analysis method shall be followed by additional checks such as FEM analysis and load tests.

**Table 2** Diameter effects of laterally loaded pile

| Pile dia. (m) | Lateral load (tons) | Deflection (P–Y model) | Deflection (3D FEM model) |
|---|---|---|---|
| 1.0 m | 31 | 25 | 27 |
| | 56 | 53 | 57 |
| | 76 | 86 | 85 |
| | 92 | 121 | 105 |
| 1.5 m | 100 | 37 | 42 |
| | 160 | 76 | 71 |
| | 200 | 110 | 99 |
| | 250 | 169 | 126 |
| 2.0 m | 100 | 18 | 24 |
| | 160 | 33 | 40 |
| | 200 | 43 | 52 |
| | 250 | 61 | 66 |

## 4 Field Load Test

Pile was tested to maximum of 137 tons/100 mm deflection whichever comes first. Point of application of the load is kept 2.0 m above ground level. Application of the loading was carried out by means of jacking the test pile against the reaction system. The readings were taken with the help of two calibrated dial gauges and the average readings were taken (Fig. 4).

## 5 Comparison of Numerical Methods and Field Test Results

The results of the numerical analysis and the load test are plotted in the following Table 3.

The deflections obtained from the numerical analysis are compared with the field records and are found to be in the order (Figs. 5 and 6).

It is observed that depth-wise variation in deflection is following similar trends between the two numerical methods, and the point of fixity is found to be the same in both the numerical methods.

**Fig. 4** Lateral load test setup

**Table 3** Summary of results (numerical analysis versus load test)

| Sl. No. | Loading (kN) | Deflection, mm (P–Y method) | Deflection, mm (FEM method) | Deflection, mm (load test) |
|---|---|---|---|---|
| 1 | 170 | 13 | 13 | 6 |
| 2 | 310 | 25 | 27 | 21 |
| 3 | 350 | 29 | 32 | 28 |
| 4 | 460 | 40 | 45 | 43 |
| 5 | 560 | 53 | 57 | 59 |
| 6 | 670 | 70 | 72 | 79 |
| 7 | 760 | 86 | 85 | 90 |
| 8 | 920 | 121 | 105 | 109 |

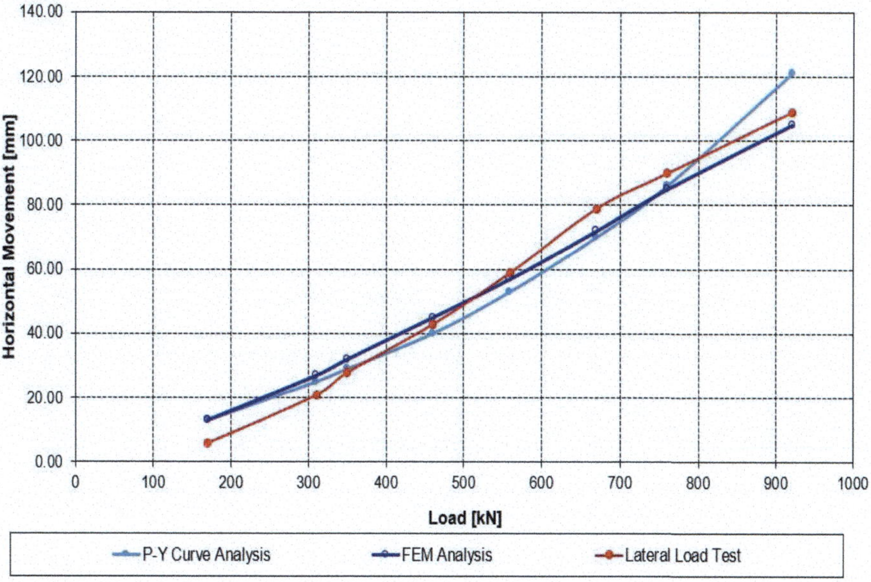


**Fig. 5** Comparison of load test results versus numerical methods

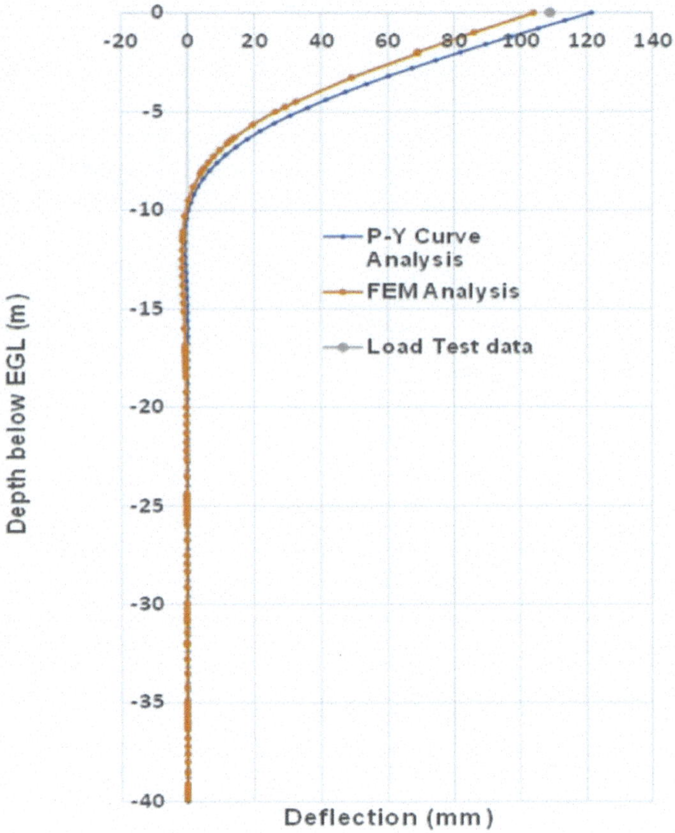


**Fig. 6** Behavior of 1 m diameter pile at 920 kN test load

## 6 Conclusion

Behavior of the laterally loaded pile is assessed by Finite Element Analysis and P–Y Method of analysis. It is observed that both methods follow similar trends and match with the load test results in most of the load cases. Programs using P–Y curves are originally developed for small diameter piles, and application of it for larger diameter piles shall be used with caution. Additional checks such as FEM analysis and load tests may be used to validate the performance.

## References

1. Reese LC, Van Impe WF (2011) Single piles and pile groups under lateral loading 2nd edn. Taylor & Francis, London. ISBN 978-0-415-46988-3
2. Tomlinson MJ. Pile design and construction practice, 5th edn. E&FN Spon, London

# Foundation for an 85 m High Bungy-Jumping Tower at Rishikesh

**Abhay Gupta, Ravi Sundaram, and Sanjay Gupta**

## 1 Introduction

The hilly terrains, an intrinsic part of the Rishikesh experience in Uttarakhand, offer a fun-filled adrenalin rush matched by no other. An 85 m high bungy-jumping tower coming up in Shivpuri, Rishikesh, is planned to be the highest bungy-jumping tower in India. This tower is set to offer an exciting dose of thrill to those ready for a courageous and sensational experience. A schematic sketch of the tower is illustrated in Fig. 1.

## 2 Project Details

### *2.1 General*

The tower itself shall be 76 m high. The proposed jump height is just over 86 m which will be highest in India once completed. The site location opens up to a panoramic view of the Ganga and the Himalayan ranges and imbibes the thrill and adventurous aura of the activity.

A. Gupta (✉)
Skeleton Consultants Pvt. Ltd., Noida 201301, India
e-mail: drabhaygupta62@gmail.com

R. Sundaram · S. Gupta
Cengrs Geotechnica Pvt. Ltd., Noida 201309, India
e-mail: ravi@cengrs.com

S. Gupta
e-mail: sanjay@cengrs.com



B. T. Jose et al. (eds.), *Proceedings of the Indian Geotechnical Conference 2022 Volume 2*, Lecture Notes in Civil Engineering 477,
https://doi.org/10.1007/978-981-97-1741-5_25

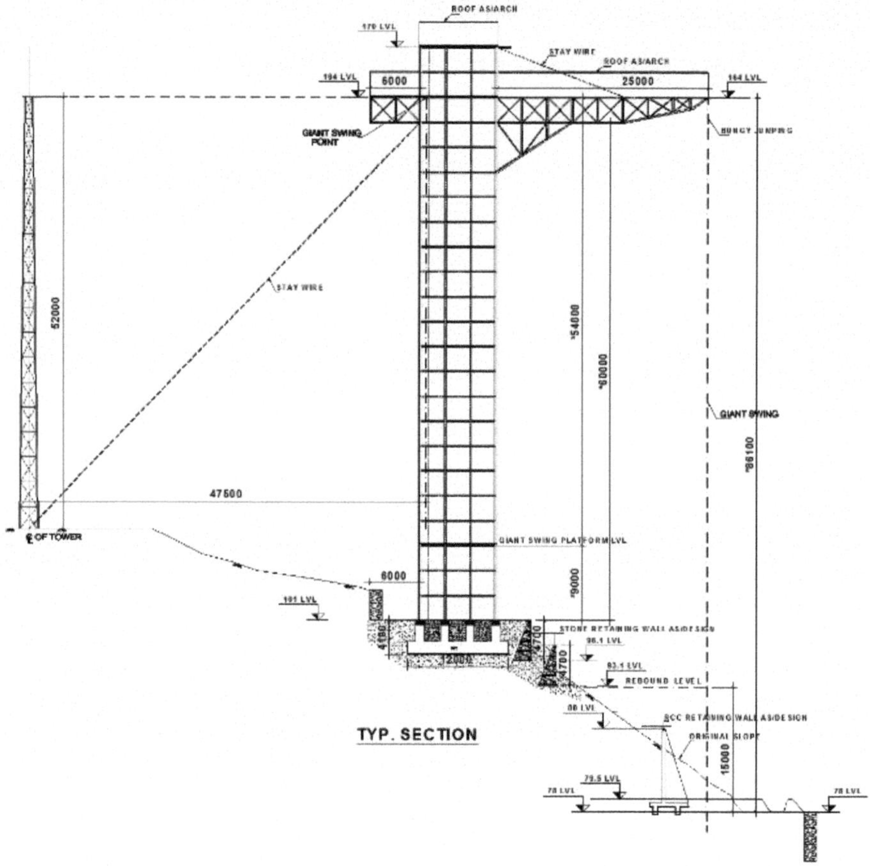


**Fig. 1** Schematic sketch of the tower

## 2.2 Location

The site is in the mountainous terrain of lesser Himalaya on the right bank of Hule River, near its confluence with the River Ganga at Shivpuri. It is accessible from NH-7 and lies at Latitude 30° 8′ 3.85″ N and Longitude 78° 23′ 3.90″ E and is about 15–20 km east of Rishikesh city. A satellite image showing the site location is illustrated in Fig. 2.

## 2.3 Project Features

The imposing steel tower will have a plan area of 9 m × 9 m and a total weight of about 175MT. It shall be located in an artificially terraced soil, covered rectangular

**Fig. 2** Satellite image showing the site location

area (about 21 m × 12 m) bound by concrete masonry walls, in moderate-to-steep, largely debris-covered slopes. The terrace is about 22 m high with relatively steep (35–40°) debris-covered and forested slope with a gentler slope below. Key details are as follows:

- The tower shall be in square shape, 9 m × 9 m in plan with a grid of 3 m × 3 m. Total height of tower is 76 m.
- There shall be total 16 columns, connected by cross-beams at every 3 m height.
- The RCC raft foundation is proposed at about 4 m below the ground.
- The bungy jump shall be performed from a cantilever boom of 25 m projection at 63 m above the ground level. The drop point shall be about 22 m below the base level of the tower.
- A boom for giant swing shall be attached on the other side of the tower of 6 m projection at 53 m above ground level and.
- A viewing deck is proposed at the top of the tower at 69 m height above the base of the tower.

## 2.4 Structural and Architectural Details

The tower is designed and constructed as RCC-steel hybrid structure. The substructure up to the plinth level shall be in RCC and the superstructure shall be constructed using braced HR steel frame of steel frame with stay wires. Steel columns have been analyzed as fixed at top of RC pedestal. Cross-bracings in vertical and horizontal

**Fig. 3** Architect's visualization of the tower and its surroundings

planes are provided at suitable locations to resist the lateral force and torsional effects due to earthquake and wind. The foundation was checked for overturning moments due to wind forces. A conceptual visualization of the tower as developed by the architect is presented in Fig. 3.

# 3 Site Conditions

## *3.1 Regional Geology*

The area represents a highly rugged topography characterized by moderate-to-steep slopes that are intervened by narrow valleys. The topography of the region appears to be controlled by structural and lithological factors [1]. The area thus exhibits high relative relief. Presence of overburden on steep slopes and high precipitation make this area prone to landslides.

The area from Rishikesh to Rudraprayag via Devprayag has a very complex geology, which varies from outer Lesser Himalayas to inner Lesser Himalayas. The entire stretch shows variations in rock type of Proterozoic to Cambrian meta-sediments. These meta-sediments are intersected by faults and thrusts at many places. The bedrock has also suffered multi-phase of folding at places [2].

The corridor from Rishikesh to Rudraprayag lies within older folded sequence over printed by Himalayan fold-thrust movement. The geological formations have suffered extensive tectonic movement and the rock formations were subjected to displacement from their original place of deposition. This transportation occurred due to large-scale thrusting of various geological formations and intensive operative compression tectonic activity in geological past resulting in numerous nappe structures.

**Fig. 4 Exposed rocks at site**. Note the shearing and disturbance to the formation

## 3.2 Site Geology

North Almoda Thrust and Saknindhar Thrust are present close vicinity of the site which are separated by distance of 1.5–2 km from each other. Site is located at nearly the centre of these two thrusts. There is high possibility that area could be part of a shear zone. Photographs in Fig. 4 illustrate the shearing and disturbance to the rocks at site.

In general, the Garhwal Himalayas is known for its fragile landscape and frequent geological hazards. Landslides are the regular threats over this region. It is a highly rugged terrain in the vicinity of the Upper Ganga River valley up to Devprayag and Lower Alakhnanda river valley up to Rudraprayag.

The area is known to experience frequent landslides, especially during every monsoon season. A map [3] showing these major thrusts (fault-lines) relative to the proposed site is presented on Fig. 5.

## 3.3 Site Description

The tower site is in the mountainous terrain of lesser Himalaya on right bank of Hule river, near its confluence with the River Ganga at Shivpuri. The tower shall be located in an artificially terraced soil, covered rectangular area (about 21 m

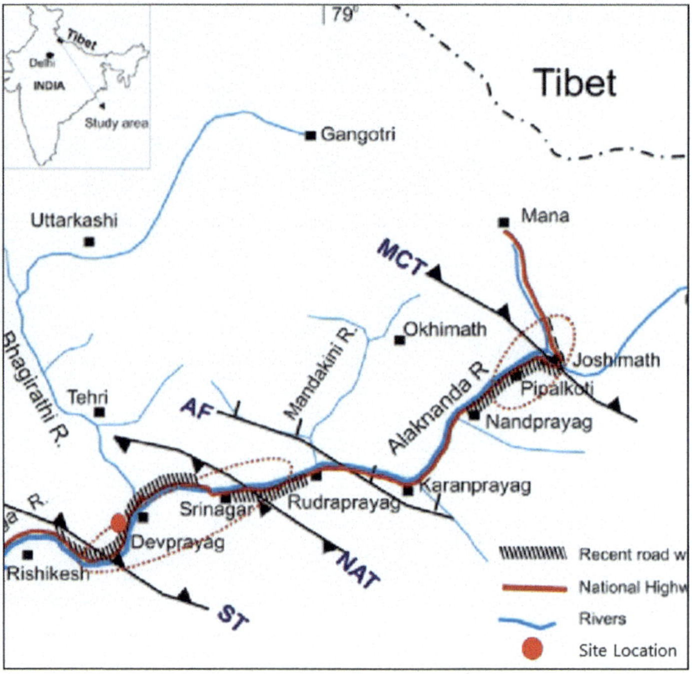


**Fig. 5 Map of the Upper Ganaga and Alakhnanda valley showing prominent fault/thrust**. These are MCT: Main Central Thrust, AF: Alaknanda Fault, NAT: North Almoda Thrust, ST: Saknindhar Thrust Sati et al. [3]

× 13 m) bound by concrete masonry walls, in moderate-to-steep, largely debris-covered slopes. The terrace is about 22 m high with relatively steep (35–40°) debris-covered-and-forested slope with a gentler slope below. Figure 6 presents photographs illustrating an overview of the site.

**Fig. 6** Site overview

Debris contains small angular fragments of shales in a silty-clayey matrix. Large dolomite boulders of up to about 3 m size lie sparsely in the debris on northern side and on the slope below the terrace. Occasional smaller trees on the slopes are tilted downslope; however, many large, fairly old trees of normal appearance exist on the debris slopes, suggesting fairly stable-slope conditions. A sketch of the site identified for the tower is shown in Fig. 7.

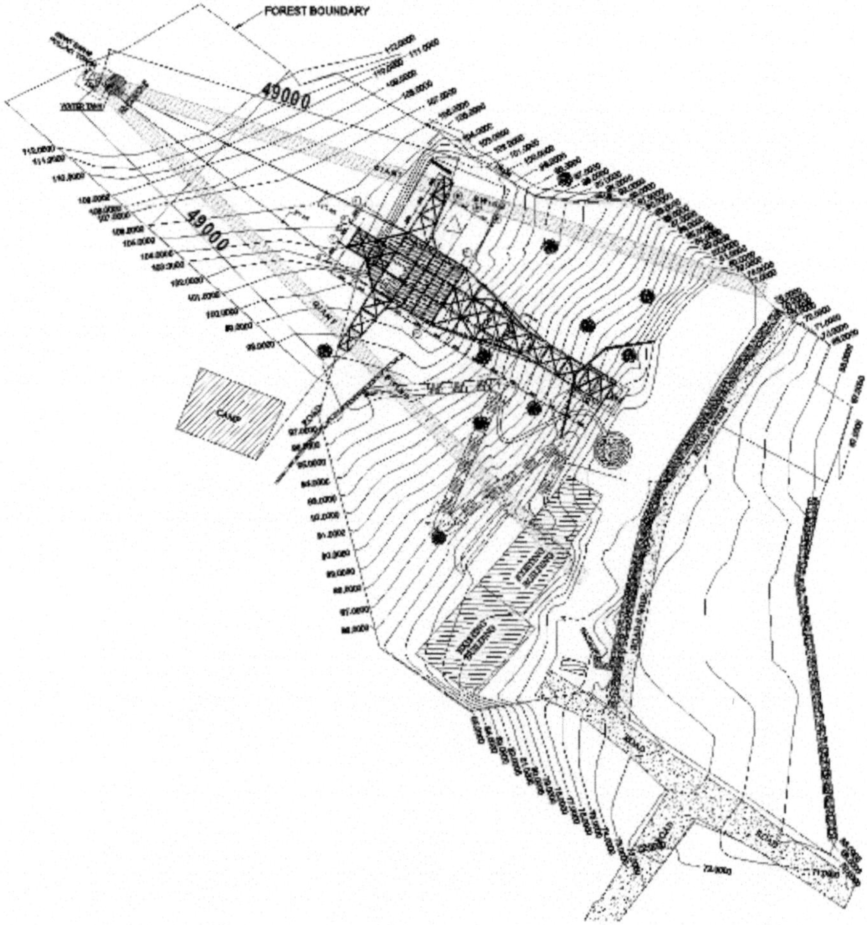


**Fig. 7** Layout plan showing tower location

## 3.4 Stratigraphy

The scope of the geotechnical investigation included a borehole and a plate load test. A photograph of the drilling in progress is illustrated in Fig. 8.

An overburden of coarse-to-fine sand intermixed with rock fragments is met at site to 7.5 m depth. It is likely that part of this may be debris material from an old landslide; however, the overburden appears to be fairly compact and stable.

Below this, completely to moderately weathered and very weak to weak purple, shale (rock) was met at site to 20.0 m with core recovery 0–43% and RQD 0%. The underlying rock classifies as completely to moderately weathered and very weak to weak grey, and sandstone (rock) was met to final explored depth of 25 m. Groundwater was met at 12.00 m depth during the period of the field investigation.

Field SPT-N values range from 35 to 45 to about 3 m depth values indicating the dense condition of strata. Below this, SPT values ranges from 55 to 67 to about 7.5 m; below this, SPT values exceed 100 to the final explored depth of 25.0 m. Groundwater was met at 12 m depth. Figure 9 presents the pictorial representation of the borelog.

**Fig. 8** Borehole drilling in progress

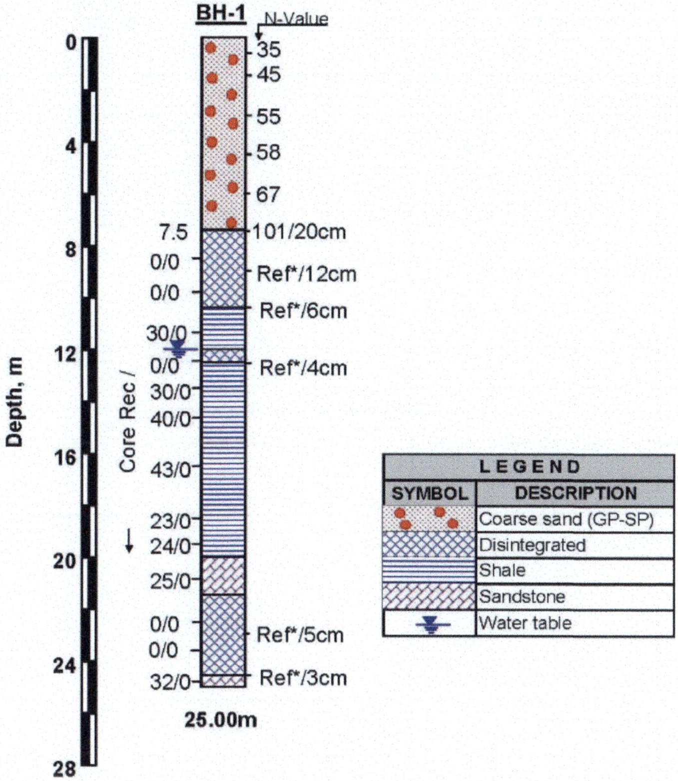


**Fig. 9** Borehole profile

Bedrock is exposed on the hillside which is thinly bedded to laminated weak shale and minor arenaceous shale/shaley arenite. Arenaceous content seems to increase southward. Rocks are often highly weathered to considerable depths. The rock formation is closely jointed and has bedding/foliation dipping westwards (WNW to SSW) at low angles.

Gentler dips of the disturbed rocks near the tower area seem the result of old slope creep. The area could be part of a shear zone as suggested by the dips of the rocks and the overturned and highly fragmented and crushed rock mass.

A plate load test was conducted on a 60 × 60 cm size test plate at 3 m depth. Figure 10 presents a photograph of the test in progress. The load-settlement curve is presented on Fig. 11.

Fig. 10 Plate load test in progress

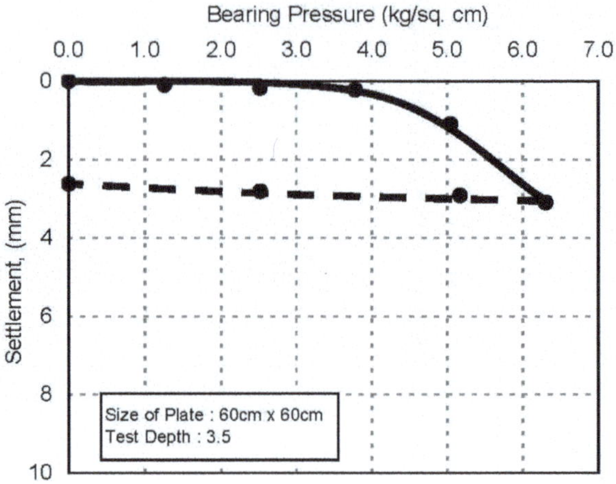


Fig. 11 Load-settlement curve—plate load test

# 4 Foundation Design

## *4.1 Net Bearing Pressure for Design*

Raft foundation of size 14 m × 15 m was planned at 4 m depth to support the steel structure weighing 175 tonnes. Analyzing the borehole data and plate load tests, the foundation is safe for a net bearing pressure of 35 $T/m^2$ for a permissible total settlement of 25 mm. The net bearing pressure due to the structural loads was about 10 $T/m^2$; hence, the structure is safe from bearing capacity and settlement standpoint.

## *4.2 Slope Stability*

**Assessment of Slopes**: In general, the hill slopes at the project site are fairly stable. Based on the overall contour map of the area, the angle of the slopes in the surrounding area has been assessed, particularly where slope angle exceeds 20 degrees. Stabilization measures such as rock bolts in critical areas, proper horticulture planning, and paving are being taken up.

**Drainage Planning**: The hill slopes are an important watershed in the area. It aids run off of rainwater from the higher reaches into the nallahs below. During monsoons, some soil erosion or denuding of vegetation could occur. This can result in a situation where the hill slope could get disturbed, causing a localized landmass to move down the slope.

Drainage of the rainwater has been carefully planned so as to divert it away from the proposed construction. Diversion of water, which may flow from the higher reaches, should be done by drains. The rainwater from the slopes shall be intercepted and drained away from the proposed construction.

All water from higher reaches have been drained off into a drainage system capable of carrying the discharge expected during the peak monsoons. The drainage is being planned such that run off does not encroach into the effective rock mass which is considered to extend 8 m beyond the outer limit of the proposed construction on all sides.

### *4.3 Slope Protection Measures*

Slope protection measures being undertaken at site for long-term stability of the slopes around the site are as follows:

1. Proper drainage is provided for the site roads. It is ensured that there is no localized ponding or stagnation of water which could seep underneath the road, which causes undermining of the road.
2. Shoulder drains are provided on either side of the roads and the water is drained away into carefully selected drainage ditches for disposal.
3. In areas where rock slopes are to be excavated, adequate precautions have been taken to ensure that the slope is stable. The slope is inspected for any possible wedges that could slide. Such wedges are removed for stability purpose.
4. A bituminous or concrete paving is provided all around the outer perimeter of the tower. This pavement extends 1.5 m beyond the outer edge of the foundations. The paving will help in limiting water ingress into the soils beneath the foundations.
5. On the uphill side, 3–5 m high retaining wall or a gabion wall is being constructed at 2–3 levels to protect the tower from rockfall from higher elevations.

**Fig. 12** Foundation construction in progress

### 4.4 *Foundation Construction*

The foundation construction is nearly over and the steel structure erection is in progress. Figure 12 presents site photographs of the foundation construction and the 16 columns.

## 5 Concluding Remarks

An 85 m high bungy-jumping tower coming up in Shivpuri, Rishikesh, will be the highest such tower in India. It is located in the mountainous terrain of lesser Himalaya on right bank of Hule River, near its confluence with the River Ganga at Shivpuri. It offers a panoramic view of the gorgeous Himalayan hill slopes to the adventurous few ready for an exciting thrill.

The RCC-steel hybrid tower, located in an artificially terraced soil, will have a plan area of 9 m × 9 m and a total weight of about 175MT. With a raft foundation at 4 m depth, the steel structure shall have 16 columns with cross-beams at every 3 m height. The jump shall be performed from a cantilever boom of 25 m projection.

An overburden of sand intermixed with rock fragments is met at site to 7.5 m depth. Below this, completely to moderately weathered and very weak to weak purple, shale (rock) was met at site to 20.0. This is underlain by completely to moderately weathered and very weak to weak grey, and sandstone (rock) was met to 25 m depth. Groundwater was met at 12 m depth.

The geotechnical investigation justified a safe net bearing pressure of 35 $T/m^2$; however, the actual loading intensity from structural considerations was only 10 $T/m^2$. In such topography, drainage planning and slope protection measures are very important; these are being meticulously implemented.

**Acknowledgements** The authors are extremely grateful to the client M/s. Bungee in Rishikesh for giving the authors an opportunity to work on this project. M/s. Skeleton Consultants Pvt. Ltd. has done the structural design for the tower and Cengrs Geotechnica Pvt. Ltd. performed the geotechnical investigation. The contribution of the architect, M/s. Studio Emergence, Mumbai, is gratefully acknowledged.

## References

1. Public Work Department Rudraprayag Uttarakhand: Preparation of Detailed Project Report, Rehabilitation and Up-gradation to 2 lane with paved shoulder configuration and strengthening of Rishikesh-Rudraprayag from km 228.0 to km 368.0 section of NH-58 in state of Uttarakhand. <0_0_12113012712111Prarup-33-GeologicalReports.pdf (forestsclearance.nic.in)>, (2013)
2. Geological Survey of India, <Geological Survey of India—Home (gsi.gov.in)>, (2000)
3. Sati SP, Sundriyal YP, Rana N, Dangwal S (2011) Recent landslides in Uttarakhand: nature's fury or human folly. Curr Sci Commentary 100(11):1617–1620

# Experimental Study of Circular Shallow Footing on the Top of Slope

**Eswara Reddy Orekanti, Venkatesh Buragadda, Pavan Kumar Arigela, B. Mounika, and K. Akshitha**

## 1 Introduction

In general, most of the building foundations are often constructed on slopes and adjacent to slopes in hilly areas and near the open excavations. Moreover, shallow foundations are generally adopted in the case of low-to-medium rise buildings. The investigation of bearing capacity of loaded slopes is very important during the construction of foundations nearer to slopes due to their more susceptible to failure [1]. The design of foundations must satisfy safe bearing capacity and allowable settlements [2]. The properties of soil and the footing characteristics play a major role on soil ultimate load-carrying capacity. In practice, a reasonable factor of safety is applied to the ultimate values to produce the allowable values of bearing capacity of foundations, depending on the uncertainty of soil behavior and the loading conditions. However, it is quite complex to model and analyze foundations which are constructed on sloping ground for their ultimate load-carrying capacity by considering many of the parameters and site conditions.

E. R. Orekanti (✉) · V. Buragadda · P. K. Arigela · B. Mounika · K. Akshitha
Department of Civil Engineering, School of Engineering, Mohan Babu University (Erstwhile Sree Vidyanikethan Engineering College), A. Rangampet, Tirupati 517102, India
e-mail: orekantieswar@gmail.com

V. Buragadda
e-mail: venkatesh.b@vidyanikethan.edu

P. K. Arigela
e-mail: arigelapavancivil@gmail.com

B. Mounika
e-mail: mounikabathuluru@gmail.com

K. Akshitha
e-mail: kakshitha9@gmail.com

 
B. T. Jose et al. (eds.), *Proceedings of the Indian Geotechnical Conference 2022 Volume 2*, Lecture Notes in Civil Engineering 477,
https://doi.org/10.1007/978-981-97-1741-5_26

As per the literature, many researchers considered various assumptions and evaluated bearing capacity of shallow foundations which are constructed on slopes which consist of cohesionless or cohesive properties [3–6]. Moreover, most of the studies were performed on sandy slopes using strip footing rather than circular or square footings [7–10]. However, very few researchers carried tests on c-φ soil using spread footing resting on the top of slope y [11, 12]. Based on the literature review, it is necessary to address the behavior of shallow footing resting on steep (>30°) sloping ground consisting of c-φ soil. Hence, the present study is an attempt to develop bearing capacity of the shallow circular foundation resting on the top of 45° sloping ground for certain c-φ soil at a setback distance through practical approach. Finally, the objective of the present research is to evaluate the bearing capacity of an isolated circular footing near the slope, specifically, to determine the contribution of $D_e/B$ (offset distance, $D_e$, to Footing width, B) on the bearing capacity.

## 2 Materials

The c-φ soil used for the experimental work was collected from surrounding village of the institute known as A. Rangampet, Tirupati, Andhra Pradesh, India. According to BIS standards, the soil physical properties were found out. The soil was collected from the field in large sacks and the state of field soil seems to have a small amount of moisture. Hence, the collected soil was dried to remove the soil natural humidity by placing in an oven for 24 h. At the end of 24 h, the dried soil particles came together and formed as many numbers of large lumps. Hence, the formed lumps were broken into fine powder form using wooden hammer and are sieved according to the experimental conditions. Table 1 shows the properties of c-φ soil used in the present laboratory model test.

**Test Tank and Footing Plate**

The test tank used for the present model tests was a rectangular tank having dimensions of 1200 mm (L) × 800 mm (B) × 800 mm (H). Two sides of the test tank were covered with non-acrylic thick wooden sheets and remaining one side consists of an acrylic sheet of 8 mm thickness. The following wooden and acrylic sheets were fixed with nut and bolt arrangement. A rigid circular footing having a size of 110 mm having a thickness of 12 mm made with mild steel was used as a model footing for the present model tests.

## 3 Experimental Program

C-φ soil was used for the present model tests and test soil was placed in the test tank and compacted at its maximum dry unit weight as mentioned in Table 1. Initially, the amount of soil required for filling the tank up to its total height with a soil slope

**Table 1** Properties of soil used in the present research

| S. No. | Property | Value |
|---|---|---|
| 1 | Specific gravity (G) | 2.66 |
| 2 | Liquid limit ($w_L$) | 42% |
| 3 | Plasticity index ($I_p$) | 23% |
| 5 | Sand | 81% |
| 6 | Silt | 8% |
| 7 | Clay | 11% |
| 8 | IS classification | SC |
| 9 | Maximum dry density | 20 kN/m$^3$ |
| 10 | Optimum moisture content | 14% |
| 11 | Cohesion (c) | 14 kN/m$^2$ |
| 12 | Angle of internal friction ($\phi$) | 28° |
| 13 | Young's modulus, E (kN/m$^2$) | 7501 kN/m$^2$ |
| 14 | Poisson's ratio, $\mu$ | 0.35 |

angle of 45° is 1.1 tons. The calculated amount of soil was placed in the test tank layer wise and compacted uniformly using tampers. A total of 22 layers of soil are compacted with a tamping rod of 3.25 kg with a height of fall of 30 cm to achieve the maximum dry unit weight of soil. The consistency of the soil compaction was determined by using core-cutter method. After preparation of the soil bed, the test tank was dismantled except on one side to create horizontal fixity as depicted in Fig. 1a. A hand-operated hydraulic jack was used for the load application on the footing. Vertical load is applied by fixing hydraulic jack within the connecting rod setup. To determine the footing displacement during loading, the dial gauges were placed on both sides of connecting rod as shown in Fig. 1b. To determine the effect of offset distance (i.e., distance from footing edge to the slope surface, $D_e$) on footing load-carrying capacity, the test was performed by maintaining the ratio of $D_e/B$ = 1.0. It means that the model footing was placed at a distance of 110 mm from the slope surface. The complete schematic diagram of the test setup is as depicted in Fig. 2.

## 3.1 *Numerical Program*

For the present research, a Plaxis 2D finite element analytical software was used to perform the numerical analysis. During the anlaysis, a denser mesh was formed with a number of 15 node triangular elements. As per the literature, 15 node triangular elements' mesh gives more accurate results as compared other node models. Hence, the deformation of the soil at failure can be accurately measured. Figure 3 presents a sample of a finite element mesh of the numerical model defined in PLAXIS.

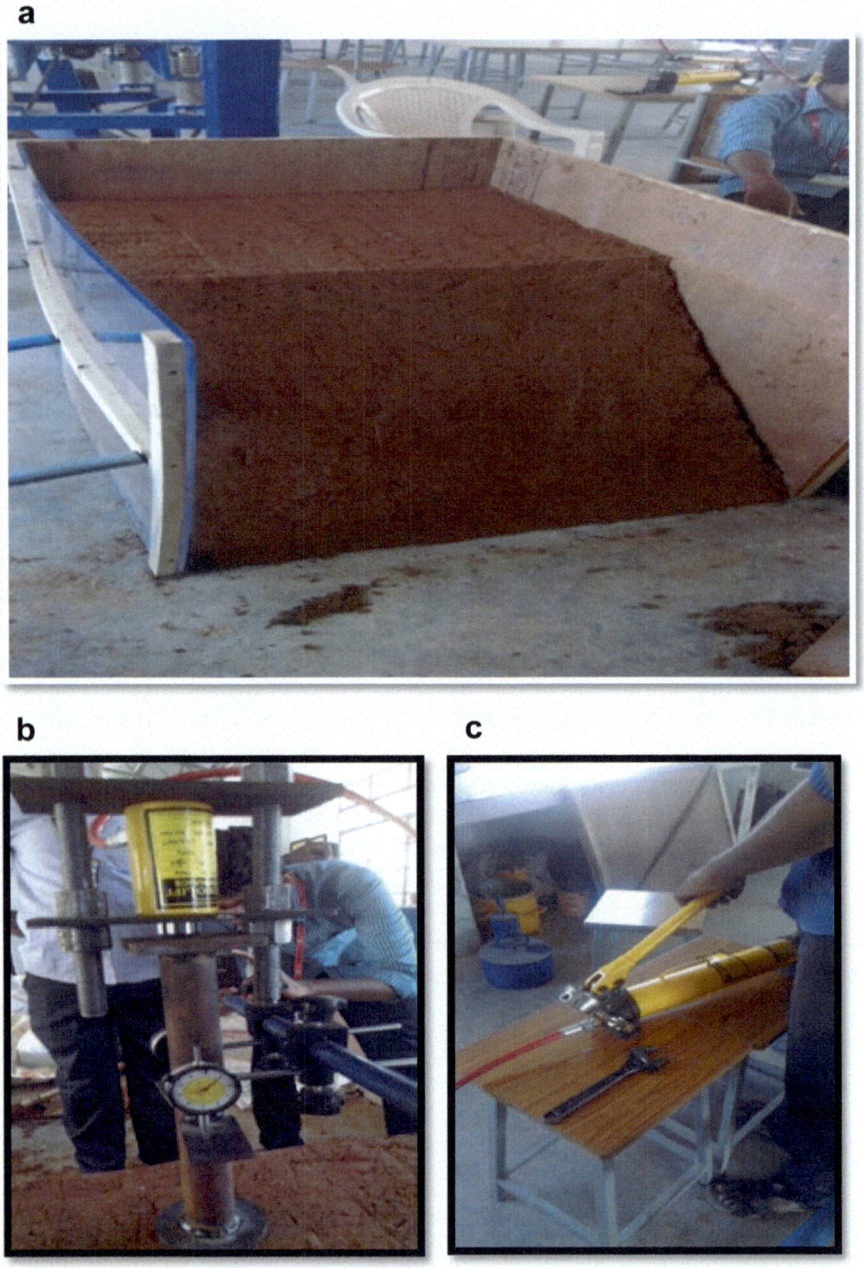

**Fig. 1** **a** Embankment placed in the test tank with an angle of 45°. **b**, **c** Total test arrangement of the footing placed on soil bed along with hydraulic jack and dial gauges

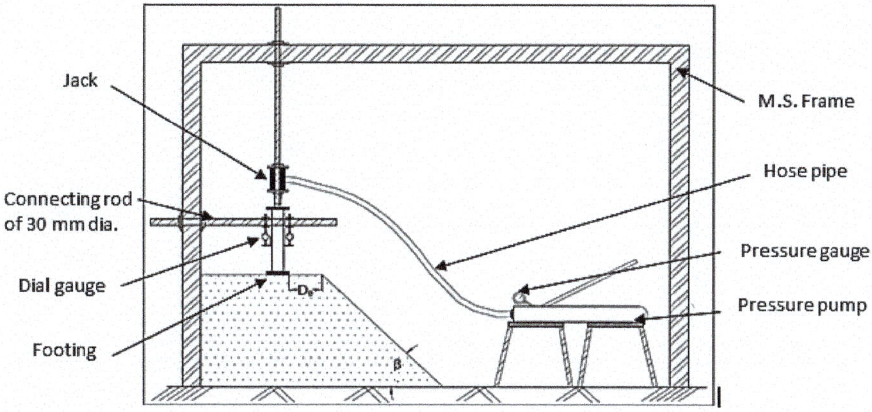


**Fig. 2** Schematic view of total test setup

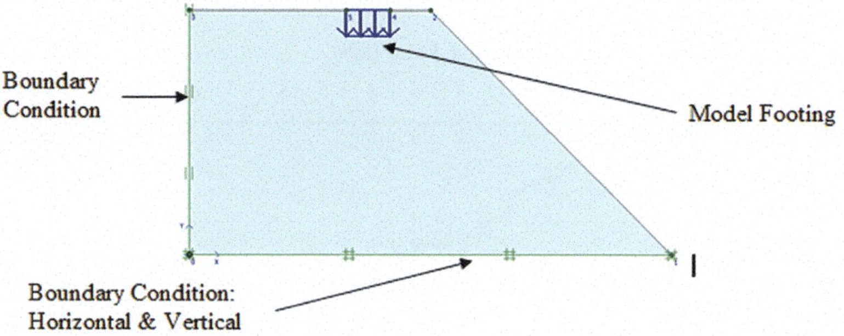


**Fig. 3** Boundary condition for the numerical model

Table 2 depicts the model soil properties used in the numerical analysis of the present research.

## 4 Results and Discussion

Figure 4 shows the load-settlement results of footing during the experimental model tests. The load-carrying capacity of the footing goes on increases with increase in footing settlement (s). However, after certain settlement, the footing load-carrying capacity diminished as marked in Fig. 4. Finally, Fig. 4 revealed that the load-carrying capacity at footing failure or ultimate load-carrying capacity is 6.20 kN (i.e., 652.4 kPa) at a corresponding settlement of 8 mm, i.e., s/D of 7.27%. Similarly, Fig. 5 depicts the load-settlement results of the footing during numerical analysis. The numerical test results also show that the footing ultimate load-carrying capacity is 6.34 kN (i.e., 667.14 kPa) at the corresponding s/D of 6.82%. Therefore, it is

**Table 2** Soil properties used in the Plaxis 2D Model

| S. No | Property | Value |
|---|---|---|
| 1 | Young's modulus, E (kN/m$^2$) | 7501 |
| 2 | Poisson's ratio, $\mu$ | 0.35 |
| 3 | Angle of dilatancy, $\Psi$ (°) | 0 |
| 4 | Material type | Undrained |
| 5 | $\gamma_{unsaturated}$ (kN/m$^3$) | 20.27 |
| 6 | $\gamma_{saturated}$ (kN/m$^3$) | 22.8 |
| 7 | Cohesion, c (kN/m$^2$) | 14 |
| 8 | Angle of internal friction, $\phi$ (°) | 28 |
| 9 | Coefficient of permeability, k (m/day) | 1.08E-04 |

evident from the results that the numerical test results are well matched with the experimental test values.

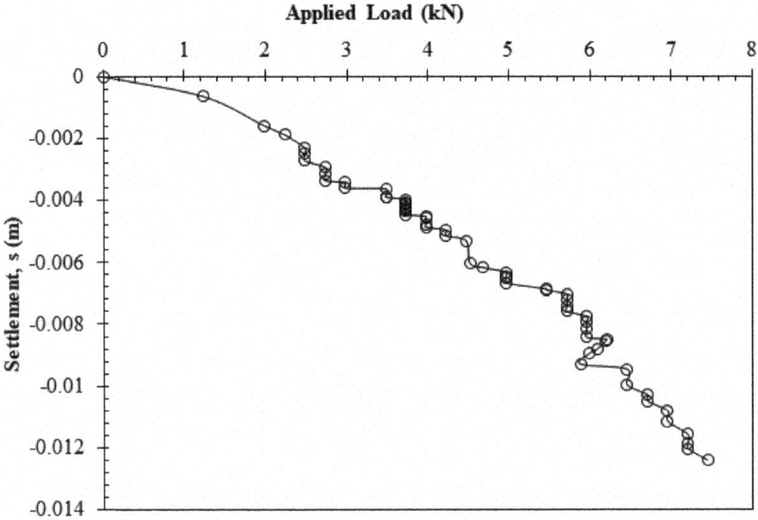


**Fig. 4** Load-settlement curves of footing during experimental model test

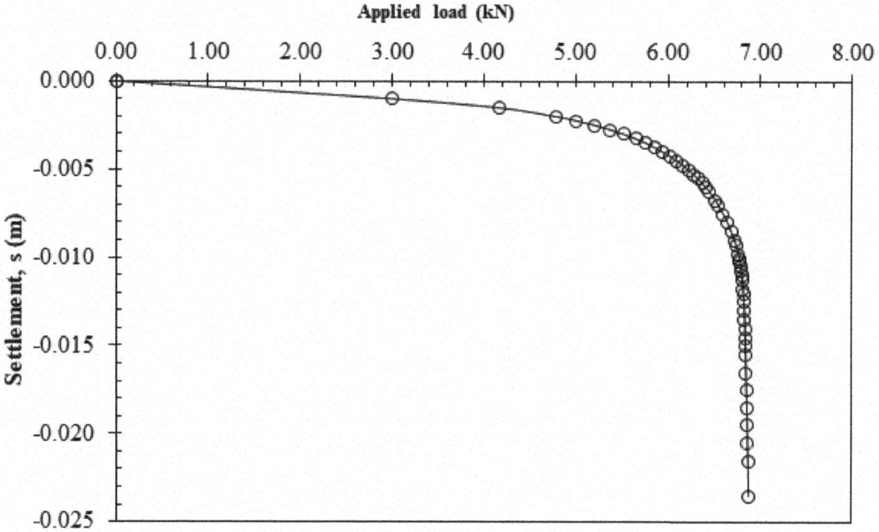


**Fig. 5** Load-settlement curve of footing during numerical analysis

## 5 Conclusion

The present model test results of the shallow footing placed at an offset distance to footing size ratio ($D_e$/B) of 1.0 show that the ultimate load is 652.4 kPa. Further, the results from the numerical analysis using Plaxis 2D, were also matched well with the experimental test results. Hence, it is possible to understand the load-settlement behavior of the shallow footing for further conditions such as larger offset distances (i.e., $D_e$/B > 1) and slope angle by performing the numerical analysis in future using the same model soil parameters as used in the present study.

## References

1. Shields D, Chandler N, Garnier J (1990) Bearing capacity of foundation in slopes. J Geotech Eng ASCE 116(3):528–537. https://doi.org/10.1061/(ASCE)0733-9410(1990)116:3(528)
2. Meyerhof GG (1957) The ultimate bearing capacity of foundations on slopes. In: Proceeding 4th international conference on soil mechanics and foundation engineering, vol 1, pp 384–386
3. Castelli F, Motta E (2010) Bearing capacity of strip footings near slopes. Geotech Geol Eng 28(2):187–198. https://doi.org/10.1007/s10706-009-9277-9
4. Narita K, Yamaguchi H (1990) Bearing capacity analysis of foundations on slopes by use of log-spiral sliding surfaces. Soils Found 30(3):144–152. https://doi.org/10.3208/sandf1972.30.3_144
5. Saran S, Sud VK, Handa SC (1989) Bearing capacity of footings adjacent to slopes. J Geotechn Eng ASCE 115(4):553–573. https://doi.org/10.1061/(ASCE)0733-9410(1989)115:4(553)
6. Shields DH, Scott JD, Bauer GE, Deschenes JH, Barsvary AK (1977) Bearing capacity of foundation near slopes. In: Proceeding 10th international conference on soil mechanics and

foundation engineering, Tokyo, Japan, Japanese society of soil mechanics and foundation engineering, vol 1, pp 715–720
7. Acharyya R, Dey A (2017) Finite element investigation of the bearing capacity of square footings resting on sloping ground. INAE Lett 2(3):97–105. https://doi.org/10.1007/s41403-017-0028-6
8. Azzam WR, ElWakil AZ (2015) Experimental and numerical studies of circular footing resting on confined granular subgrade adjacent to slope. Int J Geomech 16(1):1–15. https://doi.org/10.1061/(ASCE)GM.1943-5622.0000500
9. Castelli F, Lentini V (2012) Evaluation of the bearing capacity of footings on slopes. Int J Phys Modell Geotech 12(3):112–118. https://doi.org/10.1680/ijpmg.11.00015
10. Salih Keskin M, Laman M (2013) Model studies of bearing capacity of strip footing on sand slope. KSCE J Civ Eng 17(4):699–711. https://doi.org/10.1007/s12205-013-0406-x
11. Acharyya R, Dey A, Kumar B (2018) Finite element and ANN based prediction of bearing capacity of square footing resting on the crest of c-φ soil slope. Int J Geotech Eng 14(2):176–187. https://doi.org/10.1080/19386362.2018.1435022
12. Leshchinsky B (2005) Bearing capacity of footings placed adjacent to c'-φ' slopes. J Geotech Geoenviron Eng 141(6):1–13. https://doi.org/10.1061/(ASCE)GT.1943-5606.0001306

# Settlement in Geosynthetic Reinforced Square Footing Over Plastic Soil

**Ankur Mudgal, Bibek Jha, Raju Sarkar, Amit Kumar Srivastava, Akshit Mittal, and Nehal Jain**

## 1 Introduction

In Geotechnical engineering, failure of a foundation constructed over Plastic soil is a very prevalent issue. The foundation may lose its stability due to poor load carrying capacity of the surrounding soil, which results in higher settlements than the acceptable values and deterioration in the load carrying capacity of the soil. Many researchers have proffered many solutions to mitigate these problems, through different soil improvement techniques. The first comprehensive study on soil reinforcement was conducted by [1], wherein aluminium strips were used to reinforce the sand beds. Examined the performance of strip footings on geogrid reinforced sand bed over a soft clay slope [2]. The study showed that the use of geogrid layers in the replaced sand not only significantly improves the performance of footing but also leads to high reduction in the depth of reinforced sand layer required to achieve the allowable settlement. Carried out the laboratory model footing tests on reinforced soil bed and reported their results as a comparative study of effectiveness of geogrid and geotextile as soil reinforcement [3]. The results showed that the geogrid is more efficient than the geotextile in respect of bearing capacity of foundation on reinforced

A. Mudgal (✉) · B. Jha · N. Jain
CEG Test House and Research Center PVT. LTD, Jaipur, India
e-mail: ankur.mudgal@cegtesthouse.com

B. Jha
e-mail: bibek.jha@cegtesthouse.com

N. Jain
e-mail: nehal.jain@cegtesthouse.com

R. Sarkar · A. K. Srivastava · A. Mittal
Department of Civil Engineering, Delhi Technological University, Delhi, India
e-mail: aksrivastava@dce.ac.in



B. T. Jose et al. (eds.), *Proceedings of the Indian Geotechnical Conference 2022 Volume 2*, Lecture Notes in Civil Engineering 477,
https://doi.org/10.1007/978-981-97-1741-5_27

**Table 1** Geotechnical properties of soil

| Properties | Values | Protocols/standards |
|---|---|---|
| Specific gravity | 2.67 | IS 2720 (Part III) |
| Liquid limit (%) | 29 | IS 2720 (Part V) |
| Plastic limit (%) | 20 | IS 2720 (PartV) |
| Plasticity index (%) | 9 | IS 2720 (Part V) |
| Maximum dry unit weight (kN/m$^3$) | 17.6 | IS 2720 (Part VII) |
| Optimum moisture content (%) | 15 | IS 2720 (Part VII) |
| Angle of friction | 22° | IS 2720 (Part XI) |

sand. In in this paper, small scale laboratory tests have been performed and obtained results are presented.

## 2 Material Used

### *2.1 Soil*

The geotechnical properties of the soil were determined as per Indian standards listed in Table 1. The soil was classified as a low compressible soil.

### *2.2 Geosynthetics*

A single type of geogrid, and geotextile were used in this study. The geogrid used was bi-oriented and was made of polypropylene thermoplastic, whereas the geotextile used was woven type also made of polypropylene material. The material testing certificate of geosynthetics as provided by the manufacturer is tabulated in Table 2. The geogrid and geotextile in the study have been represented by GGR and GTX respectively, throughout the study.

## 3 Experimental Setup

### *3.1 Preparation of Test Bed*

Test bed was prepared with dimensions of 750 × 450 × 600 mm (L × B × H). Initially the soil was air dried and pulverized and then it was compacted at its maximum dry unit weight of 17.6 kN/m$^3$. Predetermined water content was thoroughly mixed to soil

**Table 2** Technical characteristics of geosynthetics (courtesy: supplier's data)

| Geosynthetic | Property | Data | Unit | Test method |
|---|---|---|---|---|
| Geogrid | Mesh type | Quadrangular apertures | – | – |
| | Polymer type | Polypropylene | – | – |
| | Aperture size | 30 × 30 (MD × CMD) | mm | – |
| | Stiffness at 0.5% strain | 550 × 350 (MD × CMD) | kN/m | ISO-10319 |
| Geotextile | Tensile strength | 475 × 384 (MD × CMD) | kN/m | IS 1969 |
| | Opening Size | 0.075 | mm | ASTM D4751 |
| | Weight of fabric | 200 | $g/m^2$ | ASTM D5261 |
| | Elongation at break | 30 × 28 (MD × CMD) | (%) | IS 1969 |

to achieve the optimum moisture content, i.e. 15%. In case of unreinforced condition, soil was compacted in three lifts, whereas for the reinforced case, the thickness of each lift was decided according to the spacing between the reinforcements. The soil was then poured into the tank and compacted using a rammer with a base of 150 mm diameter up to a marked height. During the test, the height of fall of the rammer, number of blows to be given, and the required amount of soil sample was determined to maintain the condition of uniformity of soil sample in the tank. At the end of compaction, a spirit level was used to check the alignment of the horizontal surface of prepared test bed.

### 3.2 *Layout of Geosynthetics*

The geosynthetic configurations were decided according to testing procedure described in the testing programme, i.e. at the respective u/B, b/B, h/B, ratios, and number of reinforcement layers N. All the five different dimensionless parameters (i.e. u/B, b/B, d/B, N, and h/B) were varied to ascertain the optimums in geosynthetic placements. Geometry of geosynthetics reinforced bed is shown in Fig. 1.

### 3.3 *Test Setup*

A number of laboratory footing tests were conducted on unreinforced and reinforced soil by using testing facilities developed at laboratory. The dimensions of the test set up are as follows: Length = 750 mm, width = 450 mm, and height = 600 mm. The back and sides of the tank was fabricated from 20 mm thick steel sheet braced with structural steel member whereas front side of the tank consisted of acrylic sheet of 20 mm thickness for visual observation. An angle was fixed on the face of acrylic sheet to prevent its buckling during the compaction and loading. Inner surface of

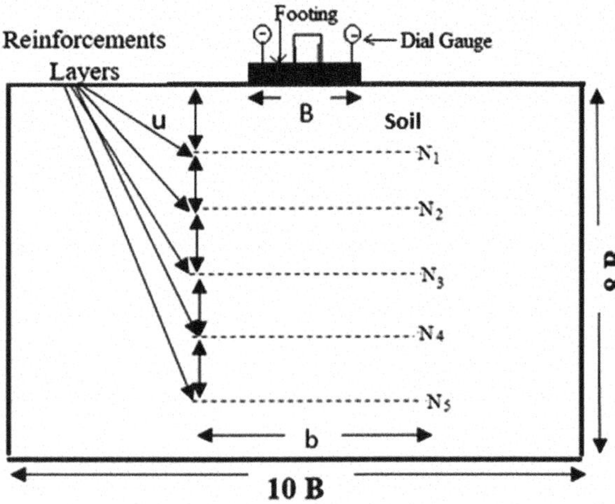


**Fig. 1** Geometry of geosynthetics reinforced bed

tank was greased to prevent the adverse effect of friction on the test results. A steel plate of size 75 × 75 mm was used as a model footing and the dimensioning of the tank were done in accordance with the footing width, so as to avert the boundary effect. Hence, dimensions of 10B, 6B, and 8B were chosen as the length, width, and height of the tank respectively, where B is the model footing width, i.e. 75 mm. The base of the footing was kept rough by gluing the sand with epoxy glue. The tank was tested in loading frame consisting of two rigid and heavy steel plate columns of thickness 150 mm attached to top head of the loading frame. A load cell of 25 kN capacity was placed at center between the footing plate and upper platen to avoid the eccentric loading. The output of the load cell was logged using a data logger in the form of pressure. Two dial gauges with accuracy of 0.001 mm were used at points diametrically opposite to the footing. Average reading obtained by both the dial gauges was considered for settlement analysis. The test bed was tested as per the provision of (ASTM 1997) up to a settlement of 20 mm, where the load increments were applied and maintained at the obtained value until the rate of settlement was less than 0.03 mm/min over three consecutive minutes. Sitting load was applied initially over the footing to fix the footing over the soil base, so as to obtain planar strain conditions. A general arrangement of the testing setup is shown in Fig. 2.

### *3.4 Testing Procedure*

The aim of the study was to investigate the effect of reinforcements in bearing capacity of reinforced soil foundation. The model tests were conducted with both the reinforcements, i.e. GGR, and GTX. The detailed experimental programme is shown in Table 3. Test series "A" was carried out on unreinforced soil footing bed to compare the results obtained by tests conducted on reinforced soil bed. Two series of tests (B

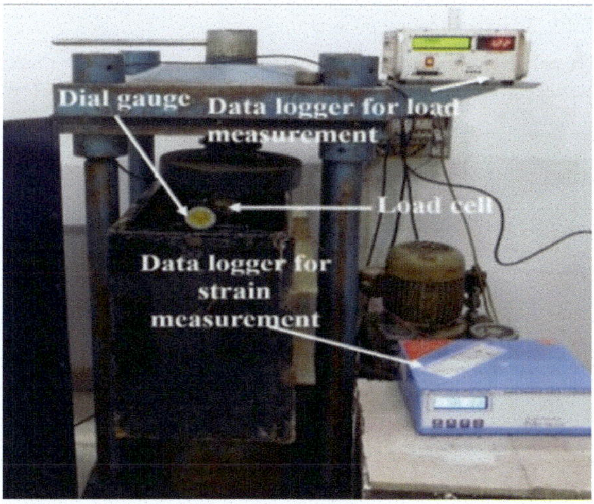


**Fig. 2** General arrangements of testing setup

and C) were conducted for both the reinforcements, i.e. GGR and GTX. Initially, at reinforcement width equal to 5B, the u/B values were varied form 0.17, 0.34, 0.51, 0.68, and 0.85, respectively. Effect of number of reinforcement layers was estimated by fixing the top layer at maxima obtained from the previous test and varying the number of reinforcements until the effect of the reinforcement becomes diminished, or becomes considerably insignificant for any further extensions in number of reinforcement layers. The vertical spacing between the reinforcement layers was also fixed at optimum u/B value. Similar testing procedure were adopted for different reinforcement widths, i.e. 4B and 6B. The effect of spacing between two corresponding reinforcements was analysed by varying the distance between two reinforcements by a factor of 0.08 B, 0.16 B, 0.24 B, 0.32 B, and 0.4 B, respectively, and fixing the top layer at the optimum obtained from the previous tests.

# 4 Results and Discussion

## *4.1 Determination of Optimum Depth of First Layer of Reinforcement*

The settlement measured from the dial gauges is considered as footing settlement and denoted as (s). The ratio of footing settlement (s) to the width of footing (B) is defined as settlement ratio (SR), indicated in percentage. A non-dimensional parameter bearing capacity ratio (BCR) was calculated at each settlement ratio to give an incisive account of the improvement in load carrying capacity due to inclusion of reinforcement in the Plastic soil bed. The BCR is defined as the ratio of the bearing pressure of a reinforced soil to that of an unreinforced soil, when evaluated at the

**Table 3** Experimental programme

| Test series | Reinforcement | *N* | *u/B* | *b/B* | h/B | No. of test | Remarks |
|---|---|---|---|---|---|---|---|
| A | Unreinforced soil | – | – | – | – | 3 | To estimate the improvements due to reinforcement |
| B | GGR | 1 | 0.17, 0.34,0.51,0.68,0.85 | 5B | – | 5 | To find out the optimum *u/B* value |
| | | 1,2,3,4,5 | Optimum depth | 5B | 0.34B | 5 | To find the effect of number of layers of geogrid |
| | | Optimum depth | 0.34 | 4B, 5B, 6B | | 15 | To check the optimum values of geogrid width and number of geogrid layers |
| | | 2 | 0.34 | 5B | 0.08 B, 0.16 B, 0.24B, 0.32 B and 0.4 B | 5 | To check the effect of vertical spacing between reinforcements |
| C | GTX | 1 | 0.17, 0.34, 0.51, 0.68,0.85 | 5B | | 5 | To find out the optimum *u/B* value |
| | | 1,2,3,4,5 | Optimum depth | 5B | 0.34B | 5 | To find the effect of number of layers of geogrid |
| | | Optimum depth | 0.34 | 4B, 5B, 6B | | 15 | To check the optimum values of geotextile width and number of geotextile layers |
| | | 2 | 0.34 | 5B | 0.08 B,0.16 B, 0.24 B, 0.32 B and 0.4 B | 5 | To check the effect of vertical spacing between reinforcements |

same settlement ratio. In this study, BCR was calculated at four different settlement ratios, i.e. UBC, 4, 8, 12, and 16%. It should be noted that the BCR is similar to an improvement factor used by many researchers in their studies [4, 5]. Another parameter settlement reduction factor (SRF) being also used in the study, SRF can be calculated as.

$$\mathrm{SRF} = 1 - \frac{\boldsymbol{(S)\mathbf{r}}}{\boldsymbol{(S)\mathbf{ur}}} \times 100\,\mathrm{for}\, s/B = \mathrm{UBC}, 4,\ 8,\ 12,\ \mathrm{and}16\%$$

where $(S)_{\mathrm{ur}}$ = Settlement of unreinforced soil and $(S)_{\mathrm{r}}$ = the settlement of reinforced soil bed at bearing pressure with respect to $(S)_{\mathrm{ur}}$. This settlement reduction factor (SRF) is similar to percentage reduction settlement (PRS) used by [6, 7] in their studies to quantify the performance improvement in settlement in terms of percentage. Many researchers have also considered settlement ratio (SR) as a parameter to compare the settlement reduction with the application of geosynthetics. Foundations are designed in accordance to the allowable bearing pressure of the soil. Thus, computation of ultimate bearing capacity (UBC) becomes substantial to correctly assess the increase in construction viability of the soil with applications of geosynthetics. To determine the optimum value of first depth of reinforcement layer, initially five tests were conducted on square model footing, supported by single layer of each geosynthetic. Figure 3a, b shows the pressure-settlement curves for GGR and GTX, respectively. As can be deciphered from the curves that bearing pressure of soil increases as the ratio of u/B increases. However the rate of increases in the bearing pressure is significant until a value of u/B = 0.34 after which bearing pressure rapidly decreases with increasing the u/B value. Figure 4a, b depict the improvement factor versus u/B for GGR and GTX, respectively. It can be observed from the graph that the BCR gradually increases as u/B value increase from 0.17 to 0.34, afterwards a decrease in BCR can be observed with increases in u/B. Reported somewhat similar findings that the bearing capacity of a square footing on a geogrid reinforced soil bed improved significantly to a depth of placement of u/B = 0.33 [8]. The probable reason of these optimum values of u/B is that when u/B < 0.34, the surcharge pressure was not sufficient to generate the friction at soil–reinforcement interface.

## 4.2 *Effect of Number of Reinforcement Layers and Effective Depth*

For ascertaining the effect of increasing number of reinforcements on the bearing capacity of the soil, the top layer of reinforcement was fixed at 25.5 mm (i.e. u/B = 0.34) and then reinforcement layer was varied by fixing the vertical spacing of 25.5 mm till the effect of reinforcement becomes insignificant. The reinforcement ratio (d), which is defined as the ratio of the total depth of the reinforcement and the width of the footing was ascertained for each reinforcement case. The reinforcement depth below the base of the footing can be expressed as

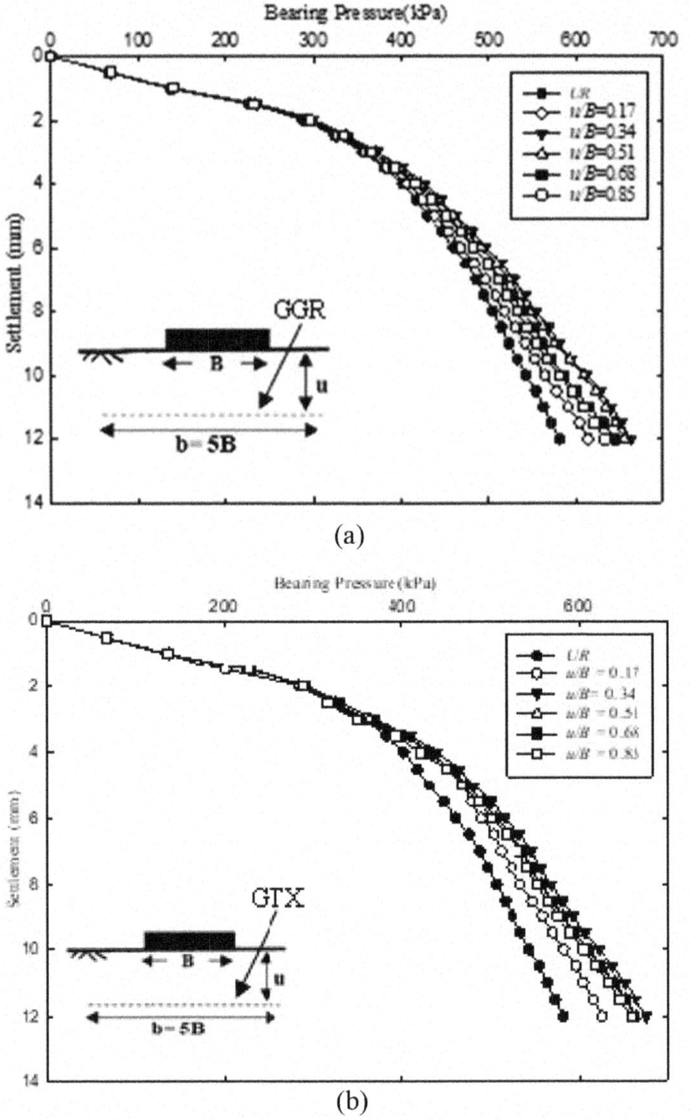


**Fig. 3** Pressure-settlement curves for **a** geogrid **b** geotextile at different u/B ratios

$$\boldsymbol{d} = \boldsymbol{u} + (\boldsymbol{N} - 1) \times \boldsymbol{h} \quad (1)$$

where $u$ = first reinforcement depth below the base of footing, $h$ = vertical spacing between two consecutive layers of reinforcement, $N$ = Number of reinforcement layers. The pressure-settlement curves were plotted for each number of reinforcement layer to compare with unreinforced one. Figure 5a, b show the pressure settlement

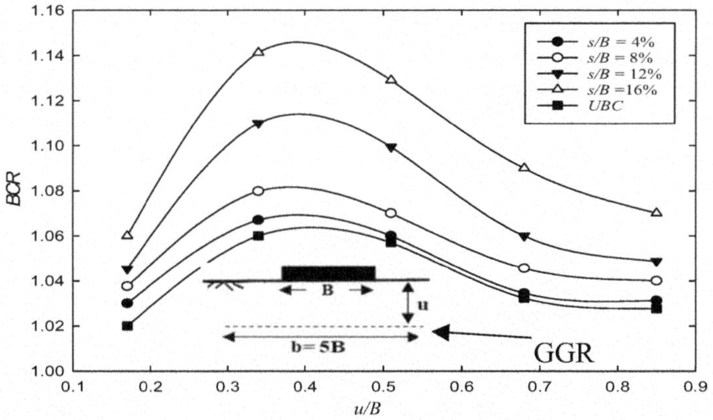


(a)

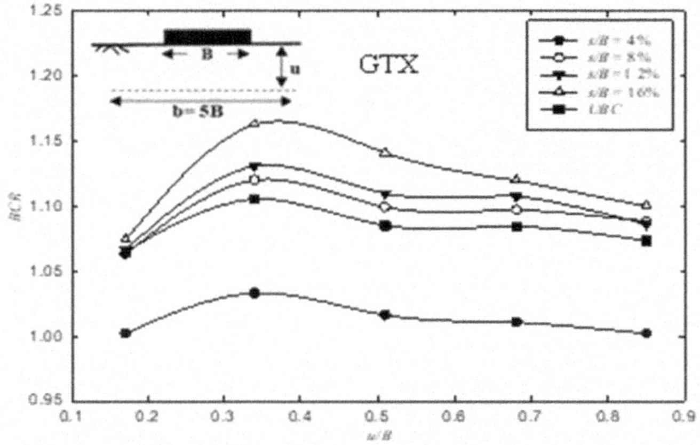


(b)

**Fig. 4** Bearing capacity ratio (BCR) for **a** geogrid **b** geotextile at different u/B ratio

for geogrid and geotextile, respectively. As expected, the value of bearing pressure increases with increment in number of reinforcement layers. However, improvement in bearing capacity becomes almost insignificant after the addition of fourth layer and third layer which are located at a depths of 1.36B and 1.02B for geogrid and geotextile, respectively. The probable reason of bearing capacity improvement is that the increase in friction at the soil reinforcement interface, which increases with the increase in reinforcement layers. Also, better interlocking between the soil and geogrids and passive earth resistance can be attributed as the reason for the development of the bearing capacity. Better interlocking between the geogrid and the soil prevented the lateral deformation of the soil. As a result of applied load, tension is mobilized in the geosynthetics which resist the shear stresses developed in the soil below the loading area and transfer them to the stable soil, thus eventually increasing the depth of the failure zone thus, results in higher bearing capacity and settlement reduction. Similar findings were also observed by [8], they reported that the inclusion of geogrid reinforcements became insignificant after a 1.33B depth of reinforcement. On the basis of their findings, they reported a maximum of four geogrid layers as optimum reinforcement in case of a strip footing.

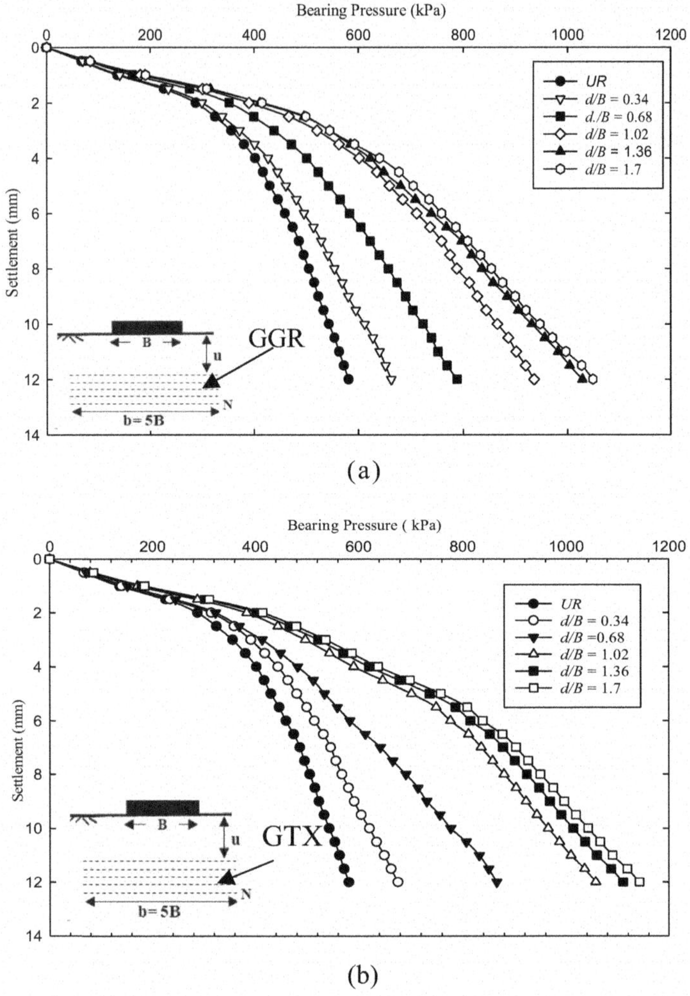


**Fig. 5** Pressure-settlement curves for **a** geogrid **b** geotextile at different reinforcement depth

### 4.3 Effect of Type of Reinforcement

Geogrid and geotextile of different stiffness are used in the present study. Technical characteristics of geosynthetics are presented in Table. 3. Reinforcements used in the study were made of same material but the tensile properties of geotextile is higher than the geogrid. In order to investigate the effect of tensile properties of reinforcement in reduction in the development of bearing capacity of soil foundation bed, the values of BCR were estimated at different settlement ratios, i.e. (4, 8, 12, and 16% and UBC). Figure 6a, b depict the variation of BCR with reinforcement depth, i.e. d/B for geogrid and geotextile, respectively. The nature of the curve may be classified in to two groups; one for settlement level s/B < 4% (lower settlement level) in which ultimate bearing capacity lies and other for s/B > 4% (higher settlement level). For the first group, geogrid impart much substantial improvement in bearing capacity than geotextile. The reason for the same can be explained that at lower settlement level, geogrid efficiently mobilized the lateral stress resistance capacity due to the confinement effect which plays a vital role in reinforcement mechanism. However for higher settlement (s/B > 4%), performance of geotextile is much better and gives more improvement than the geogrid. The reason behind this can be explained as, at certain settlement, geotextile requires higher deformation to perform on its full capacity due to its higher tensile strength. Generally, foundation constructions require to be constructed for the ultimate bearing capacity of the soil. Geogrid is better performing material than geotextile for limited settlement requirements or settlement upto ultimate bearing capacity. However, geotextiles primary function is to act as a filter or in drainage system behind retaining walls, adjacent to roads, within slopes, etc. Thus can be considered as reinforcement material where small tensile strength is required. Presented the similar findings when they compared the geogrids with geotextile at a certain settlement level [9].

### 4.4 Influence of Reinforcement Width

The effect of reinforcement width was analysed by the variation of reinforcement width as 4B, 5B, and 6B. When comparing the values of BCR from Fig. 7a–c for the same number of reinforcement layers at a different reinforcement width, it is clear that the footing performance in terms of bearing capacity improvement for both the reinforcement geogrid and geotextile are significantly improved with an increase in reinforcement width. This significant improvement continues at around five times of width of footing for both the reinforcements. Consider, for example for the geogrid case with $N = 4$ at UBC. The BCR increased from 1.44 to 1.56 when b/B ratio varied from 4 to 5. Further increment in b/B ratio from 5 to 6, the BCR was increased from 1.56 to 1.58. It is observed that a maximum rise in BCR values was observed for b/B ratio between 4 and 5 than those between 5 and 6. It is clear that satisfactory results may not be expected with increment in reinforcement width beyond 6B.

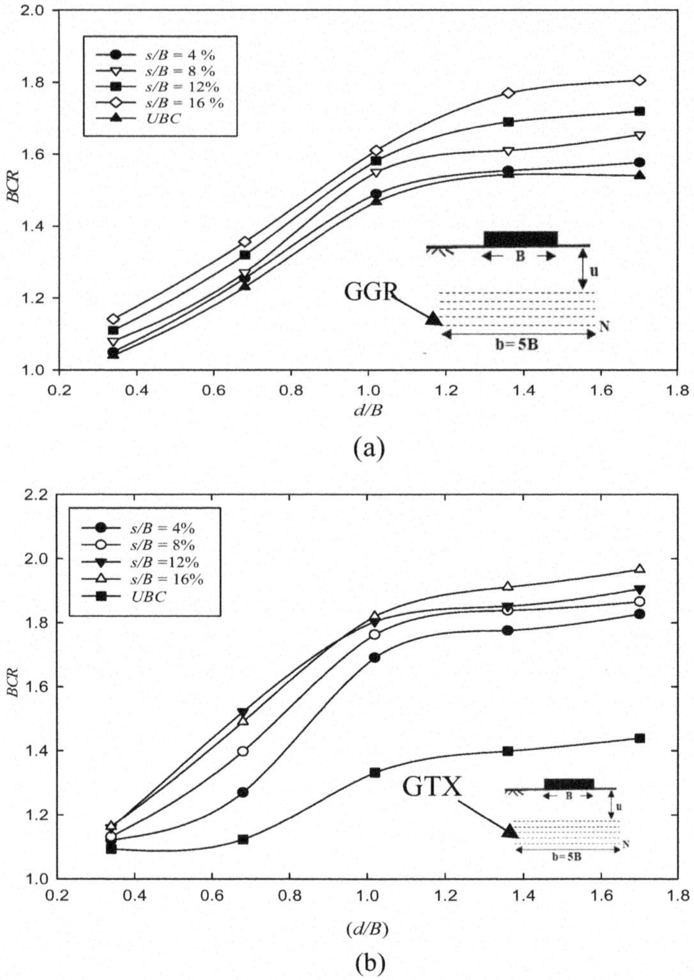


**Fig. 6** Variation of BCR versus reinforcement depth (d) for **a** geogrid **b** geotextile

Similar to this finding, [5] reported that the optimum width of geogrid to reinforce the square footing resting over the sand observed at b/B = 5–5.93. They explained that the concept of optimum width of reinforcement comes from the fact that only those portion of reinforcement is mobilized the tensile strength effectively, which lies in the shear zone below the footing. Beyond the shear zone, some more length is required as anchorage to impart pull-out resistance to the reinforcement thus, the optimum width of reinforcement is sum of length of reinforcement in anchorage zone and shear zone on the both sides. Further increment in reinforcement width beyond the optimum value will not be effective and satisfactory results cannot be expected. [8] considers the similar findings, they suggested the optimum reinforcement width equal to five times of the footing when they used the different planer geosynthetics in reinforced soil foundation. The similar results can be observed in case of geotextile.

### 4.5 *Effect of Vertical Spacing Between Reinforcements*

Figure 8 shows the pressure-settlement curves with variations of vertical spacing between two reinforcements. It can be seen that 0.16B is the optimum vertical spacing for all reinforcements. Considering, for example, GGR case at settlement ratio of s/B = 4%, the bearing capacity increases to 25.0% and then the improvement decreases to 24%, 23.0, 21.4 and 19.8% (Bearing pressure = 445.3, 441.7, 438.2, 432.5, and 426.8 kPa) for h/B ratios 0.08, 0.16, 0.0.24, 0.32, and 0.4, respectively. The optimum value of vertical spacing was obtained at 0.16B for both the geosynthetics under central loading. Similar results were suggested by [8, 10].

## 5 Regression Analysis

Regression analysis was carried out on the results obtained from the experimental analysis. The analysis was carried out on a R Integrated Development Environment (IDE) RStudio. The analysis was carried out by considering five dimensionless parameters, viz. via. u/B, h/B, N, Normalized stiffness and Normalized tensile strength and estimating the degree of significance of each parameter with the improvement factor, computed at ultimate bearing capacity of soil.

For the purpose of analysis, data set was created using the results obtained from the experimental analysis, and imported into the RStudio framework. Bearing capacity ratio was kept as the Y intercept for the purpose of analysis and functioned as the dependent variable, and all the other parameters were analysed for their influence on the improvement factor at the ultimate bearing capacity for all the reinforcements and their corresponding configurations. The relative importance of each independent parameter for computation of ultimate bearing capacity of a reinforced foundation was assessed by computing the individual t values for each of the variable. The higher the value of |t|, the greater is the variable significance. Table 4 shows the fittings obtained with different parameters.

From Table 4 it can be observed that linear model including all dimensionless parameters is the best fitted for computation. Conducting linear regression including all parameters yields the following results as reported in Table 5.

From the analysis, it can be observed that N is the most significant factor when computing the ultimate bearing capacity of reinforced foundation, with an overall t value of 23.911. Also, from the analysis of the obtained results, it can be concluded that spacing between reinforcements is more significant that the tensile modulus of the reinforcement for a range of.

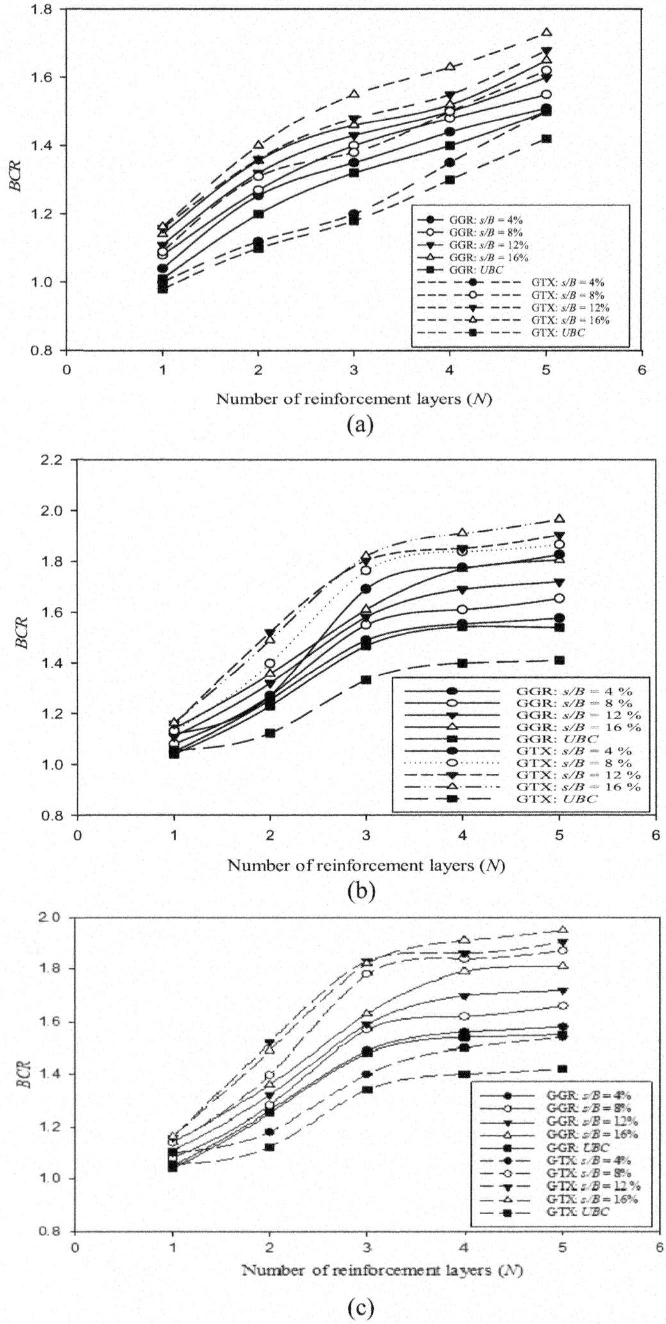


**Fig. 7** Bearing capacity ratio (BCR) for **a** width of reinforcement = 4B **b** width of reinforcement = 5B **c** width of reinforcement = 6B

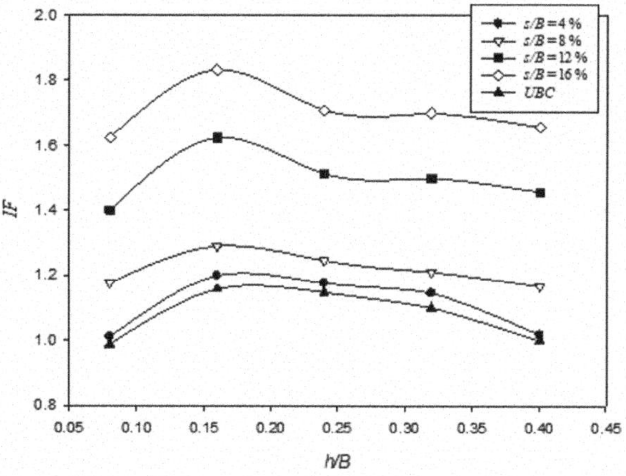


(a)

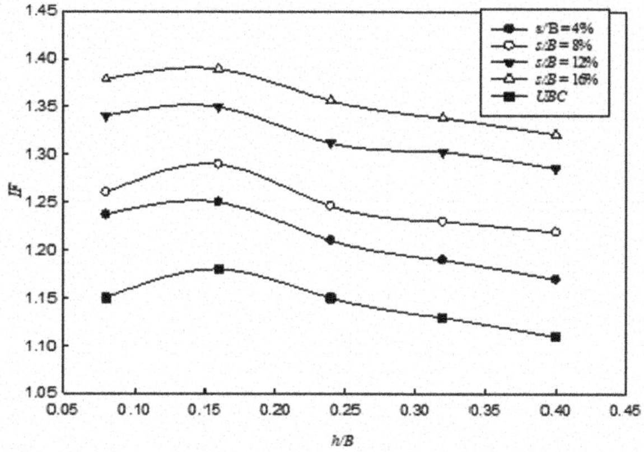


(b)

**Fig. 8** Variation of improvement factor versus h/B for **a** GGR **b** GTX

$$0.08 \leq H/B \leq 0.4$$

Which is a critical observation, as the project cost is usually associated with the spacing between reinforcements. From Table 4 it can be observed that the highest adjusted $R^2$ value is obtained when u/B, *N,* and Norm. Tensile strength are taken as the parameters for computation of the ultimate bearing capacity, even though the multiple $R^2$ value reduces (confidence level still greater than 95%).

## 6 Conclusions

On the basis of obtained results following conclusions were drawn.

**Table 4** Various possible linear models and their fittings

| Multiple $R^2$ | $R^2$ adjusted | Parameters |
|---|---|---|
| 0.9685 | 0.9524 | *u/B, h/B, N,* $N_S$, and $N_t$ |
| 0.9657 | 0.9516 | *u/B, h/B, N,* and $N_t$ |
| 0.9644 | 0.9537 | *u/B, N,* and $N_t$ |
| 0.9599 | 0.9508 | *u/B, h/B, N,* and $N_s$ |
| 0.9547 | 0.9499 | *u/B, N,* and $N_s$ |
| 0.9511 | 0.9433 | *h/B, N,* $N_S$, and $N_t$ |
| 0.5197 | 0.4416 | *u/B, h/B,* $N_S$, and $N_t$ |
| 0.4785 | 0.4315 | N |
| 0.2996 | 0.2277 | *h/B,* $N_S$, and $N_t$ |
| 0.2768 | 0.2226 | *u/B* and *h/B* |
| 0.2323 | 0.2141 | *h/B* |
| 0.1884 | 0.1448 | $N_S$ and $N_t$ |
| 0.1871 | 0.1678 | $N_t$ |
| 0.1796 | 0.1601 | *u/B* |
| 0.1117 | 0.1107 | $N_S$ |

**Table 5** Linear regression computations with all dependent variables

| Parameters | Coefficients | t value |
|---|---|---|
| *u/B* | −0.1187 | −3.009 |
| *N* | 0.1948 | 23.911 |
| *h/B* | 0.0159 | 2.175 |
| *Ns* | 0.01956 | 0.678 |
| *Nt* | −0.4798 | −1.51 |

1. The optimum depth of top most layer was found to be 0.34B times the width of square footing for both the reinforcements, i.e. geogrid and geotextile whereas optimum depth of reinforcement (d) was obtained at d/B ratio of 1.36B and 1.02B for geogrid and geotextile, respectively.
2. The soil reinforced with geotextile behave differently from the geogrid. The improvement in bearing capacity increases with increasing in reinforcement layers. Optimum number of reinforcement layers was obtained at N = 4 for geogrid reinforced soil and N = 3 for geotextile reinforced soil.
3. The width of reinforcement also played a crucial role in amassing maximum benefit from reinforcements. A substantial improvement in performance of reinforcement was found when the width of reinforcements was equal to 5 times the width of footing for geogrid and geotextile.

4. For the foundation construction point of view, Geogrid was the best performing material. Although, geotextile performed better at higher settlement ratios, geogrid provided better reinforcement for lower settlement ratios for which the structures are usually designed for.
5. Regression analysis showed that the most significant parameter for computation of the ultimate bearing capacity of the soil foundation is number of layers of reinforcements.
6. The vertical spacing between reinforcements is a more significant parameter than the Normalized stiffness and the Normalized tensile strength of the geosynthetic.
7. The numerical model for computation of the ultimate bearing capacity includes number of layers of geosynthetics, Initial layer spacing and the Normalized tensile strength.

## References

1. Binquet J, Lee KL (1975b) Bearing capacity tests on reinforced earth slabs. J Geotech Geoenviron Eng 101
2. El Sawwaf MA (2007) Behavior of strip footing on geogrid-reinforced sand over a soft clay slope. Geotext Geomembr 25(1):50–60
3. Rowe RK, Soderman KL (1987) Stabilization of very soft soils using high strength geosynthetics: the role of finite element analyses. Geotext Geomembr 6(1–3):53–80
4. Mandal, Sah (1992) Beating capacity tests on geogrid-reinforced clay. Geotext Geomembr 327–333
5. Latha GM, Somwanshi A (2009) Bearing capacity of square footings on geosynthetic reinforced sand 27:281–294
6. Thallak SG, Saride S, Dash SK (2007) Performance of surface footing on geocell-reinforced soft clay beds. Geotech Geol Eng 25:509–524
7. Tafreshi MSN, Dawson A (2010) Comparison of bearing capacity of a strip footing on sand with geocell and with planar forms of geotextile reinforcement 28:72–84
8. Chen Q, Abu-Farsakh M, Sharma R, Zhang X (2007) Laboratory investigation of behavior of foundations on geosynthetic-reinforced clayey soil. Transport Res Rec J Transport Res Board 28–38
9. Biswas A, Krishna AM, Das SK (2016) Behavior of geosynthetic reinforced soil foundation systems supported on stiff clay subgrade. Int J Geomech 1532–3641
10. Guido VA, Chang DK, Sweeney MA (1986) Comparison of geogrid and geotextile reinforced earth slabs. Can Geotech J 23(4):435–440

# Offshore Geotechnics, Marine Geology and Sub–Sea Site Investigation

# Offshore Steel Pile Capacity Evaluation Using Static and High-Strain Dynamic Tests

**Anil Cherian**

## 1 Introduction

The axial capacity assessment of steel piles has possibly been an area of uncertainty in foundation design over the past few decades. The majority of pile design continues to be dominated by axial capacity estimation in various foundation applications, where the critical issue is more likely to be the magnitude of settlement under operating conditions. In the United Arab Emirates (UAE), steel tubular driven piles are frequently used to support offshore structures [1]. In geotechnical engineering, the accurate measurement of the ultimate pile capacity is a significant mystery. The main objective of static and dynamic load testing is to confirm that the pile can withstand the design load and does not experience any excessive displacements during its design life. Most of the foundation construction standards require the performance of preliminary static tests on a certain percentage of the piles on the site. Similarly, the pile capacity can be evaluated from the dynamic load test by driving piles using suitable hammers.

For some of the offshore projects, a comparison of both test results is performed to identify the ultimate steel tubular pile capacity. The load transfer mechanism of these piles is affected by the soil disturbance caused by driving, the conditions of driving, the frictional resistance of the pile, etc. [2]. Various researchers performed predication of the ultimate capacity of piles using different methods [3–6]. The main objective of this article is to compare the static and dynamic load test results of the same steel tubular pile from the Duroob Island (N 2,715,031.617, E 706,324.730) about 140 km away from Abu Dhabi, United Arab Emirates, to verify the capacity using traditional numerical methods like Van der Veen [7] and Chin Fung Kee [8].

A. Cherian (✉)
Strainstall Middle East L.L.C. (James Fisher and Sons plc, UK), Dubai, United Arab Emirates
e-mail: dranilct@gmail.com

 
B. T. Jose et al. (eds.), *Proceedings of the Indian Geotechnical Conference 2022 Volume 2*, Lecture Notes in Civil Engineering 477,
https://doi.org/10.1007/978-981-97-1741-5_28

**Table 1** General subsurface condition

| Layer No. | Depth range (m) | Geological description |
|---|---|---|
| 1 | 0.00 to −1.10 | Medium dense, silty, very gravelly sand |
| 2 | −1.10 to −1.60 | Loose, gravelly, silty sand |
| 3 | −1.00 to −4.10 | Loose to very loose, gravelly to slightly gravelly, very silty sand |
| 4 | −4.10 to −4.53 | Very dense, silty, very gravelly sand |
| 5 | −4.53 to −6.53 | Very weak calcarenite |
| 6 | −6.53 to −19.53 | Very weak mudstone |

## 2 Geological Condition

The geologic description of the ground materials at the site and the approximate average depth at which they were encountered are summarized in Table 1. The subsurface material consists of medium to dense silty gravelly sand followed by very weak Calcarenite and Mudstone.

## 3 Methodology

For the static load test (SLT), the reaction system was used as per ASTM D1143-20 [9]. The load is applied through a 4000 kN capacity hydraulic jack, and a 5000 kN load cell has been mounted in the line of force application to measure the applied load (Fig. 1). The details of the steel tubular test pile are provided in Table 2. The test load is applied to the pile in a series of 25% increments of the design load up to 150% of the working load. Settlement of the pile was recorded with four displacement transducers, each positioned at an equal distance around the pile and connected to a data logging system and computer display [1].

In high-strain dynamic pile testing (HSDPT) or PDA, which includes a signal-matching procedure for computed and measured force and velocity at the pile head based on the stress-wave theory [10]. Goble and Rausche [11], Rausche et al. [12], Middendorp [13], and Paquet [14] have discussed the initial major developments of dynamic pile testing and their findings. The pile shafts are instrumented with strain sensors and accelerometers prior to driving. The strain sensor and the accelerometers are set up in two sets, perpendicular to one another. Dynamic pile load testing was done after driving the steel pile to the desired depth in accordance with ASTM D4945-17 [15], and a typical set up is presented in Fig. 2. Pile driving and dynamic load testing are carried out with a hydraulic hammer (IHC S 280) with a maximum nominal energy of 280 kJ. After the test, using the recorded data, an analysis using the Case Pile Wave Analysis Program (CAPWAP) was conducted to estimate the total pile capacity.

**Fig. 1** Typical set up for static load test

**Table 2** Test pile details

| Pile ID | Pile diameter/thickness (mm) | Pile length (m) | Test load (kN) |
|---|---|---|---|
| P03 (static) | 1118/20.60 | 31.00 | 1650 |
| P03 (PDA) | 1118/20.60 | 31.00 | 6460 |

The ultimate capacity of piles is generally estimated using data from axial static load testing and load-settlement behavior at a particular percentage of test loads. The available literature offers a variety of techniques for calculating pile's maximum carrying capacities [16]. According to the common methods outlined by Van der Veen [7] and Chin Fung Kee [8], the static load test results in the current study were analyzed to determine the pile's ultimate bearing capacity. For the final design check, extrapolation of the static load test results was done in order to compare with the dynamic test results.

## 4 Results and Discussion

Static and dynamic load tests were conducted on piles embedded into the weak mudstone formation. The results obtained from SLT and the PDA are presented in Figs. 3 and 4.

According to the SLT results, at a test load of 1650 kN, there was a 4.30 mm deflection. The lesser settlement value shows that the pile is not fully mobilized,

**Fig. 2** Typical set up for dynamic test

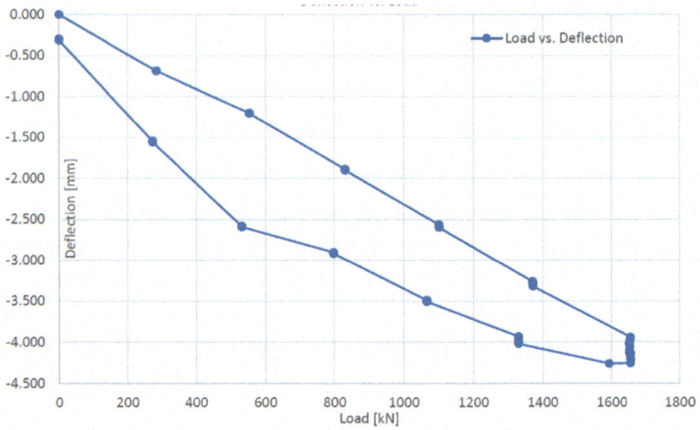


**Fig. 3** Load—deflection plot from SLT

and as a result, there is no indication of the pile toe material's bearing capacity. Furthermore, there is no evidence of any plastic soil behavior because the applied load is within the first stage of the pile capacity. During the unloading stage, the deflection-load relation shifts, exhibiting a slightly quicker reversibility at low loads with a residual deflection of less than 0.5 mm. According to the CAPWAP analysis, the settlement was 4.00 mm at a load of 1650 kN and 17.00 mm at the ultimate test

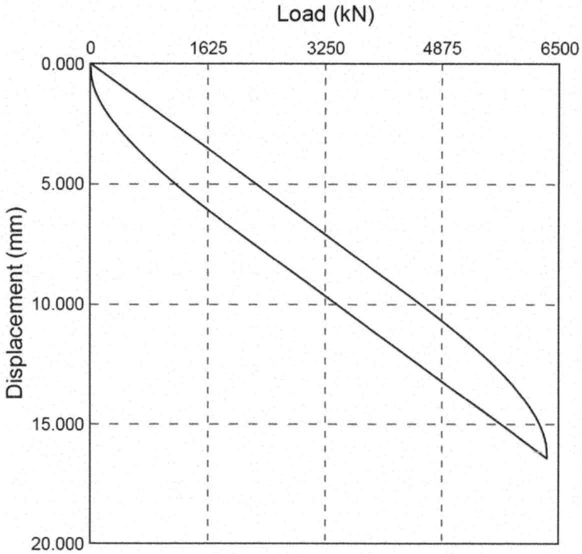


**Fig. 4** Load—displacement plot from PDA

load of 6460 kN. Based on the similar settlement values observed at 1650 kN, it was decided to perform further predictions of ultimate pile capacity from the SLT results using two common methods.

A method of analyzing the results of the load test to get a sign of the ultimate bearing capacity of soil strata/pile is adopted according to the approaches defined by Van der Veen [7] and Chin Fung Kee [8]. The settlement Δ at each loading stage is divided by the load P at that stage and plotted against Δ/P. The inverse slope of the lines gives the ultimate bearing capacity of the soil strata. Van der Veen suggests another extrapolation method whereby the hyperbolic fit is replaced by an inverse logarithmic curve. The curves lead to a vertical asymptote, which shows the ultimate capacity of the pile. However, the ultimate pile capacity is limited by a settlement equal to 10% of the pile diameter if soil failure is not reached before that. The hyperbolic and logarithmic curves were compared with the settlement values and are depicted in Fig. 5.

The pile capacity of SLT is significantly higher than the applied working and test load. The applied load indicates a linear soil behavior of the pile. The Chin method has a higher limit of ultimate pile capacity of 9500 kN, which is reached by pile settlement rather than geotechnical failure. Van der Veen's method estimates the ultimate pile capacity at 8073 kN. Based on the Chin method, the load at which the pile will reach a deflection equal to 10% of the pile diameter is 8073 kN, assuming that there is no further increment in soil resistance. This is the load obtained at an asymptotic pile deflection. Generally, the ultimate load is considered to be 80% of the load at asymptotic pile deflection and was estimated as 6458 kN. Conversely, the capacity estimated from the dynamic test using CAPWAP analysis was 6460 kN. From the comparison, it was found that both test methods are reasonably accurate and

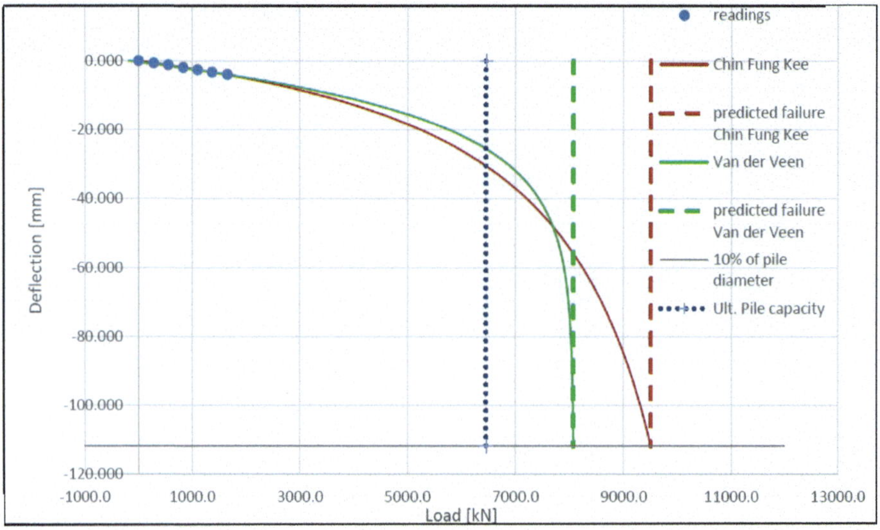


**Fig. 5** Estimated static pile capacity

a suitable estimate of the load-settlement curve can be obtained. Based on the study, it is suggested that dynamic load testing can be used as an alternative, cost-effective testing method to assess the pile capacity, especially in offshore steel tubular piles. Moreover, the dynamic load test method is successful in fast-track and restricted access sites, providing substantial cost savings to any project along with a potential quality control mechanism.

## 5 Conclusion

The present investigation of static and dynamic tests yields valuable information and illustrates which dynamic testing results can be used. It is generally believed that there are significant differences between static and dynamic tests. Comparisons between the results of static load tests performed on steel tubular piles and those of dynamic pile testing have shown the ability of HSDPT as an alternative offshore pile testing tool to produce precise estimates of the capacity. The dynamic load test technique can provide a sensible and beneficial approach to verifying the pile capacity necessary for withstanding design loads applied to the steel tubular piles. However, because this study was based on a small number of tests, further investigation is needed before routine application to reduce uncertainty in the estimation of the axial capacity of the piles.

**Acknowledgements** The author is grateful to James Fisher Strainstall for their continued support and encouragement.

## References

1. Cherian A (2022) Static behavior of strain gauge instrumented continuous flight auger (CFA) piles. Jordan J Civil Eng 16(2):229–241
2. Randolph MF, Dolwin J, Beck R (1994) Design of driven piles in sand. Geotechnique 44(3):427–448
3. Meyerhof G (1970) Safety factors in soil mechanics. Can Geotech J 7(4):349–355
4. Reese JD, O'Neil MW, Wang ST (1988) Drilled shaft tests inter-change of west belt roll road and US290 Highway, Texas. Final Report. Lymon C. and Associates. Austin
5. Kulhawy FH, O'Rourke TD, Stewart JP, Beech JF (1983) Transmission line structure foundations for uplift-compression loading, load test summaries: Appendix to EPRI Final Report EL-2870. Electric power research institute
6. O'Neill MW (2001) Side resistance in piles and drilled shafts. J Geotech Geoenviron Eng 127(1):3–16
7. Van der Veen C (1953) Experiments with model pile in groups. In: Proceedings of IV international conference on soil mechanics and foundation engineering. Paris, pp 171–178
8. Chin FK (1973) Estimation of ultimate load of piles not carried to failure. In: Proceedings of the 2nd Southeast Asia conference on soil engineering. Singapore, pp 81–90
9. ASTM D1143-20 (2020) Standard test method for piles under static axial compressive load. American Society of Testing and Materials, USA
10. Svinkin MR (2012) Engineering evaluation of static pile capacity by dynamic methods. In: Pro ceedings of 9th international conference on testing and design methods for deep founda tions, Kanazawa University, Japan, pp 179–18
11. Goble GG, Rausche F (1970) Pile load test by impact driving, Highway Research Record, pp 123–129
12. Rausche F, Goble GG, Likins G (1985) Dynamic determination of pile capacity. J Geotech Eng 111(3):367–383
13. Middendorp P (1987) Numerical Model for TNO-WAVE. Instituut TNO voor Bouwmaterialen en Bouwconstructies, TNO-IBBC, Report B1–86–75. Netherlands, pp 6–21
14. Paquet J (1988) Checking bearing capacity of dynamic loading—choice of a methodology. In: Fellenius BH (ed) Proceedings of the third international conference on the application of stress-wave theory to piles. BiTech Publishers, Canada, pp 383–398
15. ASTM D 4945–17 (2017) Standard test method for high-strain dynamic testing of deep foundations. American society of testing and materials, USA
16. Svinkin MR (2019) Sensible determination of pile capacity by dynamic methods. Geotechn Res 6(1):52–67

# Seismic Response of Spar Floating Offshore Wind Turbine

**Maria James, Sumanta Haldar, and Subhamoy Bhattacharya**

## 1 Introduction

In the present scenario, shifting to renewable energy sources is urgently needed. Although wind energy is one of the most promising forms of renewable energy, it requires good attention to make it more economical and viable. Moreover, the loading pattern and wind energy potential may get altered in the future as a consequence of climate change, necessitates at most care in the selection process [11, 14]. The availability of robust and less turbulent wind, less noise, and being far away from public visibility makes offshore wind turbines more attractive. However, the infrastructure of offshore wind farms is expensive and complex, especially in deeper water, and requires high-standard construction equipment. Water depth directly affects the cost of foundations, making it uneconomic to build fixed structures in deep water. In this situation, floating wind turbines would represent a significant advancement.

There are three fundamental concepts for floating wind turbines based on the way they attain static stability: spar, semi-submersible, and tension leg platform. Spar is a ballast stabilized structure that achieves rotational stability by keeping its center of gravity well below the center of buoyancy and translational stability from the catenary or taught mooring line. Semi-submersible is characterized by a large water plane area, while a pre-tensioned mooring line makes TLP stable. The world's first operational floating wind turbine, Hywind (Statoil), was installed in Norway

M. James (✉) · S. Haldar
IIT Bhubaneswar, Bhubaneswar, Odisha, India
e-mail: a21ce09004@iitbbs.ac.in

S. Haldar
e-mail: sumanta@iitbbs.ac.in

S. Bhattacharya
University of Surrey, Guildford, UK
e-mail: S.Bhattacharya@surrey.ac.uk



B. T. Jose et al. (eds.), *Proceedings of the Indian Geotechnical Conference 2022 Volume 2*, Lecture Notes in Civil Engineering 477,
https://doi.org/10.1007/978-981-97-1741-5_29

in 2009. In 2019, 11.4 MW of floating offshore wind was established and about which 8.4 MW are in Portugal and 3 MW from Japan. Thus, the total offshore wind reached 65.7 MW across the world by 2019 and of which 32 MW in the UK, 19 MW in Japan, 10.4 MW in Portugal, 2.3 MW in Norway, and 2 MW in France and planned to increase the total capacity to 2 GW by 2030.

Previous studies on bottom fixed offshore wind turbine show that wind turbines are prone to earthquake shaking [4, 6, 13, 23, 25, 26, 28]. However, the seismic studies on floating wind turbines are limited. Seismic waves are transferred to the floater as compressional primary (P) waves via cables and will pull the cables and transfer the movement to the floater. It can also be argued that the cables cannot transmit seismic waves due to the low mass of cables and the high damping offered by the surrounding water. Many seismic studies have been conducted on floating structures such as oil and gas platforms and tunnels, but on the seismic vulnerability of floating wind turbines is limited. Kawanishi et al. [18] studied the response of TLP under vertical earthquake shaking along with wind, wave, and current loads. The vertical component superimposes the tension in tethers, but it is within the permissible limit. The acceleration caused due to the ground shaking hardly gets transmitted into the hull. The earthquake response of the TLP under offset condition indicates that TLPs have poor resistance against wind, waves and current loads, and the responses will become more severe when vertical excitation is combined with the lateral loads. Liou et al. [20] analyzed the vertical motion, tether force, and foundation uplift of a TLP when the subsoil is excited by seismic waves. The maximum seismic response was due to the vertical earthquake ground motion in the narrow band around the fundamental frequency of TLP, which indicates the need to select a proper accelerogram for the design of TLPs. Chandrasekaran et al. [10] conducted the seismic analysis of Offshore Triangular Tension Leg Platforms and reported that TLPs would show non-linear behavior when seismic loading was combined with wave load. Retnamma and Shaji [29] conducted a seismic study on mini TLP and concluded that the earthquake response on TLP reduces as the water depth increases. The maximum reduction in response is observed for heave motion compared to pitch and surge. The dynamic response of deep-water tension leg platforms under different PGA, wave heading angles, and ground motion directions are studied by Yu et al. [34]. Downward excitation increases tension, but it was within limits.

Suroor et al. [30] stated that horizontal seismic motion does not influence the TLP due to the high flexibility of tendons in the lateral direction. Therefore, they are safe from platform and tendon forces. However, it may cause significant damage due to the slacking of tendons if horizontal shaking combines with strong currents and high sea waves (extreme environmental events). Suzuki et al. [31] conducted a seismic study of TLP by observing the responses at the nacelle level and column corner. The tendon tensions due to vertical seismic excitation are appeared to be within the design range, and no slack is observed due to seismic loading. Bhattacharya et al. [6] studied the effect of the vertical and horizontal components of seismic motion on the performance of the TLP wind turbine. They concluded that TLP-type floating wind turbines are more susceptible to vertical than horizontal ground motion. Vertical shaking causes overturning failure in TLP. Also studied was the effect of fault rupture in TLP under two conditions: (i) loss of the pretension force in the tension leg and (ii) increased tension due to the pull of the foundation under the tension leg. Loss of pretension force in TLP is most critical compared to additional tension in the tension leg since it creates an additional rotational effect at the tower top.

All the above studies are limited to tension leg platforms, and very few studies have been done on a spar floating wind turbine. Considering the high seismicity record of potential offshore wind farms and the fact that wind turbines are expensive and a slight flaw could result in a substantial loss, it is imperative to do a seismic analysis on floating wind turbines. This study focused on the seismic analysis of a spar floating wind turbine. This paper is organized into three sections. First is the evaluation of seismic performance with respect to an earthquake of high PGA. The effect of wave loading along with seismic excitation is examined further. Finally, the vulnerability of the wind turbine to earthquakes of different predominant frequencies and PGAs are evaluated. This study will aid in determining the feasibility of floating structures in seismically active regions by assessing the seismic sensitivity of spar floating wind turbines.

## 2 Numerical Model and Validation

A three-dimensional model of the platform-mooring-anchor system modeled in ABAQUS [1] is shown in Fig. 1. The soil is represented using API-based $p$–$y$, $q$–$z$, and $t$–$z$ curves [2]. The undrained shear strength, $s_u$ is assumed to be 50 kPa with a uniform variation along the depth. The submerged unit weight is 7.5 kN/m$^3$. The mooring lines are secured to the seafloor using a suction caisson. The hydrostatic stiffness is provided using springs. Quadratic hybrid beam elements (high axial stiffness compared to bending stiffness) are used to model the mooring line. At the initial stage, buoyancy loading is applied at the center of buoyancy (at 60 m from MSL), together with gravity loading. The detailed dimensions and properties of the platform-mooring-anchor system are given in Matha [22]. The model is validated by comparing the obtained natural frequency with the reported one in the literature, Matha [22] (see Table 1).

The seismic vulnerability of a spar floating wind turbine is evaluated by applying earthquake motion in two horizontal directions. Three earthquakes are chosen based on the predominant frequency and are scaled down to 0.1, 0.3, and 0.5 g. The properties of earthquakes used in this study are reported in Table 2. The typical acceleration time history of the Kocaeli earthquake before scaling is shown in Fig. 2

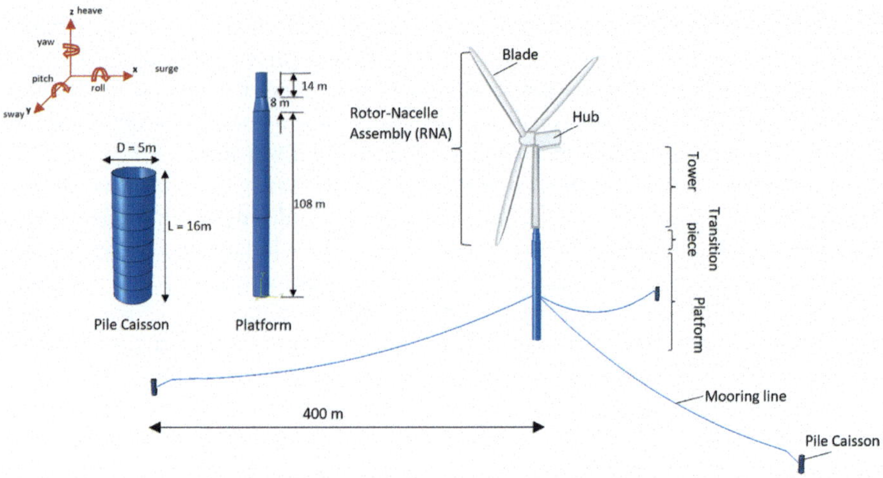


**Fig. 1** Numerical model of spar floating wind turbine developed in Abaqus CAE

**Table 1** 3D FE analysis and the reported natural frequencies of the spar floating OWT [22]

| Mode | Observed (Hz) | As reported in Matha (2009) (Hz) |
|---|---|---|
| Platform surge | 0.0079 | 0.0080 |
| Platform sway | 0.0079 | 0.0080 |
| Platform heave | 0.0329 | 0.0324 |
| Platform roll | 0.0353 | 0.0342 |
| Platform pitch | 0.0353 | 0.0343 |
| Platform yaw | 0.160 | 0.1210 |

**Table 2** Properties of earthquakes used in this study

| Earthquake | Predominant frequency of $H_1$ (Hz) | Predominant frequency of $H_2$ (Hz) | $V_{s30}$ (m/s) | Rupture distance (km) | Mag |
|---|---|---|---|---|---|
| Northridge, 1994 | 1.22 | 0.88 | 370.52 | 99.59 | |
| Kobe, 1995 | 0.59 | 2 | 312 | 0.27 | 6.9 |
| Kocaeli, 1999 | 0.195 | 6.1 | 523 | 13.49 | 7.51 |

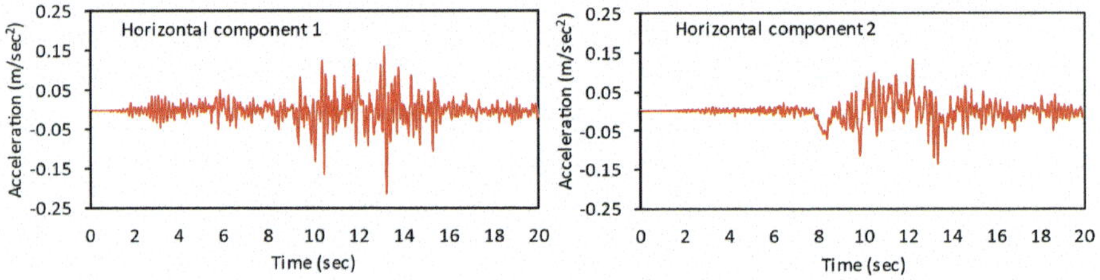


**Fig. 2** Acceleration time history of unscaled Kocaeli earthquake of 1999

## 3 Performance-Based Design

The performance is evaluated by examining the acceleration, lateral displacements, and rotational response at the nacelle. For the smooth performance of the wind turbines, DNV codes suggest limiting the combined pitch and roll rotation to 7–10°, and the displacement in surge and sway direction is restricted to 0.2 times water depth [6].

## 4 Results and Discussion

All other environmental and operational loads are initially neglected to evaluate the earthquake susceptibility in detail. Further, the performance is assessed under combined seismic, wave, and operating loads. Finally, the effect of peak base acceleration (PBA) and predominant frequency of earthquake on the spar wind turbine is evaluated and discussed in the following section.

The acceleration response in the horizontal and vertical directions is shown in Fig. 3. Here, the maximum acceleration observed at nacelle is 0.06 m/s$^2$, which is lower than the permissible value of 0.25 g. The lateral displacement and rotational response of the tower at the topmost point are presented in Fig. 4. The maximum value is 40 mm, which is relatively small compared to the restricted value for safe performance. Similarly, the rotational responses are also observed to be minimal. Due to the catenary shape and low mass of the cable, the earthquake forces are not transmitted to the floating structure. Hence from this preliminary study, floating wind turbines with catenary moorings have marginal impact during seismic shaking.

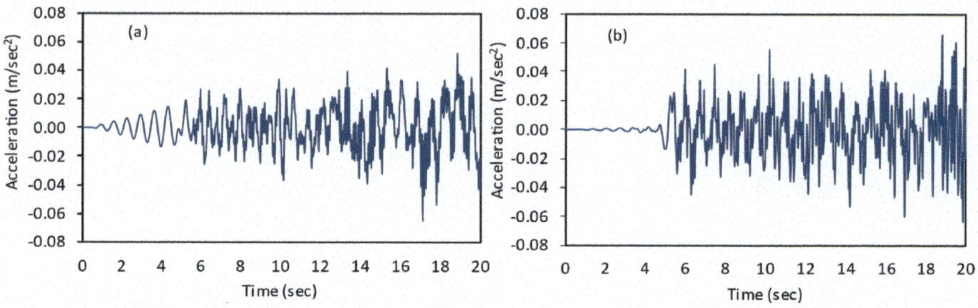


**Fig. 3** Acceleration response at nacelle in **a** surge and **b** sway direction

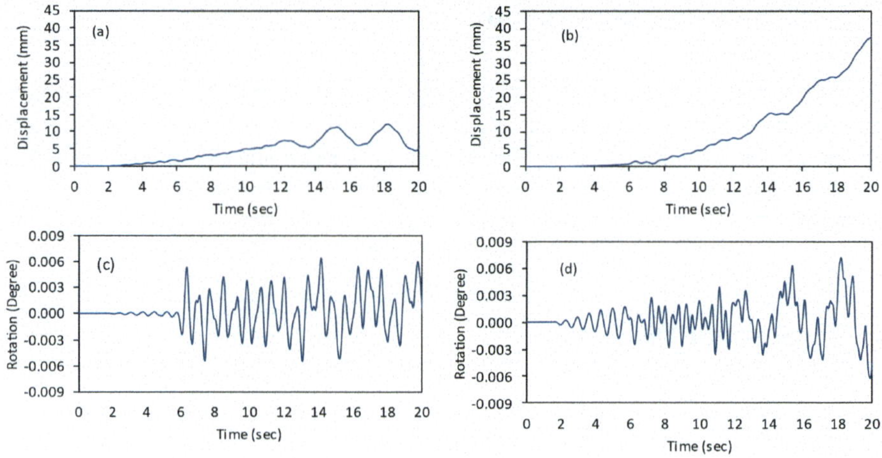


**Fig. 4** Responses at tower top during Kocaeli earthquake of 1999 scaled to 0.5 g PGA, **a** displacement in surge, **b** displacement in sway direction, **c** rotational response in roll, **d** rotational response in pitch

## 4.1 Combined Seismic and Wave Loading

The effect of wave loading along with seismic excitation is evaluated further to comprehend the impact of environmental loads during an earthquake. The wind turbine is analyzed under normal operating conditions. The Kocaeli earthquake of 1999, scaled to 0.5 g, is used as the input earthquake motion. Other operational loads such as 3P and 1P are included in this study, along with wave load. The wave height and wave period of 2.25 m and 5 s are selected, which is the wave climate of the Indian offshore at 150 m water depth (20.91°N and 69.25°E). The wave load is applied in the x direction. Therefore, only those responses are reported in Fig. 5.

The lateral displacement, the rotational response, and the acceleration at the nacelle with and without considering earthquake loads along with wave and operational loads are shown in Fig. 5. The response coincides in both cases. The design of a spar floating wind turbine is therefore governed by wave loads and not earthquake loads.

## 4.2 Frequency Study

The response of spar floating wind turbines to the predominant earthquake frequency is examined in detail. From the above studies, seismic loading shows a marginal impact on a spar floating wind turbine. However, the dominant frequency plays a significant influence in defining the seismic demand of a structure, as resonance with the structural frequency may amplify the seismic demand. Many past studies [12,

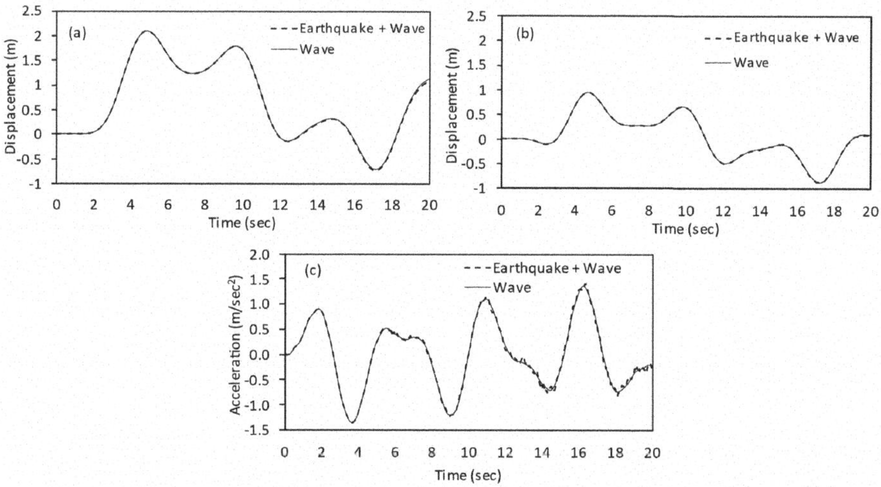


**Fig. 5** The responses at nacelle during the combined seismic, wave, and operational loading, **a** lateral displacement in surge, **b** pitch rotation, **c** acceleration in the surge

13, 23] have shown that monopile and jackets are prone to high-frequency vertical shaking due to the closeness of natural structure frequency with the predominant frequency of earthquakes in the vertical direction. But the effect of frequency content on floating wind turbines has not been studied in detail. Hence, to explore this effect, motions having three different predominant frequencies are considered. One is close to the structural frequency in the yaw direction (0.16 Hz), one far away from the structure frequency range, and the other between these two.

The various response at nacelle is shown in Figs. 6 and 7. In x axis, predominant frequency of earthquake ($f_{pr}$) is normalized with respect to the natural frequency of platform in yaw direction ($f_y$). The maximum acceleration reported is 0.067 m/s$^2$ which is much lower than the permissible limit of 0.25 g. It is observed that, as the predominant frequency increases, the responses reduce. The maximum response is recorded when the ratio of predominant frequency to structural frequency is close to 1, i.e., resonance condition. In general, the horizontal component of the earthquake is characterized by low-frequency seismic waves, whereas high-frequency content is in the vertical direction. Consequently, spar floating wind turbines are significantly more susceptible to horizontal seismic motion.

## *4.3 PGA Study*

Peak ground acceleration (PGA) is one of the primary determinants of the intensity of earthquakes. The significance of PGA on the seismic performance of spar wind turbines is analyzed at a deeper level. Figure 8 presents acceleration response at

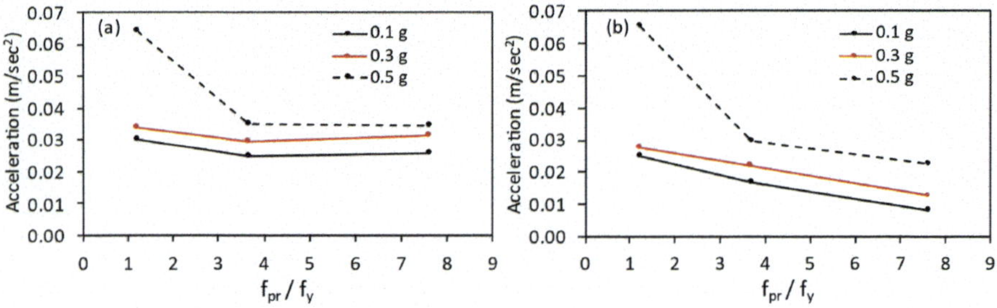


**Fig. 6** Acceleration in **a** surge and **b** sway direction at nacelle

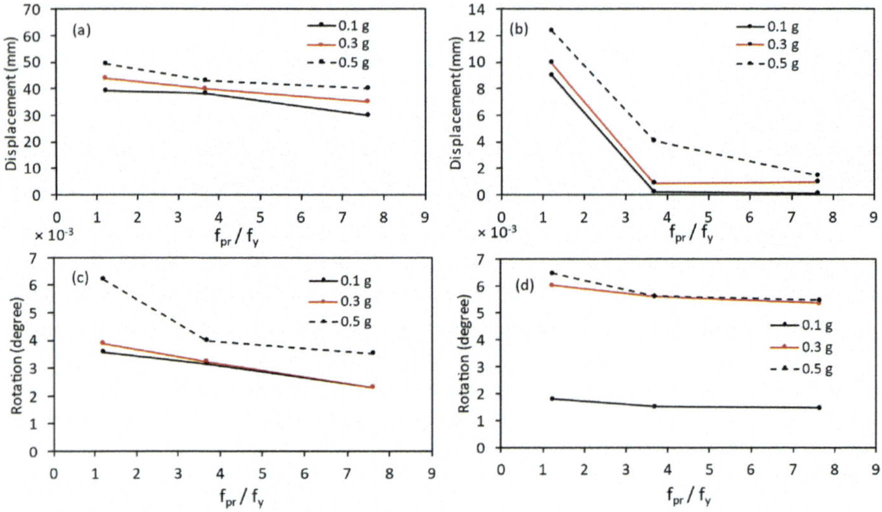


**Fig. 7** The variation of lateral displacement of the tower top in **a** surge, **b** sway direction and rotational response in **c** pitch and **d** roll direction with respect to ratio of predominant frequency of earthquake ($f_{pr}$) to yaw frequency ($f_y$)

nacelle with respect to peak base acceleration. The variation of lateral displacement of the tower top corresponding to the predominant frequency of earthquake in the surge, sway, and heave direction are shown in Fig. 9. The response increases as peak base acceleration increases. However, all the responses are observed to be within the permissible limit.

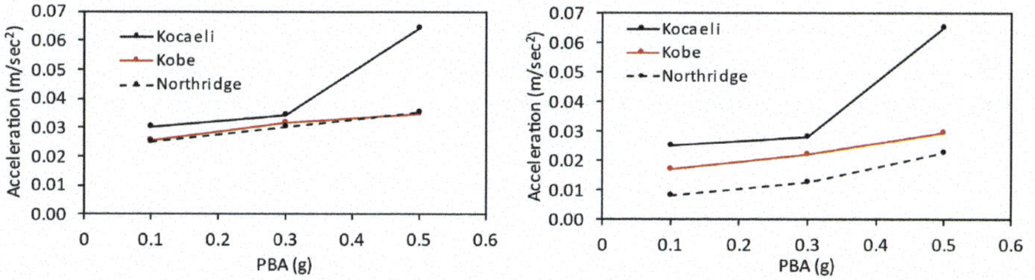


**Fig. 8** Acceleration response at nacelle with respect to various PBA in **a** surge and **b** sway direction

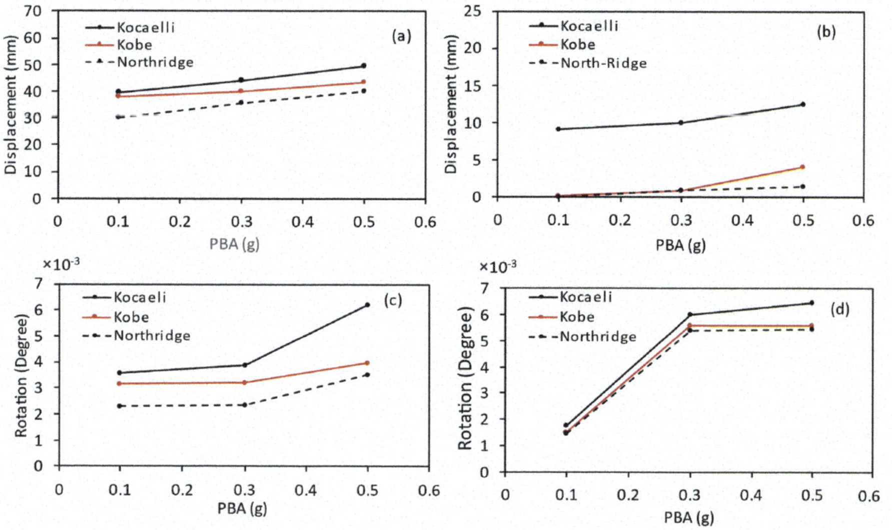


**Fig. 9** Response at nacelle with respect to various peak base acceleration, **a** displacement in surge, **b** displacement in sway, **c** pitch rotation, and **d** rotation in roll

## 5 Conclusion

The seismic susceptibility of a spar floating wind turbine is evaluated to determine the feasibility of deploying floating wind turbines in seismically active areas. The effect of high-intensity earthquake shaking, peak ground acceleration, predominant frequency, and load combination effects are evaluated in detail. Observations indicate that the response to wave loads alone and seismic plus wave loads are the same. Hence, it can be inferred that wave load will govern the design. Further, the effect of peak base acceleration and predominant frequency on the seismic design of the spar is examined. The spar wind turbines are more prone to low-frequency earthquake shaking, which generally occurs in the horizontal component of near-field earthquakes. However, in all the cases evaluated, responses are found to be minimal. Therefore, it can be inferred that spar floating wind turbines are less susceptible to earthquake dynamics due to the catenary shape, low mass of cable, and the long natural vibration period. This is a preliminary study on assessing the safety of wind turbines during ground shaking. This study's future scope is anticipated to include a more in-depth examination of the secondary risks associated with earthquakes.

## References


1. ABAQUS (2018) Abaqus user manual, version 6.10. Dassault Systemes Simulia Corp
2. API-RP-2GEO (2011) Geotechnical and foundation design considerations. American Petroleum Institute, Washington, DC, USA
3. Almheriegh MA (2021) Calculating environmental design loads for floating wind turbine. Am Acad Sci Res J Eng Technol Sci 82:52–65
4. Amani S, Prabhakaran A, Bhattacharya S (2022) Design of monopiles for offshore and nearshore wind turbines in seismically liquefiable soils: methodology and validation. Soil Dyn Earthq Eng 157:107252
5. Arablouei A, Cai W, Sari A (2016) Seismic assessment of tension-leg platforms. In: Offshore Technology Conference, OnePetro
6. Bhattacharya S, Biswal S, Aleem M, Amani S, Prabhakaran A, Prakhya G et al (2021) Seismic design of offshore wind turbines: good, bad and unknowns. Energies 14(12):3496
7. Bhattacharya S, Lombardi D, Amani S, Aleem M, Prakhya G, Adhikari S et al (2021) Physical modelling of offshore wind turbine foundations for TRL (technology readiness level) studies. J Mar Sci Eng 9(6):589
8. Bhattacharya S, De Risi R, Lombardi D, Ali A, Demirci HE, Haldar S (2021) On the seismic analysis and design of offshore wind turbines. Soil Dyn Earthq Eng 145:106692
9. Bungum H, Olesen O (2004) The 31st of August 1819 Luroy earthquake revisited. Norw J Geol 85:245–252
10. Chandrasekaran S, Jain AK, Chandak NR (2006) Seismic analysis of offshore triangular tension leg platforms. Int J Struct Stab Dyn 6:97–120
11. James M, Haldar S, Bhattacharya S (2024) Impact of climate change on the design of multi-megawatt spar floating wind turbines. Mar Struct 93:103547
12. James M, Haldar S (2021) Seismic design of large offshore wind turbine considering rocking vibration. In: Transportation, water and environmental geotechnics. Springer, Singapore, pp 411–422
13. James M, Haldar S (2022, September) Seismic vulnerability of jacket supported large offshore wind turbine considering multidirectional ground motions. In: Structures. Elsevier, vol 43, pp 407–423
14. James M, Haldar S, Varghese R, Bhattacharya S, Pakrashi V (2023) Climate change effects on offshore wind turbines. In: Wind energy engineering. Academic Press, pp 413–422
15. Jonkman J, Butterfield S, Musial W, Scott G (2009) Definition of a 5-MW reference wind turbine for offshore system development, NREL/TP-500-38060. National Renewable Energy Laboratory
16. Katsanos EI, Thöns S, Georgakis CT (2016) Wind turbines and seismic hazard: a state-of-the-art review. Wind Energy 19:2113–2133
17. Kawanishi T, Furuta H, Kato W (1987) Earthquake response of tension leg platform with gravity anchor. J Soc Naval Archtects Jpn 517z22
18. Kawanishi T, Ohashi S, Takamura H, Kobayashi H (1993) Earthquake response of the tension leg platform under unbalanced initial tension. In: The third international offshore and polar engineering conference, OnePetro
19. Kokubun K, Taniguchi T, Inoue S (2013) Effects of earthquake and tsunami on floating offshore wind turbine. In: Proceedings of international symposium on marine and offshore renewable, vol 5

20. Liou GS, Penzien J, Yeung RW (1988) Response of tension-leg platforms to vertical seismic excitations. Earthq Eng Struct Dyn 16(2):157–182
21. Ma KT, Luo Y, Kwan CTT, Wu Y (2019) Mooring system engineering for offshore structures. Gulf Professional Publishing
22. Matha D (2010) Model development and loads analysis of an offshore wind turbine on a tension leg platform with a comparison to other floating turbine concepts: April 2009 (No. NREL/SR-500-45891). National Renewable Energy Lab (NREL), Golden, CO (United States)
23. Patra SK, Haldar S (2021) Fore-aft and the side-to-side response of monopile supported offshore wind turbine in liquefiable soil. Mar Georesour Geotechnol 39(12):1411–1432
24. Patra SK, Haldar S (2021) Long-term drained and post-liquefaction cyclic behaviour of offshore wind turbine in silty sand using element tests. Arab J Sci Eng 46(5):4791–4810
25. Patra SK, Haldar S, Bhattacharya S (2022) Predicting tilting of monopile supported wind turbines during seismic liquefaction. Ocean Eng 252:111145
26. Patra SK, Haldar S (2021, June) Seismic response of monopile supported offshore wind turbine in liquefiable soil. In: Structures. Elsevier, vol 31, pp 248–265
27. Patra SK, Haldar S (2021) Response of monopile supported offshore wind turbine in liquefied soil. In: Geohazards. Springer, Singapore, pp 367–382
28. De Risi R, Bhattacharya S, Goda K (2018) Seismic performance assessment of monopile-supported offshore wind turbines using unscaled natural earthquake records. Soil Dyn Earthq Eng 109:154–172
29. Shaji SP, Jayalekshmi R. Earthquake analysis of mini tension leg platforms under random waves
30. Suroor H, Arablouei A (2019) Comparison of coupled and decoupled seismic analysis of TLP piles. Offshore Technology Conference, OnePetro
31. Suzuki K, Yamaguchi H, Akase M, Imakita A, Ishihara T, Fukumoto Y, Oyama T (2011) Initial design of tension leg platform for offshore wind farm. J Fluid Sci Technol 6(3):372–381
32. Thangam Babu PV, Reddy DV (1986) Dynamic coupled fluid-structure interaction analysis of flexible floating platforms. J Energy Resour Technol 108/207
33. Tsiapas YZ, Chaloulos YK, Bouckovalas GD, Bazaios KN (2021) Performance based design of tension leg platforms under seismic loading and seabed liquefaction: a feasibility study. Soil Dyn Earthq Eng 150:106894
34. Yu Y, Li Z, Cheng S, Zhao M, Cui Y, Yu J, Xu L (2021) Dynamic response of deepwater tension leg platform under seismic ground motion and wave loading. In: ISOPE international ocean and polar engineering conference. ISOPE, pp ISOPE-I
35. Zhang L, Shi W, Karimirad M, Michailides C, Jiang Z (2020) Second-order hydrodynamic effects on the response of three semisubmersible floating offshore wind turbines. Ocean Eng 207:107371

# Behavior of Monopile Founded Offshore Wind Turbine Under Static Load—A Parametric Study

**Sumarta Biswas and Swagata Bisoi**

## 1 Introduction

As the time progresses the amount of non-renewable energy sources are decreasing day-by-day and this type of energy is unhealthy for the environment. The process of producing non-renewable energy, high amount of greenhouse gases are emitted and these greenhouse gases are very harmful for the environment [8]. As time progresses the value and importance of renewable energy is increasing day-by-day that is good for the environment. A major part of the renewable energy is wind energy and the rate of growth of wind energy is increasing gradually. Over the last few years, the installation of offshore wind farms have grown rapidly, suitable renewable energy production field in the wind power industry [6]. Various offshore wind turbine foundations used currently are monopile, gravity, tripod, jacket and floating foundations. Of all the foundation types, the monopile foundation used about 65%. The monopile foundation is a simple foundation due to its simple shape and easy installation process [2]. The foundation is a single pile foundation made of steel with diameter of 3.5–4.5 m. The pile is driven to install up to 25 m or more depending on the seabed soil properties. The monopile foundation is the extension of the turbine tower under water and connected by transition piece into the seabed [5].

Many studies have been performed on monopile foundation. Several methods have been proposed to estimate the deformation of pile under cyclic lateral load. Achmus et al. [1] came to the conclusion that degradation stiffness model is able to account for cyclic lateral loading of piles in the determination of pile deformations. They also presented how some important factors like pile length, diameter and loading

S. Biswas (✉) · S. Bisoi
Department of Civil Engineering, Meghnad Saha Institute of Technology, Kolkata 700150, India
e-mail: sumarta_b.ce2018@msit.edu.in

S. Bisoi
e-mail: swagata.bisoi@msit.edu.in

 
B. T. Jose et al. (eds.), *Proceedings of the Indian Geotechnical Conference 2022 Volume 2*, Lecture Notes in Civil Engineering 477,
https://doi.org/10.1007/978-981-97-1741-5_30

state effects the deformation and gave the design chart, in which they considered a normalized ultimate lateral resistance of pile. According to their studies horizontally loaded offshore pile often needs the suitability of the 'zero-toe-kick' or vertical tangent criterion for its design when diameter of the monopile is large. Nowadays the renewable energy generation from offshore wind turbine are increasing to reduce the use of fossil fuel. Although wind turbine creates geotechnical challenges especially in the design section. Doherty et al. [4] consider the reliability methods to design monopile foundation and jacket structure foundations so that next generation pile design can be beneficiated. They concluded that the large diameter monopiles could be use in intermediate water depths while jacket/tripod structures in deeper waters. They also discussed about the limitations of the design methods considered for their study both for monopile foundations and for jacket piles subjected to lateral loading and axial forces, respectively. In last few year, large diameter monopile foundation usually used for wind turbines to increase its bearing capacity though there is no any mature method existed that can define how the diameter affects the bearing capacity of large diameter steel pile. Liu et al. [9] investigated the stress and deformation behavior of the soil pile system and compared it with large diameter and small diameter hollow steel piles subjected to vertical and horizontal loads in sands based on the finite element method. They proposed a method for calculating vertical bearing capacity based on soil arching method. They also projected a method to calculate lateral displacement considering shell theory for large diameter hollow steel pile. However, the soil arching method over predicted the vertical bearing capacity compared to API [2] method. The shell theory to calculate lateral displacement is 5–45% more than *p–y* curve based on API code. The characteristic of monopile foundations subjected to lateral load investigated by Lada et al. [7]. They considered the effect of pile diameter, length and load eccentricity due to lateral loading condition on monopole foundation. This article mainly focused on the analysis of large diameter monopile based on numerical model results. Meanwhile cyclic loading is also taken under consideration and they came to this conclusion that the cyclic loading leads to an accumulated rotation, thus by using the serviceability limit state (SLS) design the maximum permanent rotation must not exceed the limit value for a given wind turbine.

A parametric study has been carried out in this study considering the parameters such as diameter of the pile, soil stiffness, load amplitude and tower height. The API [2]-based static *p–y*, *t–z* curve considered to model the monopile. The tower is not considered in this study. However the tower and rotor-nacelle assemble (RNA) mass applied as a point mass on the top of the monopile has been considered in this study. The objective of the study is to verify the deflection of the monopile head for the variation of the several parameters such as diameter of the pile, load amplitude and tower height under static horizontal load. The effect of soil stiffness on the behavior of offshore wind turbine under static load has also been considered in the present study. The loading condition on offshore wind turbine is dynamic in nature due to wind and wave frequency in the field. However, the response due to dynamic load is insignificant as compared to static load when the fundamental frequency is away from resonance condition [3]. Therefore, the present study considered the static load without considering any resonance condition.

## 2 Method of Analysis

API [2]-based static *p–y* curve and *t–z* curve have been considered in the present study to verify the behavior of the monopile founded OWT. The vertical load in terms of tower mass and RNA mass has been considered as point mass and applied at the top of the pile (Fig. 1). Lateral load is considered from wind load and applied as static load. The moment is applied at 5 m above the top of monopile due to tower.

### 2.1 *P–y Curve*

*P–y* curve is generated from the following equations and Table 1 [2],

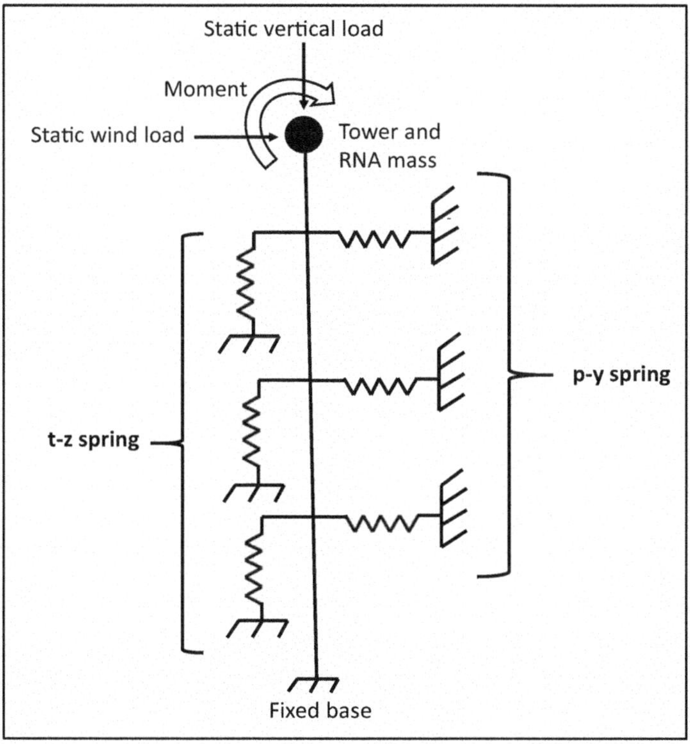


**Fig. 1** A monopile supported offshore wind turbine in clay

**Table 1** Lateral resistance versus deflection at different depths [2]

| $X > X_R$ | | $X < X_R$ | |
|---|---|---|---|
| *P/pu* | *y/yc* | *p/pu* | *y/yc* |
| 0 | 0 | 0 | 0 |
| 0.115 | 0.05 | 0.115 | 0.05 |
| 0.23 | 0.1 | 0.23 | 0.1 |
| 0.28 | 0.2 | 0.28 | 0.2 |
| 0.33 | 0.3 | 0.33 | 0.3 |
| 0.415 | 0.65 | 0.415 | 0.65 |
| 0.5 | 1 | 0.5 | 1 |
| 0.61 | 2 | 0.61 | 2 |
| 0.72 | 3 | 0.72 | 3 |
| 0–0.72[a] | 15 | 0.72 | 15 |

[a] The value is changing from 0 to 0.72 based on depth below seabed level

$$p_u = 3c + \gamma X + JcX/D \quad \text{for } X < X_R \tag{1}$$

$$p_u = 9c \quad \text{for } X \geq X_R \tag{2}$$

where

$p_u$ ultimate resistance (kPa),
$c$ undrained shear strength for undisturbed clay soil samples (kPa),
$D$ pile diameter (mm),
$\gamma$ effective unit weight of soil (MN/m$^3$),
$J$ dimensionless empirical constant = 0.375,
$X$ depth below soil surface (mm),
$X_R$ depth below the soil surface to the bottom of reduced resistance zone (mm),
$X_R$ $6D/(\gamma D/c + J)$,
$p$ actual lateral resistance (kPa),
$y$ actual lateral deflection (mm),
$y_c$ $2.5\, \varepsilon_c\, D$ (mm),
$\varepsilon_c$ strain = 0.01.

For the case where equilibrium has been reached under cyclic loading, the static *p–y* curves may be generated from Table 1.

The *p–y* curves (Fig. 2) are typically represented at different depth of soil which has been considered for the present study using Eqs. 1, 2 and Table 1. However, the node of the element and spring interval are considered at 1 m interval [3]. It is observed that as the depth increases, lateral resistance (*p* value) also increases up to $X_R$ and beyond $X_R$ it becomes constant.

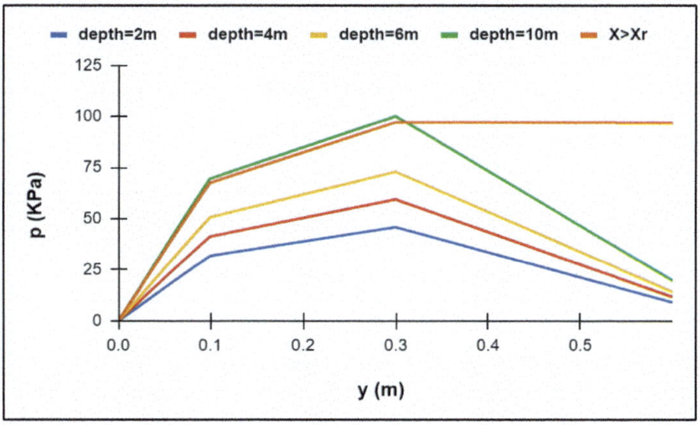


**Fig. 2** A typical representation of *p–y* curve for the variation of depth

## 2.2 T–z *Curve*

*T–z* curve is generated as follows:

$$t_{\max} = \alpha c \tag{3}$$

where

$\alpha$ a dimensionless factor ($\alpha \leq 1.0$),
$c$ undrained shear strength of the soil (kPa).

The factor, $\alpha$, can be computed by the equations:

$$\alpha = 0.5\psi - 0.5\psi \leq 1.0 \tag{4}$$

$$\alpha = 0.5\psi - 0.25\psi > 1.0 \tag{5}$$

where

$\psi$ $c/p'_o$,
$p'_o$ effective overburden pressure (kPa).

where

$z$ local pile deflection (mm),
$D$ pile diameter (mm),
$t$ mobilized soil pile adhesion (kPa),
$t_{\max}$ maximum soil pile adhesion or unit skin friction capacity (kPa).

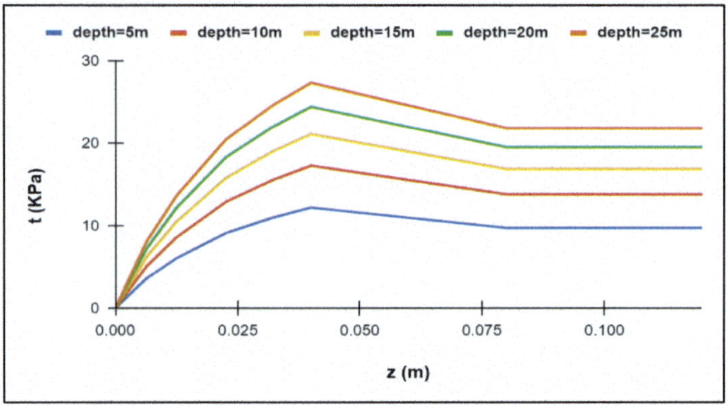


**Fig. 3** A typical representation of *t*–*z* curves at different depth

**Table 2** Frictional resistance versus deflection at different depth [2]

| *z/D* | $t/t_{max}$ |
|---|---|
| 0.0016 | 0.30 |
| 0.0031 | 0.50 |
| 0.0057 | 0.75 |
| 0.0080 | 0.90 |
| 0.0100 | 1.00 |
| 0.0200 | 0.80 |
| 0.0300 | 0.80 |

The *t*–*z* curves (Fig. 3) are typically represented at different depth of soil which has been considered for the present study using Eqs. 3, 4, 5 and Table 2. It is observed that as the depth increases, the skin friction is also increases up to the tip of the pile.

## 3 Parameters

A range of parameters has been considered for this parametric study (Table 3).

## 4 Result and Discussion

The tower and RNA mass are applied to the pile head as point mass. A static horizontal load has been applied to the pile head. After applying the load, a typical graph has been generated between the points along the length of the monopile and displacement of monopile (Fig. 4).

**Table 3** Range of parameters [3, 7]

| Name | Properties | Value |
|---|---|---|
| Tower | Tower height | 60, 65, 70 m |
| | Density of tower material | 7850 kg/m$^3$ |
| | Diameter of tower | 4 m |
| | Thickness of the tower | 0.04 m |
| | RNA mass | 80 ton |
| Monopile | Diameter of monopile | 4, 5, 6 m |
| | Length of monopile | 20 m |
| Soil | Undrained shear strength of soil | 25, 50, 100 kPa |
| | Unit weight of soil | 8 KN/m$^3$ |
| Static lateral load | Static wind load | 15, 16, 19.3 MN |

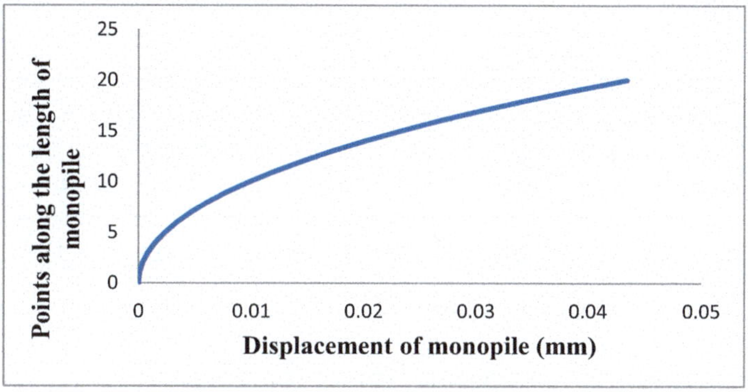


**Fig. 4** The response of the monopile head applying a static load

## *4.1 The Change in Deflection Due to the Variation of Static Horizontal Load*

The wind load which is externally acting on the wind turbine has been applied while keeping the other parameters constant. The load is applied as point load at the top of the pile. The load considered for this study are 15, 16 and 19.3 MN. The moment is applied on the pile head due to the tower height of OWT. Table 4 shows the change in deflection of the pile head due to the increased loading conditions. It is observed that the deflection of the monopile at ground level, has been increased to 6.6% when the wind load acting on the wind turbine is increased from 15 to 16 MN. The deflection of the monopile at ground level, has been increased to 20.4% when the wind load acting on the wind turbine is increased from 16 to 19.3 MN.

**Table 4** Deflection of monopile head due to change in static wind load

| Wind load (MN) | Deflection (mm) | Deflection increased (%) |
|---|---|---|
| 15 | 0.043446117 | – |
| 16 | 0.046320078 | 6.6 |
| 19.3 | 0.055784668 | 20.4 |

**Table 5** Deflection of monopile head due to change in soil stiffness

| Soil stiffness (kPa) | Deflection (mm) | Deflection reduced (%) |
|---|---|---|
| 25 | 0.043446117 | – |
| 50 | 0.043256133 | 0.4 |
| 100 | 0.042738263 | 1.2 |

## 4.2 *The Change in Deflection Due to the Variation of Soil Stiffness*

The variation in deflection of monopile head due to the change in soil stiffness keeping the other parameters constant which is presented in Table 5. The stiffness considered for this study are 25, 50 and 100 kPa. It is observed that the deflection of the monopile at ground level, has been reduced to 0.4% when the soil stiffness is increased from 25 to 50 kPa. The deflection of the monopile at ground level, has been reduced to 1.2% when the soil stiffness is increased from 50 to 100 kPa.

## 4.3 *The Change in Deflection Due to the Variation of Monopile Diameter*

The diameter of the monopile considered for this study are 4, 5 and 6 m to verify the deflection of the pile head (Table 6). The results show that the deflection of the monopile at ground level, has been reduced to 58.8% when the diameter of the monopile is increased from 4 to 5 m. The deflection of the monopile at ground level, has been reduced to 51.7% when the diameter of the monopile is increased from 5 to 6 m.

**Table 6** Deflection of pile head due to change in the diameters of monopile

| Diameter of monopile (m) | Deflection (mm) | Deflection reduced (%) |
|---|---|---|
| 4 | 0.043446117 | – |
| 5 | 0.017899612 | 58.8 |
| 6 | 0.0086436 | 51.7 |

**Table 7** Deflection of monopile head due to change in tower height

| Tower height (m) | Deflection (mm) | Deflection increased (%) |
|---|---|---|
| 60 | 0.043446117 | – |
| 65 | 0.049318506 | 13.5 |
| 70 | 0.055178652 | 11.9 |

### *4.4 The Change in Deflection Due to the Variation of Tower Height*

The tower height is considered for this study are 60, 65 and 70 m to verify the deflection of the pile head (Table 7). It is observed that the deflection of the monopile at ground level, has been increased to 13.5% when the tower height is increased from 60 to 65 m. The deflection of the monopile at ground level, has been increased to 11.9% when the tower height is increased from 65 to 70 m.

## 5 Conclusion

A parametric study is carried out for different parameters to investigate the effect of response due to horizontal static load. API [2]-based static *p–y* and *t–z* curves are considered to model the monopile. The horizontal load applied to the pile top and moment is applied at the pile head due to the tower interaction. The mass of tower and RNA is considered as point mass at the pile head. Based on the results, the following conclusions can be drawn:

- The response of the monopile decreases significantly due to increasing soil stiffness which is important to design the monopile in order to select the monopile diameter.
- Wind load largely affects the system responses. The monopile response increases with an increase in wind load.
- An increase in tower height marginally affects the monopile response. The lateral deflection increases with an increase in tower height.
- The response of the monopile decreases significantly due to increasing the monopile diameter.

## References

1. Achmus M, Abdel-Rahman K, Kuo YS (2009) Behavior of monopile foundations under cyclic lateral load. Comput Geotech 36:725–735

2. API (2005) Recommended practice for planning, designing and constructing fixed offshore platforms—working stress design, API RP2A, 22nd edn. American Petroleum Institute, Washington, D.C.
3. Bisoi S, Haldar S (2014) Dynamic analysis of offshore wind turbine in clay considering soil-monopile-tower interaction. Soil Dyn Earthq Eng 63:19–35
4. Doherty P, Casey B, Gavin K (2011) The geotechnical challenges facing the offshore wind sector. Geo-Frontiers Congress. ASCE, pp 162–171
5. Kuo YS, Achmus M, Abdel-Rahman K (2012) Minimum embedded length of cyclic horizontally loaded monopiles. J Geotech Geoenviron Eng 138:357–363
6. Kwon DK, Kareem A, Butler K (2012) Gust-front loading effects on wind turbine tower system. J Wind Eng Ind Aerodyn 104(106):109–115
7. Lada A, Ibsen LB, Nicolai G (2014) The behavior of the stiff monopile foundation subjected to the lateral loads. Aalborg University, Department of Civil Engineering. DCE Technical Memorandum No. 47
8. Li Y, Paik KJ, Xing T, Carrica PM (2012) Dynamic overset CFD simulations of wind turbine aerodynamics. Renew Energy 37:285–298
9. Liu R, Zhou L, Lian J, Ding H (2015) Behavior of monopile foundations for offshore wind farms in sand. J Waterway Port Coastal Ocean Eng 142(1):04015010–04015011

# Importance of Environmental Baseline Survey for Offshore Wind Farms and Methods for Conducting It

**Paras Vaid**

## 1 Introduction

Wind energy is a carbon–neutral source of energy that reduces air pollution emissions, preserves water resources, and helps achieve long-term energy security. It does not produce solid waste and liquid waste, on the other hand electricity which is generated from oil, coal, biomass, natural gas, or nuclear power, leaves behind enormous waste. Even though the offshore wind energy sector is not as mature as the onshore one but owing to the similarities between the two, the former has been able to become commercial-ready and efficient in a comparatively short period of time. Many countries are actively involved in the research and development of the Offshore Wind Farms (OWFs) at a large scale. Nations such as Germany, United Kingdom, Belgium, Denmark, have already achieved measurable success in the field. However, during 2018, there was a gross of 4496 MW offshore wind power installed in six different countries out of which China came out to be the largest investor of this field with 1800 MW installed capacity. In fact, China overtook, for the first time, the United Kingdom (U.K.) which had installed capacity of 1312 MW. U.K. had been the market leader during the past years [1]. Today, the wind energy may be considered as a major player in the power generation and energy market. These global developments have stimulated the development of OWFs in India which enjoys a vast coastline of over 7600 km [2].

Planning to construct a commercial-scale OWF requires knowledge of the shape of the earth, the geologic makeup of the earth, the condition of the seafloor, meteorological and oceanographic conditions, information about the physical characteristics

P. Vaid (✉)
Geotechnical Section, Institute of Production, Engineering and Ocean Technology, ONGC, Panvel, Maharashtra, India
e-mail: vaid_paras@ongc.co.in



B. T. Jose et al. (eds.), *Proceedings of the Indian Geotechnical Conference 2022 Volume 2*, Lecture Notes in Civil Engineering 477,
https://doi.org/10.1007/978-981-97-1741-5_31

of the rocks and soils that make up the land, marine habitats, and ecological environment. The above information is captured by conducting geophysical, geotechnical, metocean, and Environmental Baseline Surveys (EBS).

An EBS is a kind of survey that is used to determine the pre-existing environmental status of an area prior to the development of a project [3]. It sets up a benchmark against which the impacts of development activities can be measured.

## 2 Importance of EBS

There are major environmental concerns related to OWF development such as alterations to the food webs, risk of collisions, increase in noise levels, changes to pelagic and benthic habitats, and pollution caused by the increased vessel traffic [4]. An accurate assessment of the above is possible only if we have an established baseline against which to measure the environmental changes. For such purpose, an EBS is conducted. An efficiently planned EBS helps to strive for: protection of the existing ecological environment and development with minimum harm to marine habitat.

Globally, the hydrocarbon industry has enough experience as well as required expertise to carry out the geophysical, geotechnical, and metocean surveys for offshore hydrocarbon installations. That knowledge has been extended to OWFs with suitable modifications to account for the difference in construction, design, extent, and other characteristics as compared to offshore hydrocarbon installations. However, historically, the EBS has not been given as much importance as other surveys since it doesn't directly affect the revenues, instead, is sometimes seen as an additional expense. Furthermore, there is a lack of long-term studies of the EBS techniques adopted for OWFs since the offshore wind energy has been exploited at a large scale only over the past few decades.

## 3 Factors to Be Considered While Planning and Conducting EBS

The planning of EBS should take into account the breeding and feeding seasons of birds and marine species in addition to the seasons and routes of their migration. In fact, the construction activities related to OWFs should be undertaking only during some specific periods of time in a year to minimize the harmful effects on marine population and birds [5]. In addition, the EBS can be conducted only in suitable weather conditions and additional survey data may be collected in order to make up for the gaps in data owing to inclement weather. Furthermore, there might be other wind projects or developments occurring in the neighbourhood of the proposed project for offshore wind power development. In such a case, the cumulative effect of all the projects may be considered on the species and their habitats.

Apart from this, it is necessary to include all elements of the OWF development while planning the EBS which include the offshore wind farm site: meteorological masts, array cables, the offshore turbines and also their support structures, scour protection, offshore substations; conveying cable passageway from offshore wind farm to shore; onshore provisions: substation and onshore conveying cable course; operations station, etc. [6].

## 4 Worldwide Best Practices to Conduct EBS

The present paper tabulates the receptor, i.e. the receiving environment and the corresponding survey type required to conduct the EBS. Tables 1 and 2 have been prepared based on the Ireland and Baltic Sea environments, respectively. These two regions have been chosen specifically since the organizations who had conducted the EBS therein have been generous enough to share their comprehensive research documents freely on their websites. There may be other region-specific receptors and/or survey types which may necessitate incorporation in the EBS. Furthermore, all the survey types enlisted in Tables 1 and 2 may not necessarily be a part of each and every project. The baseline data needs and monitoring will be specific to each project depending upon its location, extent, scale, duration, and other factors. Due to a very large number of variables involved in planning the EBS and the fact the there is a huge variation in the survey type requirements from region to region, the extent of detailedness in the present paper has been kept at a broad degree.

Table 1 presents a summary of the pre-construction baseline survey methods for OWFs and has been prepared from the document released by the Government of Ireland in 2018 [7]. The document aims to serve as a suggestion on baseline marine ecological evaluation and monitoring activities for the offshore clean energy projects with a focus on Ireland's marine environment. The details relating to the survey methods enlisted in Table 1, the time/duration/extent of survey, data/statistical analyses required, practical considerations to conduct the surveys, etc., can be found in the document [7]. In addition, the post-construction survey methods have also been given in the document.

Furthermore, from 2010 to 2015, the countries of Estonia, Latvia, Finland, and Sweden collaborated for a project called MARMONI to come out with inventive ideas for the monitoring and evaluation of marine biodiversity in the Baltic Sea environment. One of the major aims of the project was to create an ingenious approach for monitoring and evaluation which is based on a set of marine biodiversity markers. Table 2, which is prepared from the document released by the project team in 2016 [8], presents the baseline survey investigation methods for OWFs for different receptors.

In addition to the biological parameters listed in Table 2, a number of non-biological parameters may also be registered during the baseline study. These include:

**Table 1** Receptors (species and habitats) and the corresponding baseline survey methods for Ireland's marine environment [7]

| S. No. | Receptors | | Pre-construction baseline survey method description |
|---|---|---|---|
| 1 | Benthos and littoral habitats | Subtidal reef | EBS methods to find the total habitation area coverage and also the approximate area coverage of each of the biotope inside that habitat<br>1. Grab sampling survey to access macro faunal communities, organic carbon content, and sediment particle size<br>2. Drop down video done by divers |
| | | Subtidal sediment | |
| | | Intertidal sediment | |
| 2 | Birds | Construction and operation | 1. Offshore (greater than 2 km from the shore): Aerial/Boat/ High definition (HD) videography transect survey, conventional distance sampling using softwares like Distance, MRSea, GLM<br>2. Remote surveillance techniques: Thermal camera, radar, telemetry, night vision, tagging<br>3. Near shore (less than 2 km from the shore): Vantage point watches<br>4. Breeding seabird survey |
| | | Cable laying | 1. Point counts of shore<br>2. Breeding seabirds survey |
| 3 | Marine mammals and basking Sharks | Cetaceans | 1. Vantage point watches<br>2. Visual survey<br>3. Acoustic survey<br>4. Line transect with distance sampling |
| | | Seals | 1. Vantage point watches<br>2. Line transect with distance sampling<br>3. Haul out counts<br>4. Reporting entanglement or collision incidences |
| | | Basking Sharks | 1. Vantage point watches<br>2. Line transect with distance sampling<br>3. Reporting entanglement or collision incidences |
| 4 | Coastal processes | Coastal erosion | 1. Wave elevation information collection and examination,<br>2. Mathematical modelling in order to anticipate changes to the wave climate<br>3. Bathymetric survey<br>4. Tidal current information collection and examination and the sediment processes<br>5. dGPS and Terrestrial LiDAR survey |
| 5 | Cultural heritage | Shipwrecks and associated remains | 1. Bathymetric survey<br>2. Mapping of the potential sub surface wrecks and/or artefacts from the geophysical survey<br>3. Side scan sonar<br>4. Magnetometer survey |
| | | Coastal archaeology and cultural heritage | 1. Desk study to identify potential archaeological and cultural sites<br>2. Boreholes or vibrocores of sediment horizons which are identified to be of archaeological interest |

(continued)

**Table 1** (continued)

| S. No. | Receptors | | Pre-construction baseline survey method description |
|---|---|---|---|
| 6 | Fisheries | Commercial fisheries | 1. Examination of present stock or cpue data<br>2. Trawl surveys to identify the species available |
| | | Shell-fish fisheries | 1. Examination of available stock/cpue data |

2. Seabed characteristics like bathymetry, morphology, grain size distribution, sediment type, loss on ignition,
3. Hydrological characteristics like waves, current regime, upwelling,
4. Physio-chemical attributes of water layer like water turbidity, salinity, oxygen, temperature regime,
5. Ice conditions like drift ice movements, composition of ice types, long-term variation in regional ice coverage, and
6. Meteorological conditions.

## 5 Challenge of Application in Indian Context and Future Scope of Study

India, being among the 17 megadiverse countries in the world has a unique marine life which is housed within a plethora of habitats [9]. Furthermore, the Indian Ocean is the warmest ocean in the world. Accordingly, specific customizations of EBS in Indian context will be a key challenge. The 'FOWIND' project, i.e. Facilitating Offshore Wind in India [2, 5, 6] and 'FOWPI' project, i.e. First Offshore Wind Project of India [10] of the Government of India have provided an enabling platform to the stakeholders to gather relevant information regarding environmental conditions in the Indian Ocean region. The FOWIND and FOWPI project reports [2, 5, 6, 10], based on Tamil Nadu and Gujarat states, are available for free online access and these highlight some relevant features of the Indian Ocean region. These include information about relevant physical and biological aspects to be considered for the EBS. They also provide some important environmental data for the Indian context which can be a useful resource for the stakeholders who may plan and conduct the EBS in the Indian Ocean region.

However, the environmental data is still highly scattered with various ministries and departments of the government of India and non-governmental organizations. This data needs to be assimilated to cover the vast scope that the EBS encompasses. There is huge scope for further study with the focus to gather pertinent data relevant to the Indian context. The data so presented shall act as an enabler for stakeholders to act upon and execute their plans to carry out EBS with more confidence. This shall ultimately give a boost to the Indian Offshore Wind power development in an ecologically sustainable manner.

**Table 2** Receptors, parameters, and corresponding baseline survey methods for Baltic Sea region [8]

| S. No. | Receptor community | Receptor | Parameters | Investigation methods |
|---|---|---|---|---|
| 1 | Pelagic communities | Phytoplankton Zooplankton | 1. Species' composition<br>2. Species' distribution<br>3. Species' abundance<br>4. Biomass | 1. Physical sampling using plankton mesh and samplers from boat or ship<br>2. Remote sensing<br>3. Ship-of-opportunity sampling (i.e. unattended measurements and sampling on ships with regular schedules) |
| 2 | Benthic communities | Infauna | 1. Species' number and composition<br>2. Species' abundance<br>3. Biomass | 1. Grab sampling<br>2. Quantitative as well as qualitative sampling by the divers<br>3. Statistical analyses |
| | | Epifauna | | 1. Sampling by divers<br>2. Statistical analyses<br>3. Video survey<br>4. Modelling |
| | | Phytoben-thos | 1. Species' number and composition<br>2. Coverage<br>3. Biomass | 1. Quantitative as well as qualitative sampling by the divers<br>2. Community analysis and modelling<br>3. Video survey |
| | | Benthic habitats | 1. Distribution as well as main characteristics of the habitats | 1. Sampling by scuba divers<br>2. Video survey<br>3. Geographic modelling |
| 3 | Fish | Demersal fish | 1. Species observed and its abundance<br>2. Dominance ratios<br>3. Biomass | 1. Otter trawl fishing<br>2. Bottom-set gill net fishing<br>3. Stomach examination of the predatory fish (to ascertain tiny fish that can't be caught using gill mesh) |

(continued)

**Table 2** (continued)

| S. No. | Receptor community | Receptor | Parameters | Investigation methods |
|---|---|---|---|---|
| | | Pelagic fish | | 1. Pelagic trawling<br>2. Hydro acoustics<br>3. Vertical gill mesh |
| | | Recruitment processes | 1. Spawning the migration routes<br>2. Spawning grounds as well as related biological processes<br>3. Nursery areas as well as the related biological processes | 1. Telemetry<br>2. Video survey<br>3. To ascertain fish eggs: Examination using artificial spawning substrates, diving surveys |
| | | | 1. Larvae of fish | 1. Physical sampling using plankton mesh and samplers from boat or ship |
| 4 | Birds | Sea birds | 1. Abundance and distribution and in the project area and neighbourhood<br>2. Age and sex distribution<br>3. Numbers of all the flying as well as swimming individuals | 1. Aerial transect surveys<br>2. Ship transect surveys<br>3. Aerial photography |
| | | Migratory birds | 1. Flight direction as well as intensities at different altitude gradients (100 m steps) upto 1000 m | 1. Visual observations<br>2. Radar survey |
| 5 | Marine mammals | Seals | 1. Use of habitat | 1. Remote sensing: The tagging and tracing of seals with measuring devices which enable to get data on dive profiles and eating journeys of marine mammals as well as the hydrographic data |

(continued)

**Table 2** (continued)

| S. No. | Receptor community | Receptor | Parameters | Investigation methods |
|---|---|---|---|---|
| | | | 1. Distribution and abundance | 1. Aerial examinations during moulting and breeding period<br>2. Aerial photography |
| | | Harbour porpoise | 1. Species' distribution and abundance | 1. Aerial digital surveys<br>2. Visual ship surveys |
| | | | 1. Habitat use | 1. Ship based and aerial surveys<br>2. Stationary acoustic detectors (POD) |

## 6 Conclusions

A well-designed EBS is essential for sustainable offshore wind power development. The survey types for different receptors tabulated in the present study are based on global experience to conduct EBS, especially the Ireland and Baltic Sea environments. This data serves as a useful resource for planning an EBS in any other region even though site-specific additions/alterations would be required. The offshore wind industry is still in its nascent stage in India and the Indian Ocean environment poses many challenges owing to its unique features. The key lies in taking guidance from the best EBS practices followed world over and then suitably modifying them to suit the regional conditions.

**Acknowledgements** The author is grateful to the support and resources provided by ONGC for carrying out this study. Views expressed are those of author only.

## References

1. Global Wind Report 2018 (2019) Global Wind Energy Council (GWEC), Brussels, Belgium. Available: http://www.indiaenvironmentportal.org.in/files/file/Global-Wind-Report-2018.pdf. Last accessed 22 July 2022
2. FOWIND Project (2015) Pre-feasibility study for offshore wind farm development in Gujarat. Available: https://mnre.gov.in/img/documents/uploads/dfa326b8328b48d4bca6f5672f9d67d3.pdf. Last accessed 23 July 2022
3. Environmental Impact Assessment for International Cooperation (2000) Overseas Environmental Cooperation Center, Ministry of Environment, Government of Japan. Available: https://www.env.go.jp/earth/coop/coop/materials/10-eiae/10-eiae.pdf. Last accessed 22 July 2022
4. Bailey H, Brookes KL, Thompson PM (2014) Assessing environmental impacts of offshore wind farms: lessons learned and recommendations for the future. Aquat Biosyst 10(8). Available:https://doi.org/10.1186/2046-9063-10-8. Last accessed 24 July 2022

5. FOWIND Project (2018) Feasibility study for offshore wind farm development in Tamil Nadu. Available: https://mnre.gov.in/img/documents/uploads/3fc822d4816d4e1093ec854144fde5d1.pdf. Last accessed 18 July 2022
6. FOWIND Project (2018) Feasibility study for offshore wind farm development in Gujarat. Available: https://gwec.net/wp-content/uploads/2018/03/FEASIBILITY-STUDY-FOR-OFFSHORE-WIND-FARM-DEVELOPMENT-IN-GUJARAT.pdf. Last accessed 22 July 2022
7. Department of Communications, Climate Action & Environment, Government of Ireland (2018) Guidance on marine baseline ecological assessments & monitoring activities for offshore renewable energy projects, Part 1. Available: https://assets.gov.ie/76530/2caa8f12-f1e7-4d76-ab34-19174ff5b9e6.pdf. Last accessed 23 July 2022
8. MARMONI Project (2016) Guidelines for environmental impact studies on marine biodiversity for offshore windfarm projects in the Baltic Sea Region. LIFE+ Nature & Biodiversity project. Innovative approaches for marine biodiversity monitoring and assessment of conservation status of nature values in the Baltic Sea. Available: http://marmoni.balticseaportal.net/wp/wp-content/uploads/2011/03/Windfarm-EIA-Guidelines_March2016.pdf. Last accessed 13 July 2022
9. Wildlife Conservation Society, India. Available: https://india.wcs.org/Programmes/Marine-program. Last accessed 21 July 2022
10. FOWPI Project. Available: https://www.cecp-eu.in/resource-center/post/fowpi-website/home. Last accessed 22 July 2022

# The Uplift Resistance of an Offshore Pipeline and the Effect of Soil Homogeneity

**Debtanu Seth, Bappaditya Manna, and J. T. Shahu**

## 1 Introduction

The shear strength within the backfill soil can vary depending on the pipe installation method and the degree of consolidation of the soil above the pipeline at the post-installation stage. However, to reduce the computation cost and simplify the analyses, the shear strength (*Su*) is often considered uniform throughout the depth of the soil bed. Again, it is also a common practice to consider linearly increasing [1] the shear strength profile of the normally consolidated or over-consolidated offshore clay beds. The consideration of the shear strength profile along the soil bed depth plays an essential role in the study of the offshore pipeline upheaval displacement behaviour.

Several studies have explored the upheaval displacement behaviour of pipelines embedded within soft-clay. The load-displacement behaviour of the pipeline during up-heaval displacement is experimentally studied by [2] and [3] under 1-g conditions. While several researchers [4] also performed multiple scaled-model tests under enhanced gravity to study the parameters mentioned above. Moreover, the soil-pipe interaction during pipe uplift was also studied by several authors [5, 6]. The load-displacement relation and interaction between the pipelines and the soil are also estimated in different analytical studies [7, 8]. In these studies, different soil parameters such as soil unit weight ($\gamma$'), pipe embedment depth (*H/D*), and pipe parameters such as pipe weight, pipe-soil interface shear and tensile capacities, pipe diameter (*D*) are varied and their effect on the resistance of soil against pipe uplift (*Pu*) is observed. However, the shear strength profile [*Su*-profile] within the clay-bed is critical in deciding the pipeline uplift displacement behaviour, and it is crucial to explore

D. Seth · B. Manna (✉) · J. T. Shahu
Department of Civil Engineering, IIT Delhi, Hauz Khas, New Delhi 110016, India
e-mail: bmanna@iitd.ac.in



B. T. Jose et al. (eds.), *Proceedings of the Indian Geotechnical Conference 2022 Volume 2*, Lecture Notes in Civil Engineering 477,
https://doi.org/10.1007/978-981-97-1741-5_32

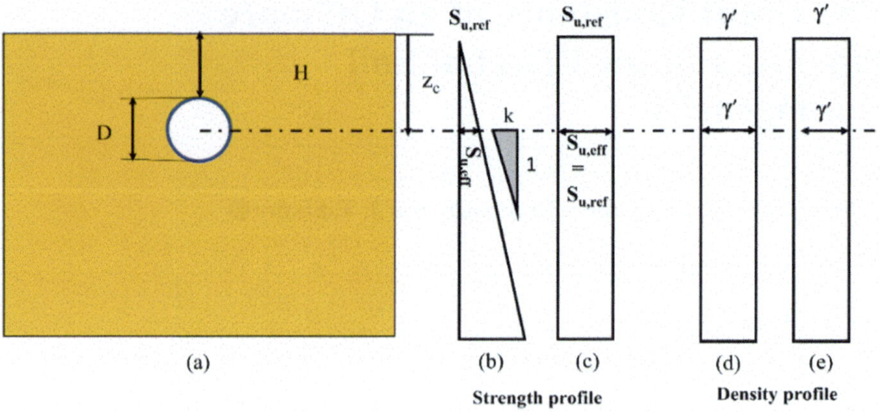


**Fig. 1** A visualization of the **a** schematic diagram of the pipe-soil system; **b**, **c** strength profile for NC and homogeneous soil, respectively; and **d**, **e** density profile for NC and homogeneous soil, respectively

the effect of variations of *Su*-profile along the depth on pipeline upheaval resistance (*Pu*).

Thus, in the current study, the variation of the *Pu* with different soil and installation parameters (*H/D*, γ') are explored for homogeneous and normally consolidated (NC) soil beds. Here homogeneous soil bed implies that the undrained shear strength (*Su*) is constant throughout the soil depth. In contrast, in the case of NC soil, the *Su* varied linearly with increasing depth of soil bed. A schematic diagram is given in Fig. 1 to describe the shear strength and unit weight distribution considered for the study. The diameter of the pipe is represented as *D* and the embedment depth is given by *H*, while *Zc* represents the depth of the centre of the pipeline. Moreover, *Su,ref* implied the shear strength at the top of the soil bed, and the *Su,eff* represented the undrained shear strength at a depth *Z*.

## 2 Methodology

In the current study, a numerical analysis is performed using a finite element package (PLAXIS 2D) to study the effect of several soil and installation parameters on the resistance against the pipeline uplift. Furthermore, the effect of soil homogeneity on those parameters is also studied. Moreover, it is considered that the soil is homogeneous along the pipe length, and the interaction between soil and pipe is similar throughout the pipe length. Thus, the pipe-soil system is modelled as a 2D plane strain problem in the current study. The further details of the analysis are discussed below.

### 2.1 Boundary Dimensions and Boundary Conditions

The boundary dimension and condition significantly affect the analysis and must be considered carefully. The criterion of boundary dimension is such that there should not be any additional stress at the boundary during the movement of the pipeline. At the same time, the boundary condition should be chosen in such a way that the movement of the soil at the model boundary simulates the movement of the soil in the actual condition. Thus, a boundary optimization study was performed to finalize the dimension of the boundary, and a boundary dimension of 7 diameters from the bottom and side of the pipeline was chosen for bottom and side boundaries, respectively. The side boundaries are considered free to move in the vertical direction and fixed in the horizontal direction. In comparison, the bottom boundary is fixed and restrained from moving in any direction.

### 2.2 Elements

In finite element studies, the whole structure is discretized into several elements. The discretized soil volume is often called a 'mesh'. In PLAXIS 2D, triangular elements of 15 and 6 nodes are available to model a soil volume. Thus, in the current study, the soil volume is discretized into several 15-node triangular elements, and the pipeline crosssection is discretized into multiple 5-node beam elements.

Moreover, the size of the elements decides the accuracy and the duration of the analysis. For smaller size elements, both accuracy and duration of the analysis are higher. While for larger element sizes, both analysis accuracy and duration decrease. Thus, a convergence study similar to [9] is performed to decide the element size, and a relative element size of 0.33 is considered [10, 11]. A similar element size is used for both homogeneous and normally consolidated soil. A typical 15-node triangular element and a 5-node beam element are shown in Fig. 2.

### 2.3 Constitutive Modelling and Model Parameters

For homogeneous and normally consolidated soil conditions, the soil volume is modelled as an elastic perfectly plastic material with Mohr–Coulomb failure criteria. Furthermore, almost similar parameters are used to represent the soil volume, except for the undrained shear strength of the soil. Some parameters are constant throughout the analysis; while others are varied to perform a parametric study. The elastic modulus to shear strength ratio ($E/Su$), Poisson's ratio ($\mu$), interface reduction factor (*Rinter*), and interface tensile capacity are kept constant throughout the analysis for both homogeneous and normally consolidated soil conditions. Moreover, for both cases, the shear strength of the soil can be represented as:

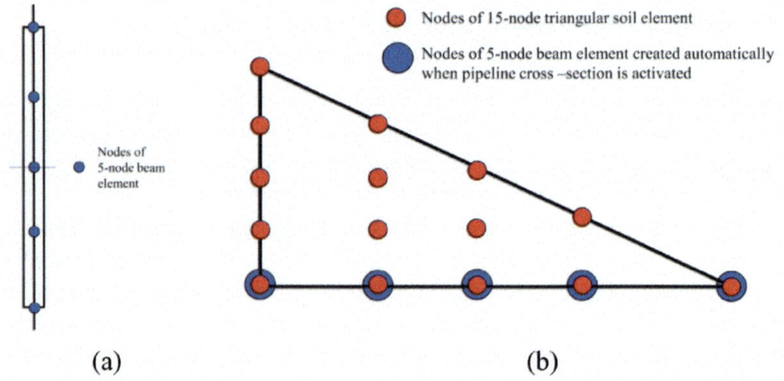


**Fig. 2** A typical diagram of a **a** 5-node beam element constituting the pipe cross-section and a **b** 15-node triangular element constituting the soil volume

$$S_{u,eff} = S_{u,eff} + k.z \tag{1}$$

where

*Su,eff* = effective undrained shear strength at z depth.

*Su,ref* = effective undrained shear strength at the surface of the soil bed.

$k$ = incremental shear strength of the soil.

On the other hand, the unit weight ($\gamma'$) of the soil and the embedment depth ratio (H/D) are varied to study the dependency of the pipeline uplift resistance on these parameters. The pipe is modelled as an elastic material with a rough outer surface. The roughness of the pipe is represented by the interface reduction factor (*Rinter*). Moreover, the buckling of the pipeline is simulated by applying nodal displacements. The details of the parameters are given in Table 1.

## 2.4 Phases of Analysis

The analysis is done in three phases: the initial, construction, and uplift. In the initial phase, the self-weight of the soil volume is calculated. While in the construction phase, the pipe element is activated, which implies the installation of the pipeline. Moreover, in the uplift phase, the nodal displacements are activated to simulate the pipe uplift.

## 2.5 Validation

The results from the current study are validated with the results obtained by [7] from their finite element analysis. The comparison between the results from the literature

**Table 1** Parameters used to model the soil and pipe volume

| Parameter | Magnitude | |
|---|---|---|
| | Homogeneous soil | NC soil |
| Undrained shear strength at the soil surface (*Su,ref*) (kN/m$^2$) | 0.01 | 10 |
| Incremental shear strength with depth (k) | 2 | 0 |
| Soil unit weight ($\gamma$') (kN/m$^3$) | 2, 5, 10, 20 | |
| Soil cover depth ratio (*H*/*D*) | 0, 1, 2, 4, 6, 8, 10, 15, 20 | |
| Elastic modulus to shear strength ratio (*E*/*Su*) | 500 | |
| Poisson's ratio ($\mu$) | 0.495 | |
| Strength reduction at the interface (*Rinter*) | 1.0 (Fully rough pipe case) | |
| Interface tensile capacity | 0 (No tension condition) | |
| Pipe diameter (m) | 1 | |
| Pipe in-plane axial stiffness (kN/m) | 7.8 × 106 | |
| Pipe out of the plane axial stiffness (kN/m) | 6.225 × 106 | |
| Pipe flexural rigidity (kN m$^2$/m) | 5265 | |

and the current study is given in Fig. 3. The validation is done for weightless soil. The tests are performed for smooth and perfectly rough outer surfaces of pipeline. Moreover, an immediate breakaway of soil from the pipeline during pipe uplift or no tension (NT) condition is considered. It can be observed that the results obtained from the current study match well with the results predicted by the numerical model in the previous studies with a 10% or lesser deviation.

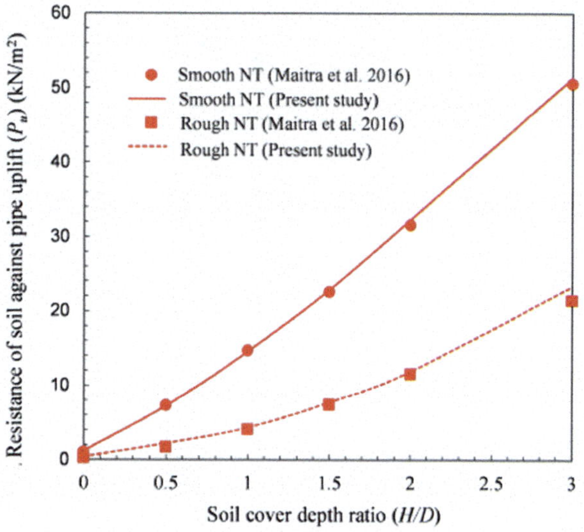


**Fig. 3** Comparison between the *Pu* predicted by the current model and the *Pu* obtained from the previous studies in the case of the weightless soil ($\gamma' = 0$)

# 3 Result and Discussion

The observation obtained from the current study and the possible explanations is discussed in this section. Initially, the effect of several soil and installation parameters on the resistance against pipe uplift (*Pu*) is discussed. Later, the effect of homogeneity is also explored. For all the cases, the no tension interface condition is considered, implying that the soil below the pipeline was immediately separated when the upheaval buckling started.

## *3.1 Upheaval Buckling of the Pipeline Embedded in Homogeneous Soil Bed*

A homogeneous soil bed refers to a soil bed where the shear strength of the soil does not vary with the depth of the soil bed. Moreover, for a pipeline buried in a homogeneous soil bed, *Pu* depends on the pipeline's embedment depth ratio (H/D), and the soil unit weight ($\gamma'$).

**Effect of Embedment Depth Ratio (H/D)**. The variation of the *Pu* with *H*/*D* is shown in Fig. 4. In Fig. 4 the vertical axis represents the resistance against pipe uplift (*Pu*), and the horizontal axis represents the pipeline embedment depth ratio (*H*/*D*).

An increasing trend of the *Pu* is observed with *H*/*D*. However, *Pu* attained a constant magnitude after reaching a peak value, and no variation of *Pu* is observed for further increase in *H*/*D*. The whole section of the plot represented two mechanisms of failure. A global failure of soil above the pipeline by the wedge failure mechanism

**Fig. 4** Variation of resistance against pipe uplift with pipe embedment depth ratio for homogeneous soil bed

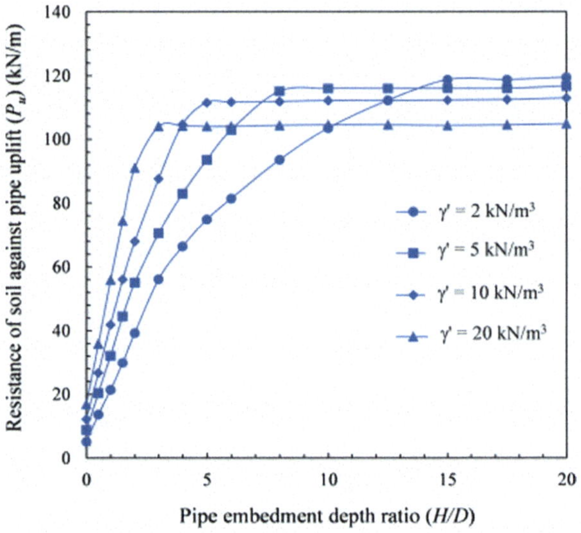

is represented by the increasing part of the plot. At the same time, a local failure of the soil around the pipeline by flow around mechanism is represented by the stabilized portion of the graph. Moreover, the global failure depends on the weight of the failure block and the shear strength at the failure boundary, which further depends on the pipe embedment depth [12]. Thus, *Pu* increased with increasing *H*/*D* initially.

However, after the *Pu* reached the peak magnitude, a transition of the failure mechanism from global failure to local failure occurred. Unlike global failure, local failure is independent of the pipeline embedment depth [10, 12]. Thus, no variation of *Pu* is observed with a further increase in the *H*/*D* after *Pu* reaches the maximum value.

It can also be observed from the figure that the *H*/*D* corresponding to which the *Pu* reached a peak value is dependent on the unit weight ($\gamma$) of the soil. With increasing $\gamma'$, *Pu* attained the peak value for lower embedment depth ratio (*H*/*D*).

**Effect of Unit Weight of Soil** ($\gamma'$). The variation of the *Pu* with $\gamma'$ is shown in Fig. 5. In Fig. 5 the vertical axis represents the resistance against pipe uplift (*Pu*), and the horizontal axis represents the pipeline unit weight of soil ($\gamma'$).

It can be observed that the resistance against the pipe uplift (*Pu*) increased with an increasing unit weight of soil ($\gamma'$). However, after reaching a peak value, *Pu* decreased with increasing $\gamma'$. It can also be observed that the *Pu* attained a similar magnitude at the post-peak condition irrespective of the pipe embedment depth ratio (*H*/*D*).

However, the unit weight of soil corresponding to which *Pu* attained peak decreased with increasing *H*/*D*. From the Fig. 5. it can be observed that *Pu* reached peak value only for *H*/*D* greater than or equal to 4. For *H*/*D* lesser than 4, the resistance against the (*Pu*) did not even reach the peak value for the selected range of soil unit weight.

**Fig. 5** Variation of resistance against pipe uplift with a unit weight of soil for homogeneous soil bed

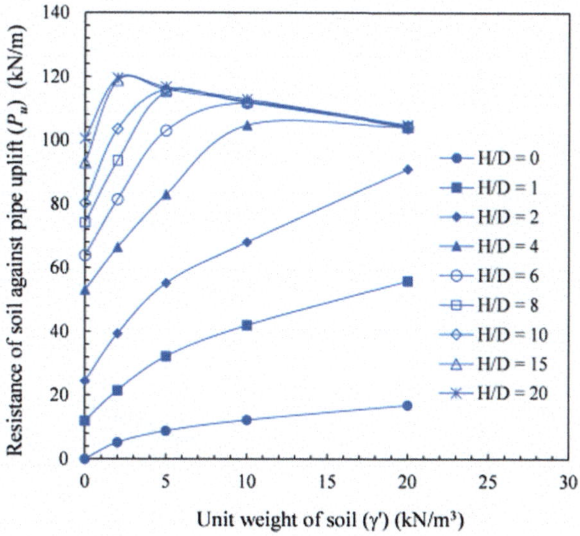

### 3.2 *Upheaval Buckling of the Pipeline Embedded in Normally Consolidated Soil Bed*

In the case of normally consolidated (NC) soil, the shear strength of the soil is considered to be increasing linearly with increasing soil depth. Similar to the previous case, in case of NC soil also, the resistance of soil against pipeline uplift is dependent on the embedment depth ratio of pipeline (*H*/*D*) and the soil unit weight ($\gamma'$).

**Effect of Embedment Depth Ratio (*H*/*D*)**. The variation of the *Pu* with *H*/*D* is shown in Fig. 6. for different magnitudes of the soil unit weight. It can be observed that the Pu increased with increasing pipe embedment depth ratio irrespective of the soil unit weight, similar to the homogeneous soil bed condition. However, *Pu* never reached the peak value for the chosen combination of the soil unit weight and the pipeline embedment depth. Thus, the post-peak behaviour is not observed in the NC condition. Moreover, only global shear failure occurred within the soil above the pipeline for the considered range of the $\gamma'$ and *H*/*D*. The transition of failure mechanism from global to local failure is not observed in case of pipelines embedded in normally consolidated soil.

**Effect of Unit Weight of Soil** ($\gamma'$). The effect of soil unit weight ($\gamma'$) on the resistance of soil against pipe uplift (*Pu*) is presented in Fig. 7. It can be observed from the figure that *Pu* increased with increasing $\gamma'$ up to a peak value. At the post-peak stage, the uplift resistance is observed to be decreasing with further increase in $\gamma'$, similar to the case of homogeneous soil bed. However, *Pu* reached the peak value for the embedment depth ratio higher than or equal to 8. For *H*/*D* lesser than 8, *Pu* is yet to reach the peak value for the considered range of soil unit weight.

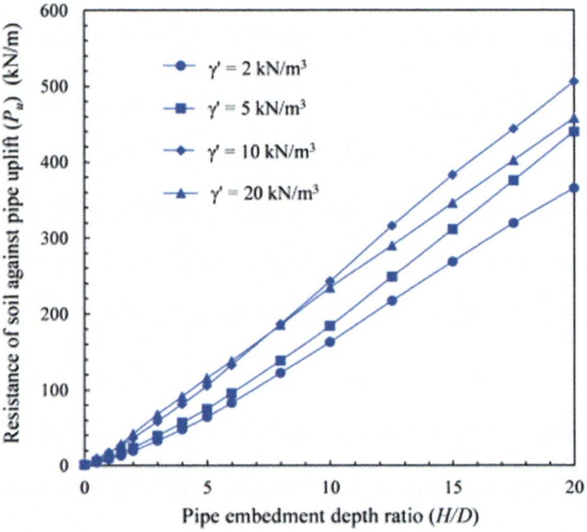


**Fig. 6** Variation of resistance against pipe uplift with pipe embedment depth ratio for normally consolidated soil bed

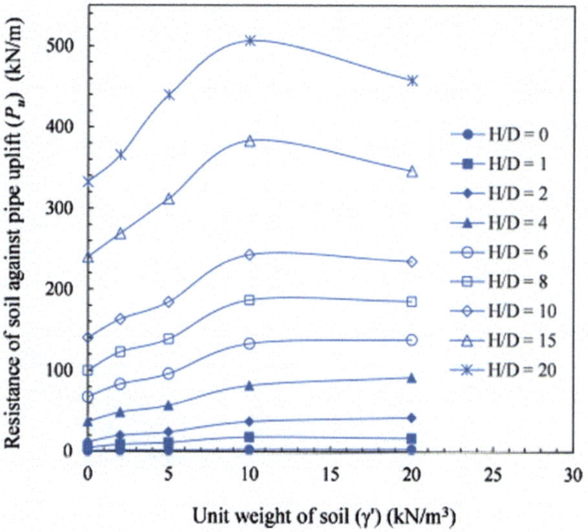


**Fig. 7** Variation of resistance against pipe uplift with a unit weight of soil for normally consolidated soil bed

## *3.3 Effect of Soil Homogeneity on the Upheaval Buckling Resistance of the Pipeline*

A comparison between *Pu* in homogeneous soil and normally consolidated soil is given in this section. The variation of the *Pu* with H/D is given in Fig. 8 for both the homogeneous and normally consolidated soil conditions. From Fig. 8, it can be observed that the resistance against the pipe uplift (*Pu*) for homogeneous soil and the normally consolidated soil are at the same range up to the embedment depth ratio of 5. However, *Pu* for NC soil increased phenomenally beyond *H*/*D* of 5.

For the particular case considered in the study, the shear strength for the homogeneous soil bed is 10 kN/m$^2$ throughout the soil bed. However, for the normally consolidated soil bed, the shear strength varied with an increment of 2 kN/m$^2$ per m depth. Hence, the *Pu* for homogeneous and NC soil is observed to be in the same range up to 5 m of depth. However, for higher embedment depth the shear strength of the NC soil increased manifold, and the *Pu* also increased extensively.

The behaviour of failure mechanisms for homogeneous and NC soil are also different. Initially, for both the homogeneous and normally consolidated cases global failure mechanism occurred, indicating a linearly increasing *Pu* with increasing pipe embedment depth. However, the *Pu* reached a peak value, and the transition of failure from global to local mode occurred for a *H*/*D* value well below 5 for homogeneous soil bed. While for the pipeline embedded in a normally consolidated soil, *Pu* did not achieve the peak value, and the transition of failure mode did not occur even for a *H*/*D* value as high as 20.

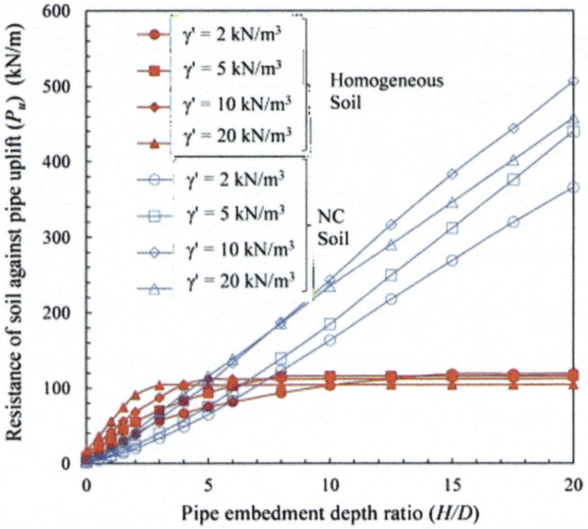


**Fig. 8** Comparison between resistance against pipe uplift within a homogeneous soil bed and a normally consolidated soil bed

## 4 Conclusion

In the current study, the soil homogeneity along the depth of the soil bed and its effect on the *Pu* is explored numerically using a finite element package called PLAXIS 2D. Two different soil conditions are considered: homogeneous and normally consolidated (NC) soil. The outer surface of the pipeline is considered to be perfectly rough, and no tension condition is assumed at the pipe-soil interface. The pipe embedment depth and the soil unit weight are varied systematically within a practical range, and the variation of the *Pu* is studied for both conditions. Moreover, the variation of *Pu* for the two conditions is compared with each other. The behaviour of the *Pu* is further described from the failure mechanism of the soil around the pipeline. The significant observations from the current study are discussed below.

- The current model is validated by the model proposed by Maitra et al. [7]., and they showed a good match.
- The *Pu* increased with increasing embedment depth for both the homogeneous and NC soil conditions. However, *Pu* for homogeneous soil conditions attained a peak value for a particular value of $H/D$. While for NC soil, the Pu did not achieve the peak value even for the considered range of the $H/D$.
- The *Pu* increased with increasing soil unit weight ($\gamma'$) until reaching the peak for homogeneous and NC soil conditions. However, for further increase in $\gamma'$, the *Pu* is observed to be decreasing.
- The *Pu* for homogeneous and NC soil conditions are compared. It can be observed that up to an embedment depth ratio of 5, both the *Pu* are in a similar range. However, for further increase in $H/D$, the *Pu* increased significantly.

The current analysis is performed using the Mohr–Coulomb model, a simplified model. Such analyses are essential for the initial understanding of the soil-pipe interaction and study of the failure mechanism of the soil. However, more advanced models should be used to understand the pipe uplift behaviour better.

## References


1. Macaro G, Utili S, Martin CM (2021) DEM simulations of transverse pipe–soil interaction on sand. Géotechnique 71(3):189–204
2. Schaminee PEL, Zorn NF, Schotman GJM (1990) Soil response for pipeline upheaval buckling analyses: full-scale laboratory tests and modelling. In: Offshore technology conference. OnePetro, Texas
3. Chen RP, Zhu B, Ni WJ (2016) Uplift tests on full-scale pipeline segment in lumpy soft clay backfill. Can Geotech J 53(4):578–588
4. Palmer AC, White DJ, Baumgard AJ, Bolton MD, Barefoot AJ, Finch M, Baldry JAS (2003) Uplift resistance of buried submarine pipelines: comparison between centrifuge modelling and full-scale tests. Geotechnique 53(10):877–883
5. Mohri Y, Fujita N, Kawabata T (2001) A simulation on uplift resistance of buried pipe by DEM. Adv Pipel Eng Constr 1–12
6. Jiang M, Zhang W, Wang J, Zhu H (2015) DEM analyses of an uplift failure mechanism with pipeline buried in cemented granular ground. Int J Geomech 15(5):04014083
7. Maitra S, Chatterjee S, Choudhury D (2016) Generalized framework to predict undrained uplift capacity of buried offshore pipelines. Can Geotech J 53(11):1841–1852
8. Wang Z, van der Heijden GHM, Tang Y (2018) Localized upheaval buckling of buried subsea pipelines. Mar Struct 60:165–185
9. Halder P, Manna B (2020) Performance evaluation of piled rafts in sand based on load-sharing mechanism using finite element model. Int J Geotechn Eng 1–18
10. Kumar P, Seth D, Manna B, Shahu JT (2021) Lateral and uplift capacity of pipeline buried in seabed of homogeneous clay. J Pipeline Syst Eng Pract 12:04021020
11. Seth D, Manna B, Kumar P, Shahu JT, Fazeres-Ferradosa T, Taveira-Pinto F, Rosa- Santos P, Carvalho H (2021) Uplift and lateral buckling failure mechanisms of offshore pipes buried in normally consolidated clay. Eng Fail Anal 121:105161
12. Seth D, Manna B, Shahu JT, Fazeres-Ferradosa T, Taveira-Pinto F, Rosa-Santos P, Pinto FVT (2021b) Offshore pipeline buried in Indian coastal clay: Buckling behaviour analysis. Ships Offshore Struct 1–16

# Effect of Linearly Increasing Cohesion on Bearing Capacity Factor of Conical Footings in c-$\phi$ Soil

**Prakarsh and Sunil Khuntia**

## 1 Introduction

Installation of the massive, inverted, conical spudcan footings into fine-grained silt material is troublesome. The industry has always classified installation circumstances as either undrained in 'clays' or drained in 'sands' so either only cohesion is considered or $\phi$ is considered at a time. Cassidy and Houlsby [1] calculated bearing capacity factors of conical footing whereas Kumar et al. [2] took linearly increasing cohesion soil profile for under-reamed piles and Griffiths and Yu [3] took it for slopes. In this study, both parameters are taken simultaneously so that the analysis can be extended to silt and partially saturated states where both parameters are present simultaneously. The soil taken in the present study was assumed to obey Mohr–Coulomb's failure criterion. This study presents you with the Bearing Capacity Factors (*BCF*) that can be used for soils where cohesion increment with depth is considered.

Spudcans are susceptible to various types of soil conditions as jack-up platforms are mobile platforms. They cannot be designed separately for each type of condition therefore the design of the spudcan has to be done while keeping the variability of soil condition over a large volume of soil in mind.

### *1.1 Jack-Up Rigs*

Offshore drilling in shallow to medium depth water is done in the ocean using mobile drilling platforms known as jack-up rigs. A floatable drilling platform (self-propelled

Prakarsh · S. Khuntia (✉)
National Institute of Technology Rourkela, Rourkela, Odisha, India
e-mail: sunilkhuntia.nitrkl@gmail.com



B. T. Jose et al. (eds.), *Proceedings of the Indian Geotechnical Conference 2022 Volume 2*, Lecture Notes in Civil Engineering 477,
https://doi.org/10.1007/978-981-97-1741-5_33

or towed by tugs) with legs that can be moved up and down is known as a mobile jack-up rig. These platforms are generally triangular or square. The depth at which they work is around 100–150 m below the water surface. The platform has three legs with three conical footings at the bottom known as spudcans. These foundations penetrate the seabed and thus provide stability to the platform above. In the preloading stage, the spudcan can penetrate to a depth of three times the diameter of the spudcans in soft clay. The diameter of the spudcan is around 10–20 m. The legs are raised when the platform has to be relocated. The legs are lowered and jacked into the seabed to create a foundation, and the once-floating hull raises itself with the help of the settled legs to become an elevated working structure. Before the operations begin, the footings are tested to 100% of their capacity to ensure their performance during a storm or adverse conditions.

### 1.2 Soil Profile

The soil profile for offshore engineering varies immensely. In many cases, the soil is in layers. These soil layers' shear strengths might vary a lot from each other. Albeit, in this study, the soil profile has been taken to uniform where the cohesion of the soft clay increases linearly with depth.

## 2 Methodology

### 2.1 Problem Definition

The diagram of the problem of the conical footing is shown in Fig. 1, where $D$ and $\beta$ represent the footing diameter and the cone apex angle, respectively. Since the geometry of this problem is a conical shape, the problem can be modelled under axisymmetric conditions. The vertical pressure applied on the conical footing is denoted by $Q_u$.

The soil is defined as weightless material to neglect the effect of unit weight on the undrained bearing capacity solutions. As the in-situ vertical effective stress in the ground results in an increase in the strength of soil with depth. The linear functions of undrained cohesion increasing with depth can be expressed as in Eq. 1.

$$c(z) = c_o + \rho z \tag{1}$$

where $\rho$ = linear cohesion gradient, $m = \rho D/c_o$ = cohesion gradient ratio, $c_o$ = cohesion on ground surface.

Since the soil is taken as a weightless material the Terzaghi equation for shallow footing is given below

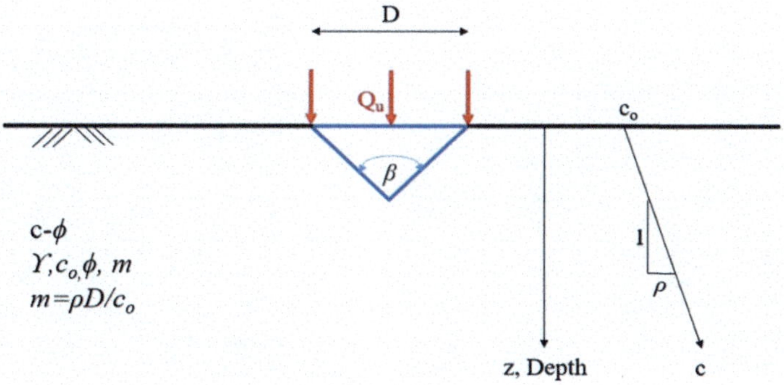


**Fig. 1** Problem definition of a conical footing on soil with LIC

$$Q_u = c_o N_c + q N_q + 0.5 \gamma B N_\gamma \tag{2}$$

For weightless soil and no surcharge condition, Eq. 2 will be reduced to

$$Q_u = c_o N_c \tag{3}$$

as $\gamma = 0$. Subsequently, the considerable variables are reduced to three dimensionless input variables after using the dimensionless technique by Butterfield [4]

$$N_c = \left(\frac{Q_u}{c_o}\right) \propto \mathrm{f}(\beta, m, \phi) \tag{4}$$

where $N_c$ is the Bearing Capacity Factor. The chosen values of those three dimensionless variables are $\beta = 60°$, $90°$, $120°$,$135°$, $150°$, and $170°$ to represent a wide range of cone apex angles; $m = 1$, 2.5, 5,7.5, 10 and $\phi = 0°$, $5°$, $10°$, $15°$, $20°$, $25°$, $30°$.

### 2.1.1 Procedure

The solutions to the problem have been obtained by finite element limit analysis using OPTUM G3 [5]. The mixed bound-based numerical computation has been used to obtain a bracketed solution between the lower and upper bound. The model of a conical footing simulated is shown in Fig. 2a, b. The symmetry of the problem enables us to perform the analysis by modelling the problem in axisymmetric conditions. The axis of symmetry is situated on the left of the model. The left and right boundaries of the model are restrained to have only vertical displacements. The bottom boundary is restrained in both directions, i.e. horizontal as well as vertical. The top boundary is a free surface with no restraints. The footings and soil have a unit weight of zero. The footing has been assigned with rigid material property whereas the soil has been assigned as a Mohr–Coulomb material. In the M-C material modelling, the cohesion

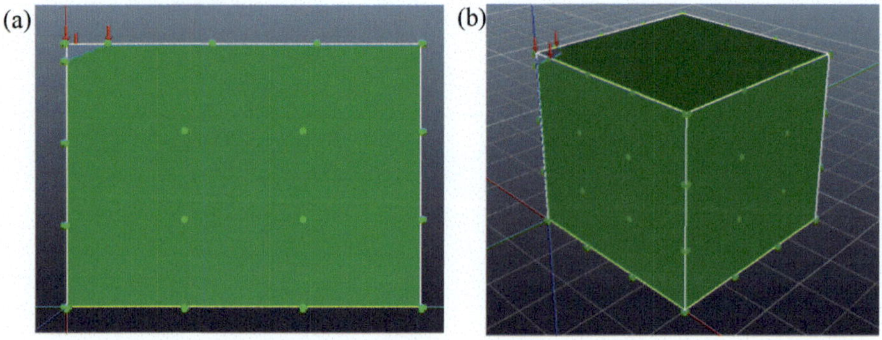


**Fig. 2** Stimulation of cone footing model in **a** 2D, **b** 3D view

has been set to increase with depth and the value of $\phi$ is fixed. At the top surface of the conical footing, the vertical uniform multiplier load $Q_u$ is applied over the area. The interface between soil and footing is assumed as rough.

The domain size is set to be large enough such that the yielding of material remains inside the domain. Automatic mesh adaptivity has been used to refine the mesh in the zone where the yielding of material takes place to avoid extra computational time. In this study, 5 adaptive iterations were used with starting 5000 number of elements and the maximum number of elements was kept at 15,000, consequently, the mesh used can be seen in Fig. 3a, b.

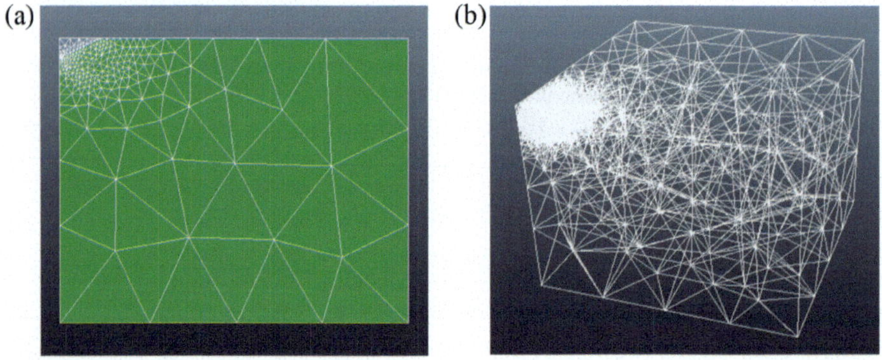


**Fig. 3** Mesh used for conical footing model in **a** 2D, and **b** 3D

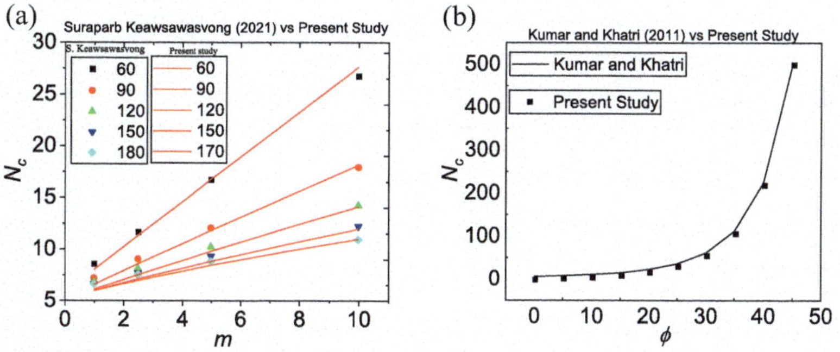


**Fig. 4** Validation of $N_c$ values of the present study with results available in the literature **a** Keawsawasvong [6]; **b** Kumar and Khatri [7]

## 3 Results and Discussion

All the solutions for Bearing Capacity Factor ($N_c$) have been obtained with the ranges of $\beta = 60°$, 90°, 120°, 135°, 150° to 170°, $m = 0$–10, and $\phi = 0°$–30°. The effects of $\phi$, $\beta$, and $m$ on the bearing capacity factor $N_c$ are demonstrated in Figs. 5, 6, and 7, respectively.

### 3.1 *Verification of the Present Model Study*

Before the results and discussion, the results were verified with the results of Suraparb Keawsawasvong [6] and Kumar and Khatri [7]. The study shows $N_c$ is a function of cone apex angle, strength gradient ratio, and anisotropic strength ratio. In the present study, the anisotropic strength ratio was taken as 1.0 making it an isotropic condition but a non-homogenous case with an increase in cohesion with respect to depth. The comparison has been shown in Fig. 4. The AUS model has been used in combination with finite element limit analysis [6] whereas the present study used the Mohr–Coulomb model with mixed bound analysis which is close to the average solutions from UB and LB solutions. Moreover, the results for $\beta = 180°$ (i.e. a circular footing), $m = 0$, and $c_o = 1$ for different values of $\phi$ has been validated with literature [6]. From Fig. 4a, b, it can be observed that the present solution is very much close to the results available in the literature.

### 3.2 *Effect of Angle of Internal Friction ($\Phi$)*

The variation of the Bearing Capacity Factor ($N_c$) with the angle of internal friction of soil ($\phi$) for different cone apex angles ($\beta$) and cohesion gradient ($m$) has been

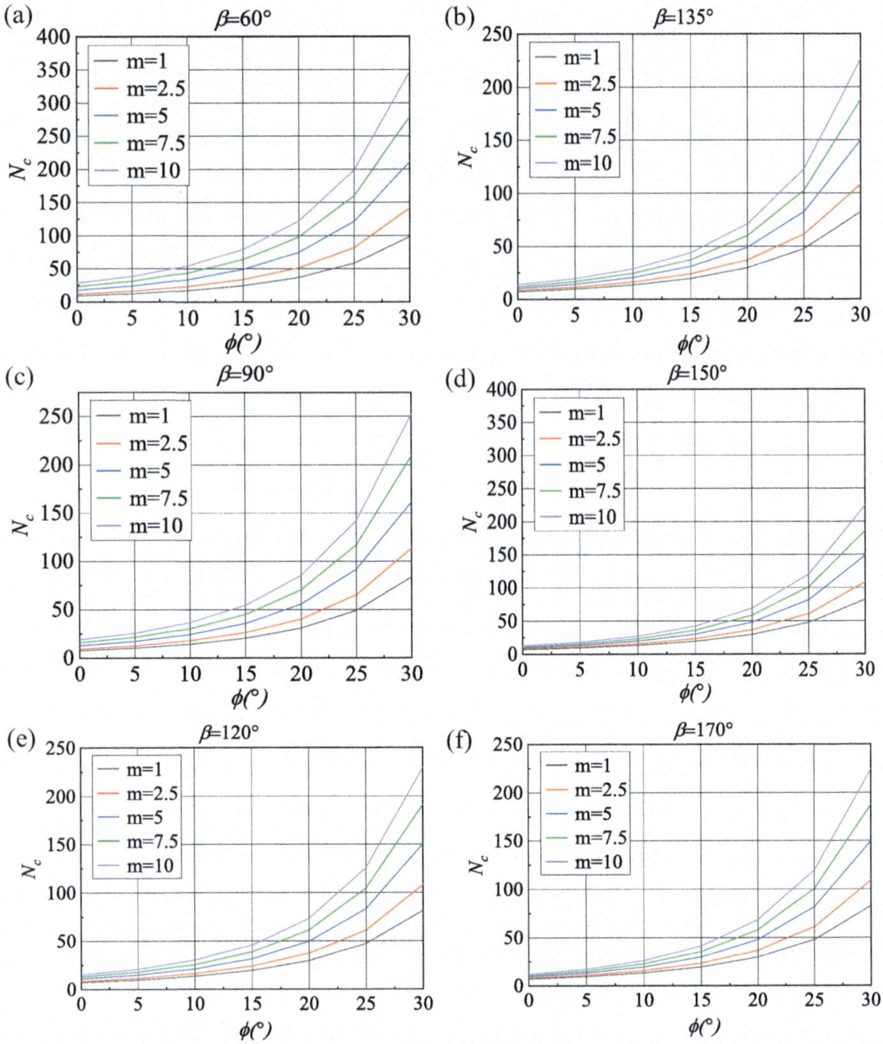


**Fig. 5** Variation of $N_c$ with $\phi$ for $m$ and **a** $\beta = 60°$, **b** $\beta = 90°$, **c** $\beta = 120°$, **d** $\beta = 135°$, **e** $\beta = 150°$, and **f** $\beta = 170°$

presented in Fig. 5. It has been observed that the value of $N_c$ increases with the increase in the $\phi$ values. The rate of increase in $N_c$ value for higher $\phi$ and $m$ values.

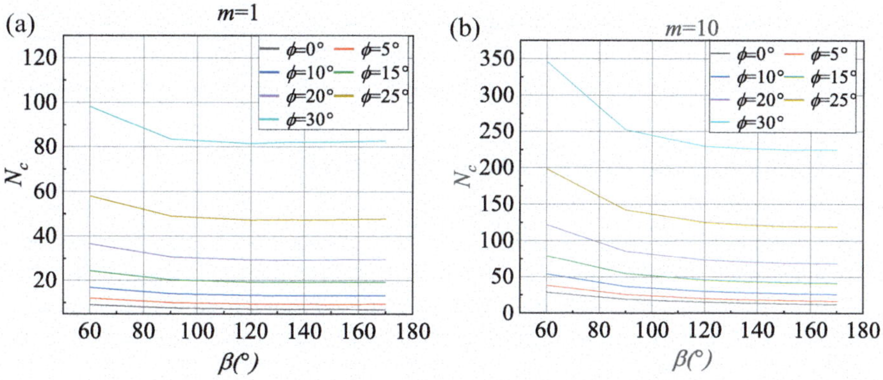


**Fig. 6** Variation of $N_c$ with $\beta$ for **a** $m = 1$, and **b** $m = 10$

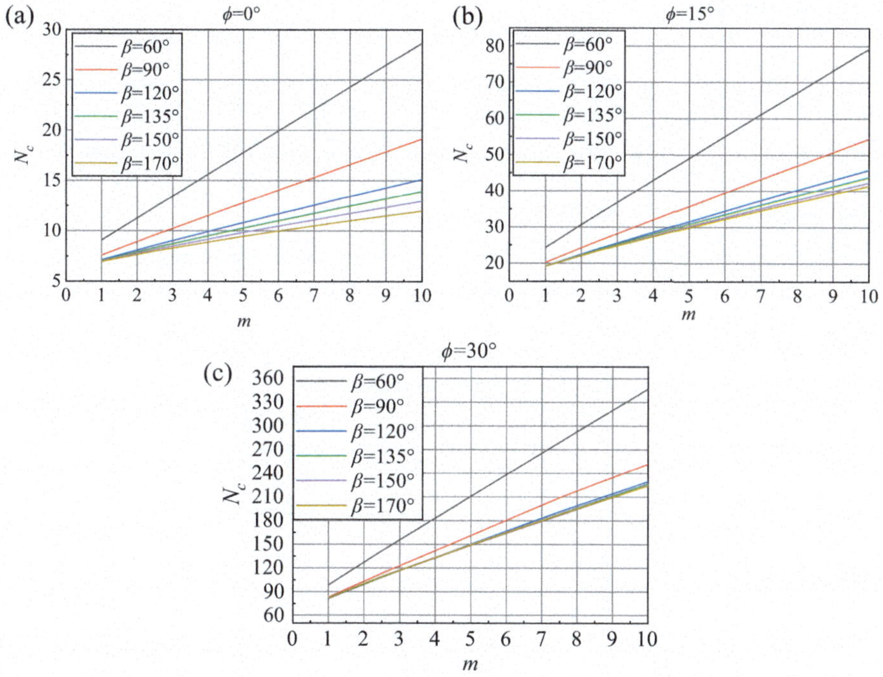


**Fig. 7** Variation of $N_c$ with $m$ for **a** $\phi = 0°$, **b** $\phi = 15°$ and **c** $\phi = 30°$

### *3.3 Effect of Cone Apex Angle* (**B**)

To examine the effect of the cone apex angle of the footing with respect to the angle of internal friction ($\phi$), the results have been obtained for a typical value of $m$ equal to 1 and 10 as shown in Fig. 6a, b, respectively. It can be observed that there is a sharp reduction in $N_c$ value for the case of low apex angle of conical footing. With the increase of apex angle ($\beta$), the rate of decrement of $N_c$ value reduces.

### 3.4 *Effect of Cohesion Gradient Ratio* (M)

The variation of the Bearing Capacity Factor ($N_c$) versus the cohesion gradient ratio ($m$) can be seen in Fig. 7a–c for the cases of $\phi = 0°$, 15°, and 30°, respectively, with $\beta = 60°$, 90°, 120°, 135°, 150°, and 170°. It is found that a high $m$ value makes a higher $N_c$ value. The $N_c$ value linearly increases with the increase in $m$ value for any value of $\beta$.

## 4 Conclusions

In the present study, the Bearing Capacity Factor ($N_c$) of conical footings on clays with linearly increasing cohesion is presented using the mixed bound finite element limit analysis. The magnitude of $N_c$ is significantly dependent on the cohesion gradient ratio ($m$), cone apex angle ($\beta$), and internal friction angle ($\phi$). The following conclusions have been made from the present study:

1. The Bearing Capacity Factor ($N_c$) value increases with the increase in values of internal friction angle ($\phi$) and cohesion gradient ratio ($m$) but decreases with the increase in cone apex angle ($\beta$).
2. The rate of increase of $N_c$ value is more for increasing values of $\phi$ and $m$, and the rate of decrease is less for increasing values of $\beta$.
3. The mixed bound finite element limit analysis is capable of determining the closer solution for any three-dimensional geotechnical problems.

## References


1. Cassidy MJ, Houlsby GT (2002) Vertical bearing capacity factors for conical footings on sand. Géotechnique 52(9):687–692
2. Kumar A, Khatri VN, Gupta SK (2020) Effect of linearly increasing cohesion on the compression and uplift capacity of the under-reamed pile in clay. SN Appl Sci 2(2):1–17
3. Griffiths DV, Yu X (2015) Another look at the stability of slopes with linearly increasing undrained strength. Géotechnique 65(10):824–830
4. Butterfield R (1999) Dimensional analysis for geotechnical engineers. Geotechnique 49(3):357–366
5. Krabbenhoft K, Lyamin A, Krabbenhoft J (2020) Optum computational engineering (OptumG3). Available from https://www.optumce.com

6. Keawsawasvong S (2021) Bearing capacity of conical footings on clays considering combined effects of anisotropy and non-homogeneity. Ships Offshore Struct 1–12
7. Kumar J, Khatri VN (2011) Bearing capacity factors of circular foundations for a general c–ϕ soil using lower bound finite elements limit analysis. Int J Numer Anal Meth Geomech 35(3):393–405

The manufacturer's authorised representative in the EU is Springer Nature Customer Service Centre GmbH, Europaplatz 3, 69115 Heidelberg, Germany. If you have any concerns regarding our products, please contact ProductSafety@springernature.com

Printed and bound by CPI Group (UK) Ltd, Croydon, CR0 4YY
01/10/2026
02225935-0012